Lecture Notes in Artificial Intellige

Edited by J. G. Carbonell and J. Siekmann

T0238278

Subseries of Lecture Notes in Computer Science

Raoul Medina   Sergei Obiedkov (Eds.)

Formal
Concept Analysis

6th International Conference, ICFCA 2008
Montreal, Canada, February 25-28, 2008
Proceedings

 Springer

Series Editors

Jaime G. Carbonell, Carnegie Mellon University, Pittsburgh, PA, USA
Jörg Siekmann, University of Saarland, Saarbrücken, Germany

Volume Editors

Raoul Medina
L.I.M.O.S. Université Blaise Pascal
Clermont-Ferrand, France
E-mail: Medina@isima.fr

Sergei Obiedkov
Higher School of Economics
Moscow, Russia
E-mail: sergei.obj@gmail.com

Library of Congress Control Number: 2008920681

CR Subject Classification (1998): I.2, G.2.1-2, F.4.1-2, D.2.4, H.3

LNCS Sublibrary: SL 7 – Artificial Intelligence

ISSN   0302-9743
ISBN-10   3-540-78136-6 Springer Berlin Heidelberg New York
ISBN-13   978-3-540-78136-3 Springer Berlin Heidelberg New York

Springer is a part of Springer Science+Business Media

springer.com

© Springer-Verlag Berlin Heidelberg 2008
Printed in Germany

Typesetting: Camera-ready by author, data conversion by Scientific Publishing Services, Chennai, India
Printed on acid-free paper   SPIN: 12227778   06/3180   5 4 3 2 1 0

Preface

Formal Concept Analysis (FCA) is a mathematical theory of concepts and conceptual hierarchy leading to methods for conceptually analyzing data and knowledge. The theory itself strongly relies on order and lattice theory, which has been studied by mathematicians over decades. FCA proved itself highly relevant in several applications from the beginning, and, over the last years, the range of applications has kept growing. The main reason for this comes from the fact that our modern society has turned into an "information" society. After years and years of using computers, companies realized they had stored gigantic amounts of data. Then, they realized that this data, just rough information for them, might become a real treasure if turned into knowledge. FCA is particularly well suited for this purpose. From relational data, FCA can extract implications, dependencies, concepts and hierarchies of concepts, and thus capture part of the knowledge hidden in the data.

The ICFCA conference series gathers researchers from all over the world, being the main forum to present new results in FCA and related fields. These results range from theoretical novelties to advances in FCA-related algorithmic issues, as well as application domains of FCA. ICFCA 2008 was in the same vein as its predecessors: high-quality papers and presentations, the place of real debate and exchange of ideas. ICFCA 2008 contributed to strengthening the links between theory and applications.

The high quality of the presentations was the result of the remarkable work of the authors and the reviewers. We wish to thank the reviewers for all their valuable comments, which helped the authors to improve their presentations. Selecting the papers was a tough job in these conditions. We are grateful to the members of the Program Committee and of the Editorial Board for giving their valuable advice.

The Conference Chair of ICFCA 2008, held at the Université du Québec à Montréal (Canada), was Robert Godin. He did a tremendous amount of work before and during the conference to make it a real success. He was helped in this task by his tireless and cheerful colleagues: Rokia Missaoui, Petko Valtchev, and Mehdi Adda. They contributed considerably to the success of the conference as well as to its friendly atmosphere. In the name of all of the participants, we wish to express here our warmest thanks to them.

February 2008   Raoul Medina
                                                                Sergei Obiedkov

Organization

Executive Committee

Organizing Conference Chair

Robert Godin   LATECE, Université du Québec à Montréal
(UQAM), Montréal, Canada

Conference Organization Committee

Rokia Missaoui   Université du Québec en Outaouais (UQO),
Gatineau, Canada

Petko Valtchev   Université du Québec à Montréal (UQAM),
Montréal, Canada

Mehdi Adda   Université de Montréal (UdeM), Montréal,
Canada

Program and Conference Proceedings

Program Chairs

Raoul Medina   LIMOS, Université Clermont-Ferrand 2, France

Sergei Obiedkov   Higher School of Economics, Russia

Editorial Board

Peter Eklund   University of Wollongong, Australia

Bernhard Ganter   Technische Universität Dresden, Germany

Robert Godin   LATECE, Université du Québec à Montréal
(UQAM), Montréal, Canada

Sergei O. Kuznetsov   Higher School of Economics, Moscow, Russia

Rokia Missaoui   Université du Québec en Outaouais (UQO),
Gatineau, Canada

Uta Priss   Napier University, Edinburgh, UK

Stefan E. Schmidt   Technische Universität Dresden, Germany

Gregor Snelting   Universität Passau, Germany

Gerd Stumme   Universität Kassel, Germany

Rudolf Wille   Technische Universität Darmstadt, Germany

Karl Erich Wolff   University of Applied Sciences, Darmstadt,
Germany

Program Committee

Mike Bain	University of New South Wales, Sydney, Australia
Peter Becker	Australia
Radim Belohlavek	Binghamton University - State University of New York, USA
Sadok Ben Yahia	Faculty of Sciences of Tunis, Tunisia
Claudio Carpineto	Fondazione Ugo Bordoni, Italy
Frithjof Dau	University of Wollongong, Australia
Vincent Duquenne	ECP6-CNRS, Université Paris 6, France
Sébastien Ferré	Université de Rennes 1, France
Linton Freeman	UCI, California, USA
Alain Gély	LITA, Université Paul Verlaine, Metz, France
Joachim Hereth Correia	Technische Universität Dresden, Germany
Derrick G. Kourie	University of Pretoria, South Africa
Marzena Kryszkiewicz	Warsaw University of Technology, Poland
Leonard Kwuida	Universität Bern, Switzerland
Wilfried Lex	Universität Clausthal, Germany
Christian Lindig	Universität des Saarlandes, Germany
Engelbert Mephu Nguifo	IUT de Lens - Université d'Artois, France
Lhouari Nourine	LIMOS, Université Clermont Ferrand 2, France
Jean-Marc Petit	LIRIS, INSA Lyon, France
Alex Pogel	New Mexico State University, Las Cruces, USA
Sandor Radeleczki	University of Miskolc, Hungary
Camille Roth	University of Surrey, UK
Sebastian Rudolph	Universität Karlsruhe, Germany
Jurg Schmid	Universität Bern, Switzerland
Selma Strahringer	Cologne University of Applied Sciences, Germany
Petko Valtchev	Université du Québec à Montréal (UQAM), Montréal, Canada
Serhyi Yevtushenko	Luxoft, Ukraine
Mohammed J. Zaki	Rensselaer Polytechnic Institute, New York, USA

Sponsoring Institutions

Centre de Recherche en Informatique de Montréal (CRIM)
Département d'informatique, Université du Québec à Montréal (UQAM)
Faculté des Sciences, Université du Québec à Montréal (UQAM)
Le Fonds Québécois de la Recherche sur la Nature et les Technologies du Québec (FQRNT)

Laboratoire de Recherche en Génie Logiciel (GÉLOG), École de Technologie
  Supérieure, Université du Québec
Mathematics of Information Technology And Complex Systems (MITACS),
  Networks of Centres of Excellence (NCE), Canada
Ministère du Développement Économique, de l'Innovation et de l'Exportation
  du Québec (MDEIE)

Table of Contents

Communicative Rationality, Logic, and Mathematics*

Rudolf Wille

Technische Universität Darmstadt, Fachbereich Mathematik,
Schloßgartenstr. 7, D–64289 Darmstadt
wille@mathematik.tu-darmstadt.de

Abstract. In this article the following thesis is explained and substantiated: *Sense and meaning of mathematics finally lie in the fact that mathematics is able to support the rational communication of humans.* The essence of the argumentation is that the effective support becomes possible by the close connection between mathematics and logic (in the sense of Peirce's late philosophy) by which, in his turn, the communicative rationality (in the sense of Habermas' theory of communicative action) can be activated. How such a support may be concretely performed shall be illustrated by the development of a retrieval system for the library of the Center of Interdisciplinary Technology Research at Darmstadt University of Technology.

Contents

1   The Impact of Mathematicts by Logic: An Example

Sense and meaning of mathematics finally lie in the fact that mathematics is able to support the rational communication of humans.

To explain and to substantiate that thesis is the concern of this article (cf. [Wi02b], [Wi02c]). First, an example shall make clear how mathematics could have a lasting effect on the rational communication by transforming mathematical forms of thinking into logical thinking. Already in [Wi01b], the close connection of *logical and mathematical thinking* has been seen as the central reason that mathematics is able to effectively support the rational thinking and acting. Logical thinking as expression of human reason grasps the *actual reality* in the basic forms of thinking: concept, judgment, and conclusion, while, according to Ch. S. Peirce ([Pe92], p.121), mathematical thinking abstracts from logical thinking to make accessible a cosmos of forms of *potential reality*. Therefore

---

* This article is an English version of the german publications [Wi02b] and [Wi02c].

R. Medina and S. Obiedkov (Eds.): ICFCA 2008, LNAI 4933, pp. 1–13, 2008.

it is possible for mathematics as a historically, socially, and culturally determined formation of mathematical thinking, respectively, to support humans in
their logical thinking and with it in their rational communication. How such
a support could be established shall first be illustrated by the development of
a retrieval system for the library of the Center of Interdisciplinary Technology
Research (ZIT) at Darmstadt University of Technology.

In this project the mathematical theory of *formal concept analysis* [GW99],
which can be understood as *applied lattice theory* (cf. [Wi07a]), was activated.
Formal concept analysis is founded on a mathematization of concept, the most
simple basic form of human thinking (cf. [Ka88], [Se01]). This mathematization
is based on the understanding that a concept is determined by a concept extent
and a concept intent; the *concept extent* consists of all objects falling under
the concept and the *concept intent* consists of all attributes which apply to all
those objects. For phrasing this concept understanding in the set language, the
mathematical notion of a *formal context* has been introduced as a set structure
(G, M, I) for which G and M are sets and I is a binary relation with $I \subseteq G \times M$;
the elements of G are called *objects* and the elements of M are called *attributes*,
and gIm is read: the object g *has* the attribute m. A *formal concept* of (G, M, I) is
now defined as a pair (A, B) with $A \subseteq G$, $B \subseteq M$, $A = \{g \in G \mid \forall m \in B : gIm\}$,
and $B = \{m \in M \mid \forall g \in A : gIm\}$. The *subconcept–superconcept–relation* is
mathematized by $(A, B) \leq (C, D) :\Leftrightarrow A \subseteq C \ (\Leftrightarrow B \supseteq D)$. With respect to this
order relation, the set of all formal concepts of (G, M, I) form a complete lattice,
the so-called *concept lattice* of the formal context (G, M, I).

Members of the Darmstadt "Research Group on Formal Concept Analysis"
started the development of the *retrieval system* for the ZIT-library in 1991,
which was finished in 1996 (see [RW00]). It needed several experiments until a
successful approach was found for the project. For instance, common retrieval
methods turned out to be unsatisfactory because of the broad interdisciplinarity
of the documents in the library. Therefore, a specific *normed vocabulary* was developed for a satisfactory content extraction of the documents. In the average,
32 catchwords from the normed vocabulary were assigned to each document,
which yielded a very good substitute for an abstract for each document. The
assignments, stored in a relational database, gave rise to a large cross table with
1554 documents as objects indicating the rows of the table and with 337 catchwords as attributes indicating the columns of the table. In that table the crosses
indicate which catchword is assigned to which document. Mathematically, such
a cross table is understood as a formal context whose concept lattice represents
a hierarchy of concepts.

The line diagram in Fig. 1 represents a concept lattice which was the result
of a conceptual search in the ZIT-library. Its five *attributes* are named "division
of work", "rationalization", "automation", "mechanization", and "production".
The concept having the intent which consists of the first four of the listed attributes has three objects named by the *document titles* "work and technics
in a social process", "information technology: a luddite analysis", and "on the
handling of machines"; most object names are only replaced by the number of

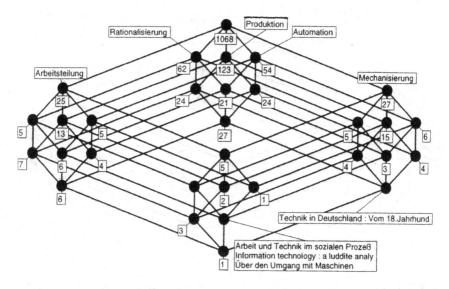

Fig. 1. Conceptual search structure of the theme "change of production"

objects of the corresponding concept to make the diagram better readable (the implementation allows to make the hidden names visable by clicking on the circles of the concepts, respectively).

The line diagram in Fig. 1 is understood as the *conceptual search structure* named "change of production" and determined by the listed five attributes. In general, 137 named *conceptual search structures*, also understood as *"conceptual views"*, have been created and characterized by a small number of catchwords. These views cover a great spectrum of themes which allow, using the program TOSCANA (see [KSVW94],[BH05]), flexible navigations with the prepared system of search structures. For example, the marked line diagram in Fig. 2 shows the resulting structure of the conceptual search structure of the theme "important industrial countries" reduced by the two catchwords "rationalization" and "automation" of the conceptual search structure "change of production".

As in many other projects of applications - members of the Darmstadt Research Group on Formal Concept Analysis have already performed more than 200 of such projects - , in the discussed project too, mathematical and non-mathematical thinking have met in *logical thinking*. Just the reality of a library with its special needs guided to an understanding of the mathematical structures and relationships of formal concept analysis as logical structures and relations. In this way the mathematical operations and connections in formal contexts became logically transparent in the "language" of content representations within cross tables without special difficulties. Analogously, the "language" of conceptual representations in line diagrams could be logically activated for content interpretations.

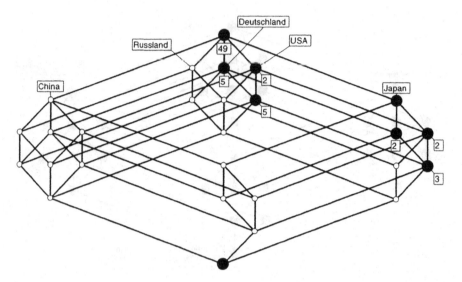

Fig. 2. Conceptual search structure of the theme "important industrial contries" reduced by the two catchwords "rationalization" and "automation"

The key idea for the development of the retrieval system was to use line diagrams of concept lattices as conceptual search structures for *a thematic search of literature*. Those line diagrams show, respectively, a thematically constructed distribution of the total stock of the literature by which the searcher can learn to make his requests more precisely. Therewith the line diagrams impart a differentiation in the contents of the stock of literature, by which the searcher can learn to specify more precisely his wishes and to focus further the proceedings of the search process. The conceptual search structures are means to activate further expert knowledge beyond the catchwords.

Each conceptual search structures represents a *local theory* based on a concept-logical structure. All these local theories are aggregated by the total connection of the underlying data table to a *global theory* which may also represented logically by a line diagram and mathematically by the corresponding global concept lattice. But it often turns out that the global concept lattice cannot be represented by a readable line diagram. Therefore smaler theories are approached which are aggerated by a few local theories; the example in Fig. 2 combines the two local theories coded in the conceptual search structures "important industrial countries" and "change of production".

The full development of the retrieval system for the ZIT-library was a process in which not only members of the Darmstadt "Research Group on Formal Concept Analysis" were involved, but also members of the ZIT, specific experts, and quite a number of students who, in particular, elaborated and tested the 137 conceptual search structures and their applications to the given stock of literature. This project could only be successfully mastered because of a high level of rational communication between the collaborators which benefited from the many experiences in working with methods of formal concept analysis over many years.

2 Communicative Rationality and Logic

After the exemplary report on a project in which mathematics could effectively support the rational communication between quite different collaborators, the previously presented thesis shall now be more explicitly explained and founded on a general level. It will be elucidated that the rational communication is dependent on *communicative rationality* (in the sense of J. Habermas), which corresponds to a *communicative logic* (in the sense of Ch. S. Peirce) and subsequently to a *communicative mathematics* (in the sense of *generalistic science*). With this connection it can be argued that mathematics can activate, via the close linkage to logic, communicative rationality by which mathematics become able to effectively support the rational communication of human beings.

Rational communication can only succeed if the participating persons are aiming at an intersubjective communication about the considered circumstances. According to J. Habermas, it is necessary for this communication that the participants "overcome their in the first place only subjective views and, owing to the common ground of reasonably motivated convictions, ascertain at the same time the unity of the objective world and the intersubjectivity of their continuity of life" ([Ha81]; Bd.1, p.28). In this connection a part of the commonly shared lifeworld has to be made conscious; "*lifeworld*" is here understood in the sense of Habermas as consisting of the culturally and socially constituted and reproduced stocks of knowledge, patterns of interpretations, as well as forms of action, self-evident conceptions, and convictions of a specific social community (see also [Wi01a]). The purpose-oriented activation of the *lifeworld background* is essential for the reflecting interpretation of situation and world references and the acquisition of intersubjective consents about such interpretations. *Real facts* and their availabilities are not simply given, but they must be interpreted and explained in mutual agreement.

For the *rational communication*, a communicative practice is therefore constitutive; this practice is aimed at the achievement, maintenance, and renewal of consent which rests on the intersubjective acknowledgment of *criticizable claims to validity*. According to Habermas, the criticizable claims to validity are not only *constitutive speech acts*, but also connected with *norm-governed actions* and *expressive attitudes*. Such actions should have the character of appropriate, in their context understandable declarations; in particular (cf. [Ha81]; Bd.1, p.35),

- the constitutive speech acts contain the claim to truth of relationships to facts in the *objective world*,
- the norm-governed actions contain the claim to rightness of relationships to legitimate normative contexts in the common *social world*,

- the expressive attitutes contain the claim to truthfulness of relationships to privileged accessible experiences in the own *subjective world*, respectively.

The reconstruction of the rationality of action, oriented toward reaching understanding, guided J. Habermas to the concept of *"communicative rationality"* which - in imitation of the characterization of rationality in the "Enzyklopdie Philosophie und Wissenschaftstheorie" [Ge93] - can be characterized as follows:

Communicative rationality is the ability
- *to intersubjectively ascertain oneself of the commonly shared life-world,*
- *to reflexively achieve consent about the situation and world references,*
- *to discursively redeem the criticizable claims to validity in refering to truth, rightness, and truthfulness.*

With this the connection of the rational communication with the communicative rationality can be described as follows: *Rational communication* can only be successful if the participants of the communication intersubjectively ascertain themselves of the lifeworld as background of their intents, if they reflexively and consensually communicate about the relevant situation and world references, and if they discursively redeem the claims to validity which are connected with their intents.

For being able to activate *communicative rationality* for rational communication, the participants of communications have to make available lifeworld interpretations, world references, and situation definitions as well as claims to validity and arguments in forms of human thinking. For this a more intense understanding of the structure, the forms, and laws of thinking is promotive and even necessary. According to the "Duden - das große Wörterbuch der deutschen Sprache" ([Du95], p.2145), the task of logic is to work out and to make available such an understanding. A philosopher who worked on this task intensively and convincingly, in particular, is the american scholar Ch. S. Peirce. In his Cambridge Conferences Lectures "Reasoning and the Logic of Things" (1898) he characterized the logic as follows ([Pe92], p.116):

Logic is the science of thought, not merely of thought as a psychical phenomenon but of thought in general, its general laws and kinds."

As subdiscipline of the scientific philosophy, the *logic* in Peirce's understanding belongs to the positive sciences ([Pe98], p.144), in contrast to mathematics which is characterized by Peirce as conditional-hypothetical science. While mathematics develops hypothetical forms of thinking for potential realities, *logic as positive science* examines general forms of thinking related to actual realities. However logic manifoldly adapts mathematical forms of thinking (as, for instance, the numbers) to apprehend actual realities with them.

For Peirce the central *maxime of logic* is the "pragmatism" invented by him ([Pe91], p.337ff.); for his pragmatism it is basic to acknowledge the inseparable

connection between rational knowledge and rational purpose. This acknowleg-
ment is joined with a *sense-critical understanding of truth* by which nothing is
able to be logically true without a purpose according to which it could be named
in this way (cf. [Ap75], p.175). For the beginning of the *search of truth* Peirce
sees "only one state of mind from which one can depart, namely exactly the
state of mind in which one is actually located at the time in which one is de-
parting - a state in which one is loaded with an immense mass of already formed
knowledge of which one cannot get rid of, even if one wants" ([Pe91]; p.434).
These individual resp. collective *self-evidences* and *convictions* are viewed to be
subjectively resp. intersubjectively true as long as some well-grounded doubt is
turned against them. For Peirce only those doubts make research meaningful,
namely such a research which removes the occasions of the doubts and reaches
new convictions viewed to be true ([Pe91], p.157f). Because of the anchoring of
the convictions in the evolutional self-understandings of humans, Peirce warns
against superficial logical reasoning which does not take into account the human
instinct and feeling ([Pe92], p.110).

As foundation for all forms of logical thinking Peirce has developed his theory
of the *categories* of firstness, secondness, and thirdness which are defined as
follows (cf. [Pe92], p.147f): *Firstness* is the mode in which anything would be for
itself, irrespective of anything else. *Secondness* is the mode in which something
is related to something else, irrespective of any third. *Thirdness* is the mode in
which the representation of the relation between something and something else
is considered. For instance, Peirce understands the phenomenon of a sign in the
sense of

1. Firstness as an "icon" where the sign is only thought by itself,
2. Secondness as an "index" where the sign related to what it signs is considered,
3. Thirdness as a "symbol" where a sign is interpreted as a connection between
   sign and that what it signs ([Pe91], p.362ff.).

From his three categories Peirce deduced *three kinds of logical reasonings*: the abduction, the induction, and the deduction. The *abduction* creates out of the hori-
zon of self-evidences a hypothesis as a First. The *induction* confirms a hypothesis by facts as Second. The *deduction* concludes a hypothesis by valid premises
based on logical laws as a Third. This means: "The deduction proves that something *must* be the case; the induction shows that something is *actually* effective;
the abduction simply assumes that something *may be* the case" ([Pe91], p.400).
Each reasoning falls always under the property of *principal criticizability* which
is characteristic for the logical thinking of humans. Peirce has expressed this in
a specific way by his *"First Rule of Logic"*: Logical reasoning tends to correct
itself, not only its conclusions but also its premises ([Pe92], p.165).

To sum up: the close connection of Peirce's logic to Habermas' communicative
rationality can be elucidated by the following determination of a *"communicative
logic"* which reflects to a high degree the comunicative part of the understanding
of logic in Peirce's late-philosophy:

Communicative rationality is the part of logic which works toward the ability
  — to intersubjectively ascertain oneself of the socially and culturally grown self-evidences, convictions, and intensions in logical thinking,
  — to achieve first of all sense-critical reflected consents about basic and situative references of logic to the real world, and
  — to reveal for the communicative-rational argumentation logical forms of thinking, in particular forms of abductive, inductive, and deductive reasoning.

Since the communicative logic, by its close connection to the communicative rationality, is able to obtain an intense *communicative-rational understanding* of the structure, the forms, and the laws of thinking, one can also succeed with the communicative logic to effectively support the rational communication of humans. An even better support may be expected if, more than by Peirce, the conceptual nature of human thinking, as elaborated by the psychologists J. Piaget ([Pi73]) and Th. B. Seiler ([Se01]) is more respected. This would mean in particular that one should activate, as basic forms of *communicative-logical thinking*,

  — the concepts as basic forms of thinking,
  — the judgments as assertional connections of concepts, and
  — the conclusions as logical inferences between judgments.

3   Communicative Mathematics

After the close connection of rational communication, communicative rationality, and logic has been revealed, it has to be explained in addition how the close connection of logical and mathematical thinking can be made explicit. For obtaining the necessary orientation for this task, the notion of *"communicative mathematics"* shall be introduced in correspondence to the notion of communicative logic.

Communicative mathematics is the part of mathematics which works toward the ability
  — to intersubjectively ascertain oneself of the socially and culturally grown self-evidences, convictions, and intensions in logical thinking,
  — to achieve first of all sense-critical reflected consents about basic and situative references of mathematics to the real world, and
  — to closely join mathematical with logical forms of thinking with the aim to reveal mathematical thinking for the communicative-rational argumentation.

This determination of communicative mathematics is closely related to ideas of generalistic science (and therefore also to generalistic mathematics) which have mainly been developed at the Darmstadt University of Technology in the last twenty years (cf. [Wi07b]). *Generalistic science* is understood as the part of a scientific discipline, respectively, which is concerned to examine the self-image of the discipline, its relationship to the world as well as the questions about sense, meaning, and connection of disciplinary activities. The task is, in particular, to generally impart suitably restructured disciplinary sciences. According to H. von Hentig, one can and must perform the necessary restructuring of disciplines by patterns which are taken from the general forms of perception, thought, and action of our civilization "to make them better learnable, available, and more generally criticizable (i.e. also beyond the disciplinary competence) ([He74]; p.33f.). As by generalistic science, communicative mathematics can particularly redeem the increasingly required transdisciplinarity. In this connection a form of resesearch is called *"transdisciplinary"* if it can be used by disciplines so that their way of thinking is rationally understandable, disposable, and activailable beyond their borders, especially for being able to contribute to solve problems which cannot be mastered by purely disciplinary methods [Wi02a]. How experiences with communicative mathematics show, it is just the transdisciplinary competence emerging out of communicative mathematics which mathematics makes effective in rational communication. For making this more understandable, the notion of communicative mathematics shall be explicated more comprehensively:

Communicative mathematis proceeds on the assumption that the self-evidences, convictions, and intentions in mathematical thinking arise and live on by the *communicative practice*. This communicative-rational process decisively contributes to the forming of what mathematicians understood by mathematics. However many mathematicians are scarcely aware of that *lifeworld background* of their understanding of mathematics. This has, above all, for the impart of mathematics the negative consequence that it is often widely overestimated what of an *explanation of mathematical circumstances* can find at all an echo in the thinking of learners. In the more than 200 application projects which have been performed by members of the "Research Group on Formal Concept Analysis" at the TU Darmstadt, we could always again make the experience that the involved non-mathematicians could hardly grasp and activate the used mathematical concepts and results; however, they could, as a rule, thoroughly handle the corresponding logical forms of thinking because of the connections to actual realities, respectively ([Wi01b], p.151ff.). This makes clear how important it is to ascertain oneself - in the sense of communicative mathematics - of the grown self-evidences, convictions, and intensions which stand behind the forms of mathematical thinking; all that allows to achieve the general *impart of mathematical knowledge* and the therefore necessary transformation into logical forms of thinking. Useful for this are systematic studies as, for instance, those studies which Ph. Kitcher carried out in his book "The Nature of Mathematical Knowledge" [Ki84] (in particular from a historical point of view).

Communicative mathematics proceeds from the view that basic and situative references to the world are constitutive and meaningful. To achieve a sense-critical reflected consent in the community of scientists is therefore a central task of communicative mathematics. Frequently it is underestimated how difficult it is to make sense-critically understandable the relationships between abstract mathematics and real facts and circumstances. Therefore *mathematizations* of real connections remain to often ineffective or are uncritically applied, which leads in quite a number of instances to hardly useful, even misleading results. As a negative example it shall only be mentioned here the mathematization of psychical "factors" by coordinate axles of euclidean spaces using "factor analysis" which, according to the reputed statistician L. Guttman, has not produced any established empirical knowledge despite multifarious applications over 70 years of research [Gu77] (cf. also [Wi95], Exkurs 1). The extensive research on *"representational measurement theory"* has been developed completely in the sense of communicative mathematics; this research elaborates as its basic results meaningful mathematical structures of measurement for appropriate interpretations of empirical data [K+71]. More generally as measurement theory, *"formal concept analysis"* proceeds by mathematically representing concept structures inherent in data (based on a contextual mathematization of concept and concept hierarchy); furthermore, those concept structures can be successfully visualized by diagrams [GW99]. By this approach the logical thinking of the persons concerned are activated to achieve consent about the material content of the data and to reach with further (mathematical and logical) means purposeful interpretations of the data. Formal concept analysis has been developed in connection with the efforts for a generalistic science which, in particular, has led to a meaningful piece of communicative mathematics.

Communicative mathematics proceeds from the view that mathematical thinking can effectively support the communicative-rational argumentation via the corresponding logical thinking. Therefore a further task of communicative mathematics lies in joining mathematical and logical forms of thinking with each other so that the mathematics can be made usable for the communicative-rational argumentation via suitable logical forms of thinking. For systematically reaching a close connection between mathematical and logical thinking, a natural approach is to mathematize the basic forms of logical thinking: *concepts*, *judgments*, and *conclusions*, and - based on that - to mathematically abstract further logical forms of thinking so that many mathematical theories can be activated as well as possible. For this the first step is done by the already well developed elaboration of a *"contextual logic"* (see [Wi97], [Pr98], [Wi00], [Pr00], [Wi02d], [Da03], [Wi04], [DK05], [Kl05], [Wi07c]) for which

- the mathematical doctrine of concepts is borrowed from formal concept analysis,
- the mathematical doctrine of judgments (building up on the doctrine of formal concepts) is developed by mathematizing J. Sowa's theory of conceptual graphs (see [So84]),
- the mathematical doctrine of conclusions is taken up from Peirce's reasoning with existential graphs as well as the usual reasoning of mathematical logic.

Even if further steps have already been made to substantially include further mathematical disciplines as *algebra*, *geometry*, and *topology*, much remains to be done for the development of communicative mathematics. In any case, all those connections must descriptively impart the mathematical forms of thinking in such a manner that not only deductive, but also inductive and abductive reasoning as well as actions of thinking in the sense of Peirce's "First Rule of Logic" are supported.

On the example in the first section it shall be briefly explained how communicative mathematics may have an effect in the three described aspects. The *self-image* of the contemporary mathematics that it needs well-defined set-theoretical basic forms for the mathematical comprehension of conceptions of thinking has led the mathematization of concept to introduce the mathematical notion of "formal context" as basic set-theoretical structure and in connection with it the mathematical notion of "formal concept" and "subconcept–superconcept–relation". The *conviction* that structures formed by formal concepts are rewarding for the examination of mathematical concepts recommends to move the mathematical structure of the concept lattice of a formal context in the center of the consideration; for this it had however first to be proved that indeed the set of all formal concepts of a formal context forms a (complete) lattice with respect to the subconcept–superconcept–relation. The *intention* of the contempory research to make understandable the meaningful mathematical structures to a large extent as much as possible has, among other developents, produced the theory of "additive" line diagrams of concept lattices (see [GW99], p.75) which has also guided the design of the lattice diagrams shown in the first section of this article.

For the development of formal concept analysis as applied lattice theory, the applications in data analysis and conceptual knowledge processing have generally been constitutive and meaningful; how fruitful this purpose-orientation of mathematics was and is, that can be particularly seen by the performed TOSCANA-projects as, for instance, the project discussed in section 1. In those application projects of formal concept analysis, mathematical thinking closely joins with logical thinking which supports the communicative-rational argumentation and, in particular, an abductive thinking.

To sum up it can be argued: Mathematics as product of human thinking and communicating finds its sense in the fact that mathematics is meaningful for humans. Because of its rational abstract nature, mathematics realizes its meaningfulness by the effective support of rational communication of humans. This support is made possible by the close connection of mathematical and logical thinking, namely a logical thinking which activates the communicative rationality in human thinking and acting. Mathematics can therefore gain sense and meaning if mathematicians and users of mathematics make an effort to obtain a deeper understanding of communicative logic and communicative mathematics.

References

[Ap75]   Apel, K.-O.: Der Denkweg von Charles Sanders Peirce. Suhrkamp-Taschenbuch Wissenschaft 141, Frankfurt (1975)

[BH05]   Becker, P., Hereth Correia, J.: The ToscanaJ Suite for implementing conceptual information systems. In: Ganter, B., Stumme, G., Wille, R. (eds.) Formal Concept Analysis. LNCS (LNAI), vol. 3626, pp. 324–348. Springer, Heidelberg (2005)

[Da03]   Dau, F. (ed.): The Logic System of Concept Graphs with Negation. LNCS (LNAI), vol. 2892. Springer, Heidelberg (2003)

[DK05]   Dau, F., Klinger, J.: From Formal Concept Analysis to Contextual Logic. In: Ganter, B., Stumme, G., Wille, R. (eds.) Formal Concept Analysis. LNCS (LNAI), vol. 3626, pp. 81–100. Springer, Heidelberg (2005)

[Du95]   Duden - Das große Wörterbuch der deutschen Sprache in 8 Bänden. 2. Aufl. Dudenverlag, Mannheim, 1993–1995, S.2145

[GW99]   Ganter, B., Wille, R.: Formale concept analysis: mathematical foundation. Springer, Heidelberg (1999)

[Ge93]   Gethmann, C.F.: Rationalität. In: Mittelstraß, J. (Hrsg.) Enzyklopädie Philosophie und Wissenschaftstheorie. Bd.3. Metzler, Stuttgart, pp. 468–481 (1993)

[Gu77]   Guttman, L.: What is not what in statistics. The Statistician 26, 81–107 (1977)

[Ha81]   Habermas, J.: Theorie des kommunikativen Handelns. 2 Bände. Suhrkamp, Frankfurt (1981)

[He74]   von Hentig, H.: Magier oder Magister? Über die Einheit der Wissenschaft im Verständigungsprozeß. Suhrkamp-Taschenbuch 207, Frankfurt (1974)

[Ka88]   Kant, I.: Logic. Dover Edition, Mineola (1988)

[Ki84]   Kitcher, P.: The nature of mathematical knowledge. Oxford University Press, Oxford (1984)

[Kl05]   Klinger, J.: The logic system of protoconcept graphs. Dissertation, TU Darmstadt 2005. Shaker Verlag, Aachen (2005)

[KSVW94] Kollewe, W., et al.: TOSCANA - ein Werkzeug zur begrifflichen Analyse und Erkundung von Daten. In: Wille, R., Zickwolff, M. (Hrsg.) Begriffliche Wissensverarbeitung - Grundfragen und Aufgaben. B.I.-Wissenschaftsverlag, Mannheim, pp. 267–288 (1994)

[K+71]   Krantz, D., et al.: Foundations of measurement, vol. 1, 2, 3. Academic Press, San Diego (1971, 1989, 1990)

[Pe91]   Ch., S.: Peirce: Schriften zum Pragmatismus und Pragmatizismus. Herausgegeben von K.-O. Apel. Suhrkamp-Taschenbuch Wissenschaft 945, Frankfurt (1991)

[Pe92]   Peirce, C.S.: Reasoning and the logic of things. In: Ketner, K.L. (ed.), with an introduction by K.L. Ketner and H. Putnam, Havard University Press, Cambridge (1992)

[Pe98]   Peirce, C.S.: The Essential Peirce. vol.2 (1893–1913). Edited by the Peirce Edition Project. Indiana University Press, Bloomington (1998)

[Pi73]   Piaget, J.: Einführung in die genetische Erkenntnistheorie. Suhrkamp-Taschenbuch Wissenschaft 6, Frankfurt (1973)

[RW00]   Rock, T., Wille, R.: Ein TOSCANA-Erkundungssystem zur Literatursuche. In: Stumme, G., Wille, R. (eds.) Begriffliche Wissensverarbeitung: Methoden und Anwendungen, pp. 239–253. Springer, Heidelberg (2000)

[Pr98]   Prediger, S.: Kontextuelle Urteilslogik mit Begriffsgraphen. Ein Beitrag zur Restrukturierung der mathematischen Logik. Dissertation, TU Darmstadt 1998. Shaker Verlag, Aachen (1998)

[Pr00]   Prediger, S.: Mathematische Logik in der Wissensverarbeitung: Historisch-philosophische Gründe für eine Kontextuelle Logik. Mathematische Semesterberichte 47, 165–191 (2000)

[Se01]   Seiler, T.B.: Begreifen und Verstehen. Ein Buch über Begriffe und Bedeutungen. Verlag Allgemeine Wissenschaft, Mühltal (2001)

[So84]   Sowa, J.F.: Conceptual structures: information processing in mind and machine. Addison-Wesley, Reading (1984)

[Wi95]   Wille, R.: Begriffsdenken: Von der griechischen Philosophie bis zur Künstlichen Intelligenz heute. Diltheykastanie, Ludwig-Georgs-Gymnasium, Darmstadt, 77–109 (1995)

[Wi97]   Wille, R.: Conceptual graphs and formal concept analysis. In: Delugach, H.S., et al. (eds.) ICCS 1997. LNCS, vol. 1257, pp. 290–303. Springer, Heidelberg (1997)

[Wi00]   Wille, R.: Contextual logic summary. In: Stumme, G. (ed.) Working with conceptual structures: Contributions to ICCS 2000, pp. 265–276. Shaker-Verlag, Aachen (2000)

[Wi01a]   Wille, R.: Lebenswelt und Mathematik. In: Hauskeller, C., Liebert, W., Ludwig, H. (Hrsg.) Wissenschaft verantworten: soziale und ethische Orientierung in der Technischen Zivilisation, pp. 51–68. Agenda-Verlag, Münster (2001)

[Wi01b]   Wille, R.: Mensch und Mathematik: Logisches und mathematisches Denken. In: Lengnink, K., Prediger, S., Siebel, F. (Hrsg.) Mathematik und Mensch: Sichtweisen der Allgemeinen Mathematik, pp. 139–158. Verlag Allgemeine Wissenschaft, Mühltal (2001)

[Wi02a]   Wille, R.: Transdisziplinarität und Allgemeine Wissenschaft. In: Krebs, H., et al. (Hrsg.) Perspektiven Interdisziplinärer Technikforschung: Konzepte, Analysen, Erfahrungen, pp. 73–84. Agenda-Verlag, Münster (2002)

[Wi02b]   Wille, R.: Kommunikative Rationalität und Mathematik. In: Prediger, S., Siebel, F., Lengnink, K. (Hrsg.) Mathematik und Kommunikation, pp. 181–195. Verlag Allgemeine Wissenschaft, Mühltal (2002)

[Wi02c]   Wille, R.: Kommunikative Rationalität, Logik und Mathematik. Math. Semesterberichte 49, 167–183 (2002)

[Wi02d]   Wille, R.: Existential concept graphs of power context families. In: Priss, U., Corbett, D.R., Angelova, G. (eds.) ICCS 2002. LNCS (LNAI), vol. 2393, pp. 382–395. Springer, Heidelberg (2002)

[Wi04]   Wille, R.: Implicational concept graphs. In: Wolff, K.E., Pfeiffer, H.D., Delugach, H.S. (eds.) ICCS 2004. LNCS (LNAI), vol. 3127, pp. 52–61. Springer, Heidelberg (2004)

[Wi07a]   Wille, R.: Formal concept analysis as applied lattice theory. In: Bělohlávek, R., Ben Yahia, S., Mephu Nguifo, E. (eds.) Concept lattices and their applications. CLA 2006, Springer, Heidelberg (to appear, 2007)

[Wi07b]   Wille, R.: Generalistic mathematics as mathematics for the general public. Contributions to general algebra. Heyn Verlag, Klagenfurt (to appear, 2007)

[Wi07c]   Wille, R.: Formal concept analysis and contextual logic. In: Hitzler, P., Schärfe, H. (eds.) Conceptual structures in practice (to appear)

Actionability and Formal Concepts: A Data Mining Perspective

Jean-François Boulicaut[1] and Jérémy Besson[1,2]

[1] INSA-Lyon, LIRIS CNRS UMR5205, F-69621 Villeurbanne cedex, France
[2] UMR INRA/INSERM 1235, F-69372 Lyon cedex 08, France
{Firstname.Name}@insa-lyon.fr

Abstract. The last few years, we have studied different set pattern mining techniques from binary data. It includes the computation of formal concepts to support various knowledge discovery processes. For instance, when considering post-genomics, we can exploit Boolean data sets that encode a relation between some genes and the proteins that may regulate them. In such a context, it appears interesting to exploit the analogy between a putative transcriptional module (i.e., a typically important hypothesis for gene regulation understanding) and a formal concept that holds within such data. In this paper, we assume that knowledge nuggets can be captured by collections of formal concepts and we discuss the challenging issue of mining/selecting actionable patterns from these collections, i.e., looking for relevant patterns that really support knowledge discovery. Therefore, a major issue concerns the computation of complete collections of formal concepts that satisfy user-defined constraints. This is useful not only to avoid the computation of too small patterns that might be due to noise (e.g., using size constraints on both their intents and extents) but also to introduce some fault-tolerance. We discuss the pros and the cons of some recent proposals in that direction.

1 Introduction

Many application domains can provide possibly huge boolean matrices whose rows denote objects and columns denote attributes (see Table 1 for toy examples). Mining such binary data, or formal contexts in the terminology of Formal Concept Analysis (FCA) [1], has been studied extensively. Indeed, popular data mining techniques have been designed for set pattern extraction (e.g., mining frequent itemsets or association rules, mining frequent closed itemsets or other condensed representations of frequent patterns [2,3]). We are interested in bi-set mining, i.e., the computation of local patterns that are sets of objects and sets of attributes being somehow "associated". Clearly, a formal concept is an interesting type of bi-set that satisfy a local constraint: its attribute set (or intent) is the maximal set of attributes that are true for each object of its associated supporting set of objects (or extent). Here, locality refers to the fact that checking whether a bi-set is a formal concept or not can be performed independently of the other patterns holding in the data. An example of a formal concept in

R. Medina and S. Obiedkov (Eds.): ICFCA 2008, LNAI 4933, pp. 14–31, 2008.

$\mathbf{r}_1$ from Table 1 is $(\{o_1,\ o_2,\ o_3,\ o_4\}, \{p_1,\ p_2\})$. Notice that this paper does not consider FCA as such and that, for instance, we are not really interested in the underlying concept lattice itself. Instead, we consider collections of formal concepts as collections of patterns. Also, we do not use formal concepts as condensed representations for collections of association rules (see [4] for a recent survey covering such issues).

Table 1. $\mathbf{r}_1$ (left) - $\mathbf{r}_2$ (right)

	p_1	p_2	p_3	p_4
o_1	1	1	0	0
o_2	1	1	0	0
o_3	1	1	0	0
o_4	1	1	1	1
o_5	0	0	1	1
o_6	0	0	1	1

	p_1	p_2	p_3	p_4
o_1	1	1	0	0
o_2	1	0	1	0
o_3	1	1	0	1
o_4	1	1	1	1
o_5	0	0	1	0
o_6	0	0	1	1

Let us introduce a couple of motivating applications for our perspective on formal concept mining (see, e.g., [5]). The objects can denote biological samples and the attribute can denote boolean gene expression properties, e.g., the fact that a given gene is over-expressed in a given sample. In such a case, the boolean properties have to be derived from the continuous values measured by, e.g., the microarray technology, and a formal concept provides an hypothesis on a maximal group of genes that have the same expression property in a given group of biological samples. A second example would be to consider that some transcription factors (i.e., the proteins which regulate gene expression) are the studied objects for which we record whether they can bind or not on the promoter sequence of some studied genes. Here again, a formal concept can be interpreted as an hypothesis on a maximal set of genes whose co-expression might be explained by its associated set of transcription factors. Clearly, one of the motivations for collecting gene expression data is indeed to be able to discover such hypothesis that correspond, from a biological perspective, to putative synexpression groups, transcription modules, regulation pathways, etc.

In this paper, we are interested in the various application domains for which, given a binary data set, one can consider that its formal concepts are a priori interesting statements about the data. In theory, computing formal concepts is exponential in the smallest dimension of the data matrix (i.e., the number of objects or the number of attributes). An important question concerns the tractability of their computation for practical applications. Given the major effort of the last decade, it turns out that computing collections of formal concepts that hold in large binary matrices can be feasible. Researchers have designed algorithms that compute complete collections of formal concepts [6]. Since these patterns are built on closed sets, the extensive research on (frequent) closed set extraction has inspired constraint-based mining of formal concepts (see, e.g., [7]): every formal concept which furthermore satisfies a size constraint on one of its components (e.g., a minimal size for its intent or its extent) can be extracted

efficiently. This is however not really satisfactory when considering that our ultimate goal is to mine actionable patterns, i.e., relevant formal concepts that can
indeed be interpreted by human experts to catalyze knowledge discovery. Real
data sets can hold hundreds of thousands of formal concepts: it is clear that looking for actionable ones among the many spurious or irrelevant ones is extremely
hard or even impossible. In fact, it is interesting to look at one fundamental
limitation of Knowledge Discovery processes based on formal concepts. Within
such local patterns, the strength of the association of the two set components
is often too strong in real-life data. Indeed, errors of measurement and boolean
encoding techniques can lead to "erroneous" zero or one values. Unexpected zero
values give rise to a combinatorial explosion of the number of formal concepts
because interesting patterns are split into less relevant ones. For example, let us
consider the data from Table 1. Assume that r_1 is a reliable representation of a
phenomenon but that data collection and preprocessing lead to r_2 instead (i.e.,
some noise has been introduced), the number of formal concepts in r_2 is approximately twice larger than in r_1. While this concerns zero values that may be one
values, we can also consider what happens in the reverse situation: the intuition
is that when some zero values have been encoded as one values by error, many
"small" formal concepts may hold. Therefore, we need to avoid computing too
small patterns but also we have to somehow relax that no exception (zero value)
can be accepted, i.e., what we call fault-tolerance. For instance, a bi-set like
$(\{o_1, o_2, o_3, o_4\}, \{p_1, p_2\})$ is not a formal concept in r_2 but it may be considered
as a relevant pattern: its objects and attributes are strongly associated (only one
zero value) and, furthermore, its "outside" objects and attributes contain more
than one zero value.

Our contribution here is to consider how some data mining researchers have
designed more or less pragmatic methods to address these problems. We avoid to
produce the technical details that are available from the referenced papers. We
will discuss the DMINER proposal for constraint-based mining of formal concepts
in the challenging case where, for instance, we want to "push" size constraints
on both dimensions [8,5]. We will also consider different approaches for designing
fault-tolerant patterns based on formal concepts [9,10,11]. The survey paper [12]
is a discussion on the needed trade-off between extraction feasibility, completeness, relevancy, and ease of interpretation of such fault-tolerant pattern types.
Notice also that [12] contains empirical results on both synthetic and real data.

Section 2 discusses the DMINER solution for constraint-based mining of formal
concepts. In Section 3, we consider the obstacles and present some available
solutions for actionable pattern discovery based on formal concepts. Section 4
briefly concludes on some current open issues in that area.

2   Formal Concept Mining

Let $\mathcal{O}$ denotes a set of objects and $\mathcal{P}$ denotes a set attributes (or properties). In
Table 1, $\mathcal{O} = \{o_1, \cdots, o_6\}$ and $\mathcal{P} = \{p_1, \cdots, p_4\}$. A data set is the materialization
of a binary relation $r \subseteq \mathcal{O} \times \mathcal{P}$. We write $(o_i, p_j) \in r$ to denote that property

j holds for object i. Boolean matrices like $\mathbf{r}_1$ and $\mathbf{r}_2$ in Table 1 are classical representations for such relations.

Formal concepts can be considered as bi-sets, i.e., couples of sets (X, Y) from $2^{\mathcal{O}} \times 2^{\mathcal{P}}$ that satisfy a constraint denoted $\mathcal{C}_{FC}(X, Y)$. The definition of such a constraint might be expressed in terms of Galois operators and closures. In this paper, let us specify it in terms of the conjunction of a density constraint (first conjunct) and a relevancy constraint (second and third conjuncts), following the presentation from [11].

Definition 1 (Formal Concept). *A bi-set $(X, Y) \in 2^{\mathcal{O}} \times 2^{\mathcal{P}}$ is a formal concept in $\mathbf{r}$ if it satisfies the constraint $\mathcal{C}_{FC}(X, Y) \equiv (\forall x \in X, \forall y \in Y(x, y) \in \mathbf{r}) \wedge (\forall x \in \mathcal{O} \setminus X, \exists y \in Y \ s.t. \ (x, y) \notin \mathbf{r}) \wedge (\forall y \in \mathcal{P} \setminus Y, \exists x \in X \ s.t. \ (x, y) \notin \mathbf{r}).*

Informally, it means that for a formal concept (X, Y), if we perform permutations of rows and columns such that all the elements from X (resp. Y) are contiguous, we observe a maximal rectangle of true values (no zero value inside, at least one zero value outside). Another useful analogy to understand the semantics of formal concepts is to consider them as bi-cliques in the bi-partite graphs represented by the boolean matrix. Computing every formal concept that holds within a boolean matrix is NP-hard. As soon as none of the dimensions is small, this extraction task is not feasible. Furthermore, when the computation is tractable, we often get a huge amount of formal concepts (e.g., millions) even in rather small data sets. As we mentioned in our introduction, this is definitely not acceptable for actionable pattern discovery. Constraint-based mining is a partial but impressive solution to both problems, i.e., computational complexity and relevancy. The idea is that the analyst can often exploit some background knowledge to specify declarative constraints that may hold for the extracted formal concepts of interest. It happens that some of these user-defined constraints can be exploited (say "pushed deeply") by the mining algorithm to prune efficiently the search space. For example, we may need patterns with a minimal number of elements on both dimensions (a counterpart of the popular minimal frequency constraint for itemset mining [13]) and/or patterns covering at least a given number of elements of $\mathbf{r}$ (intuitively, a minimum area constraint for the associated "rectangle"). As a result, when considering knowledge discovery processes based on formal concepts, we are generally computing collections of bi-sets that satisfy not only $\mathcal{C}_{FC}$ but also a user-defined constraint $\mathcal{C}_{UD}$:

$$\{(X, Y) \in 2^{\mathcal{O}} \times 2^{\mathcal{P}} \mid \mathcal{C}_{FC}(X, Y) \wedge \mathcal{C}_{UD}(X, Y)\}$$

Figure 1 provides examples of well-known user-defined constraints where α and β denote some thresholds, $a \in \mathcal{O}$, $b \in \mathcal{P}$, $E \subseteq \mathcal{O}$, and $E' \subseteq \mathcal{P}$ are constraint parameters. For instance, $\mathcal{C}^1_{size} \wedge \mathcal{C}^2_{size}$ or $\mathcal{C}_{area}$ are two different constraints to specify that patterns have to be "large enough". Also, $\mathcal{C}_{mean}$ is just one example of a constraint which enforces that the average of an external positive value associated to each element of the extent is greater than a given threshold.

Not every constraint can be processed efficiently. We have a special interest for monotonic and anti-monotonic constraints (see Definition 2) that have nice

$\mathcal{C}_{UD}(X,Y)$		
$\mathcal{C}^1_{size}$	$\equiv$	$|X| > \alpha$
$\mathcal{C}^2_{size}$	$\equiv$	$|Y| > \alpha$
$\mathcal{C}_{area}$	$\equiv$	$|X| \times |Y| > \alpha$
$\mathcal{C}_{mean}$	$\equiv$	$\sum_{i \in X} Val^+(i)/|X| > \alpha$
$\mathcal{C}_{member}$	$\equiv$	$a \in X \wedge b \in Y$
$\mathcal{C}_{inter}$	$\equiv$	$|X \cap E| > \alpha \wedge |Y \cap E'| < \beta$

Fig. 1. Examples of interesting constraints on bi-sets

properties when the search space is organized as a lattice structure thanks to a specialization relation.

Definition 2 (Monotonic constraints). *A constraint $\mathcal{C}$ is anti-monotonic w.r.t. the specialization order $\preceq$ on E iff $\forall a, b \in E$ s.t. $a \preceq b$ then $\neg \mathcal{C}(a) \Rightarrow \neg \mathcal{C}(b)$. $\mathcal{C}$ is monotonic w.r.t. $\preceq$ iff $\forall a, b \in E$ s.t. $a \preceq b$ then $\neg \mathcal{C}(b) \Rightarrow \neg \mathcal{C}(a)$.*

Following a path in an enumeration tree for candidate patterns, an anti-monotonic constraint is satisfied for all the patterns before a specific pattern and not satisfied afterwards. A popular example is the anti-monotonicity of a minimal frequency constraint which specifies that the size of the extent has to be greater than a given threshold (i.e., $\mathcal{C}^1_{size}$). This constraint can efficiently reduce the search-space and remove spurious patterns whose extent is too small. However, in the many applications where the data set is large on both dimensions and when the density in terms of true values is high, the only way to achieve tractability seems to be an increase of the minimal size threshold for the extent. Doing so, we clearly loose a priori interesting formal concepts (the larger the extent, the smaller the intent, and vice versa). Therefore, we may want to use other constraints like, for example, minimal size constraint on both the intents and the extents, i.e., conjunctions $\mathcal{C}^1_{size} \wedge \mathcal{C}^2_{size}$. Unfortunately, using the standard enumeration on formal concepts (enumeration of the intent and computation of the extend), most algorithms can only exploit anti-monotonic constraints on the intent and monotonic constraints on the extent, i.e., a conjunction like $|X| > \alpha \wedge |Y| < \beta$ (say $\mathcal{C}^1_{size} \wedge \neg \mathcal{C}^2_{size}$). Furthermore, it is not that simple to exploit constraints like $\mathcal{C}_{area}$, $\mathcal{C}_{mean}$, $\mathcal{C}_{member}$, and $\mathcal{C}_{inter}$. Even though they can be used to capture important expectation from the analysts, these constraints are neither anti-monotonic nor monotonic.

The DMINER algorithm is a depth-first search algorithm inspired by both Ganter's algorithm [14] and DUALMINER [15]. The principle of Ganter's algorithm enables to identify from an extracted formal concept the smallest formal concept that may follow it. Doing so, we can avoid the generation of many sets that are not closed, i.e., which can not correspond to formal concepts. On the other hand, for efficiency purposes, we have to follow the order related to the standard set inclusion. Given the data from Table 2, it is far more efficient to generate the formal concept $(\{o_1, o_2, o_3\}, \{p_1, p_2, p_3, p_4\})$ from $(\{o_1, o_2, o_3, o_5\}, \{p_1, p_3\})$ than from $(\{o_2, o_3, o_4\}, \{p_9, p_{10}\})$. The pattern $(\{o_1, o_2, o_3, o_5\}, \{p_1, p_3\})$ already contains a lot of information that can be used to

Table 2. A boolean context r_3

	Attributes									
	p_1	p_2	p_3	p_4	p_5	p_6	p_7	p_8	p_9	p_{10}
o_1	1	1	1	1	0	1	1	0	0	0
o_2	1	1	1	1	0	0	0	0	1	1
o_3	1	1	1	1	0	0	0	0	1	1
o_4	0	0	0	0	1	1	1	1	1	1
o_5	1	0	1	0	1	1	1	1	0	0

generate $(\{o_1, o_2, o_3\}, \{p_1, p_2, p_3, p_4\})$, e.g., we know that $(\{o_1, o_2, o_3\}, \{p_1, p_3\})$ holds in r_3 because it is "included" in $(\{o_1, o_2, o_3, o_5\}, \{p_1, p_3\})$ which is a formal concept.

It means that we do not have to scan all the data corresponding to the new patterns to be extracted. In our running example, we only need to scan the data corresponding to $(\{o_1, o_2, o_3\}, \{p_2, p_4\})$. It is even more crucial when large data sets are considered. Thus, we adopt a binary enumeration of the smallest set ($\mathcal{O}$ or $\mathcal{P}$) and the other set is computed by means of the Galois connection. This enumeration combines both Ganter's principle and a prefix-based enumeration.

It is important to exploit constraints not only to increase the relevancy of the computed patterns but also to increase computational efficiency. Most of the available algorithms can push monotonic and/or anti-monotonic constraints according to set inclusion on one dimension. We however argued in the previous section that this is not enough. Unlike most of the formal concept mining algorithms, DMINER does not consider that each candidate is only represented by means of two sets, i.e., the intent and the extent. The enumerated set, let us say the intent, is split into two sets, the first one representing the set of elements that belong to any formal concept extracted from the current candidate and the second one containing the elements that still have to be enumerated (see the inspiring principle in [15]). The two sets are the bottom and the top of a lattice which represents the current search space. For example, we may have a candidate $(o_1o_2o_3o_5, (p_1, p_1p_2p_3))$ where the intent is represented by two sets $\{p_1\}$ and $\{p_1p_2p_3\}$ instead of only one set $\{p_1p_2p_3\}$. By this way during the enumeration we always know precisely which search-space is related to the current candidate and thus increase the number of constraints the algorithm can handle. In our example the candidate is supported by the extent $\{o_1, o_2, o_3, o_5\}$ and it represents all the attribute sets Y (intents) such that $\{p1\} \subseteq_L Y \subseteq_L \{p_1, p_2, p_3\}$ where $(X, Y) \subseteq_L (X', Y') \Rightarrow X \subset X' \wedge Y \subset Y'$, i.e., the attribut sets $\{p_1\}$, $\{p_1, p_2\}$, $\{p_1, p_3\}$ and $\{p_1, p_2, p_3\}$ in our example. Notice that a candidate of the form $(O, (X, X))$ denotes the bi-set (O, X).

This pattern representation enables to push a larger class of constraints than only the anti-monotonic constraints. Indeed, each candidate denotes a search space in the form of an attribute lattice with its associated object set. For example, let us consider candidate $C = (o_1o_2o_3o_5, (p_1, p_1p_2p_3p_4))$ in r_3, each formal concept derived from C which contains p_1 contains at most the attributes from $\{p_1, p_2, p_3, p_4\}$ and its associated object set is included in $\{o_1, o_2, o_3, o_5\}$. It enables to push difficult constraints like $\mathcal{C}_{area}$ and $\mathcal{C}_{mean}$ which are neither monotonic nor anti-monotonic. Indeed, since we have an attribute lattice, we can compute bounds for some constraints. For instance, we can see that the area of C is between $|\{o_1, o_2, o_3, o_5\}| \times |\{p_1\}| = 4$ and $|\{o_1, o_2, o_3, o_5\}| \times |\{p_1, p_2, p_3, p_4\}| = 16$. If we are looking for formal concepts with an area of size at least 17 or of size at most 3, then pattern C can be pruned safely, i.e., it can not lead to acceptable formal concepts.

We adopted simple arrays as data structure to store the sets of objects and attributes. A time complexity analysis shows that it is as efficient as the other more complex data structures used in depth-first search algorithms. Finally, to check whether a set X is closed, DMINER does not have to scan all the sets from $\mathcal{P} \setminus X$. Indeed, only attributes which have been removed from the search space by enumeration have to be checked. Instead of going into much details, let us provide two DMINER executions (see Figure 2 and Figure 3) on the data sets given in Table 3. In Figure 2, the algorithm starts with the candidate $(o_1 o_2 o_3, (\emptyset, p_1 p_2))$ representing the set of all possible patterns of $\mathbf{r}_4$. Then the attribute p_1 is selected to proceed the enumeration. Two new candidates $(o_1 o_2 o_3, (\emptyset, p_2))$ and $(o_1 o_2 o_3, (p_1, p_1 p_2))$ are generated. After enumerating the attribute p_2, four formal concepts are extracted $(o_1 o_2 o_3, \emptyset)$, $(o_1 o_2, p_2)$, $(o_2 o_3, p_1)$ and $(o_2, p_1 p_2)$. Figure 3 provides an other example of DMINER execution.

To investigate the efficiency of DMINER, we studied its complexity using the time delay, i.e., the complexity to go from one solution to the next one [16]. DMINER time delay is in the worst case equal to $O(n^2 m)$ where n is the size of the enumerated set and m is the size of the other one. This complexity is the same as

Table 3. Extraction contexts $\mathbf{r}_4$ (left) and $\mathbf{r}_5$ (right)

	p_1	p_2
o_1	0	1
o_2	1	1
o_3	1	0

	p_1	p_2	p_3
o_1	0	0	1
o_2	0	1	0
o_3	1	0	1

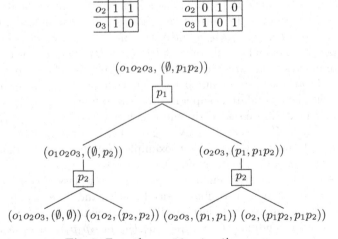

Fig. 2. Formal concept extraction on $\mathbf{r}_4$

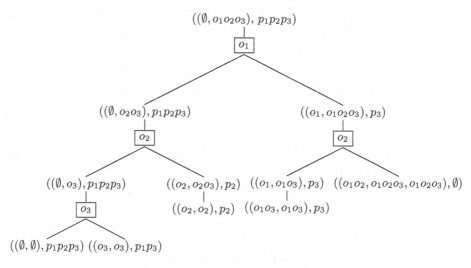

Fig. 3. Formal concept extraction on r_5

for Ganter's algorithm [14]. The time delay in average is $O(n-\log_2(K)+1)O(mn)$ where K is the number of formal concepts in the data set. It is between $O(nm)$ and $O(n^2m)$ according to K. This is an interesting result when considering the use of DMINER on data sets in which many formal concepts hold.

To refer to one concrete example, let us recall the application described in [5]. It concerns the analysis of a data set that records (a) the existence of putative binding sites of 94 transcription factors on the promoter sequences of 304 genes (selection of human genes), and (b) the over-expression property of these same genes in 10 biological simples (individuals). In other terms, the boolean context implies 104 objects (94 transcription factors and 10 biological situations, more precisely 5 for healthy individuals and 5 for diabetic patients) and 304 genes. Formal concept discovery from such a boolean context is already hard. Notice also that the obtained boolean context was rather dense in terms of true values (17% of the cells containing a true value). In such a situation, even efficient algorithms for mining frequent closed sets can turn to be intractable. This data set is particular: there are very few frequent formal concepts with a relative frequency threshold above 0.1 on genes (5 534 patterns) and then the number of formal concepts increases very fast. Without any constraint, we get more than five million formal concepts within a few minutes. In this context, extracting actionable formal concepts needs for a very low frequency threshold, otherwise almost no formal concept can be computed. Notice that actionable patterns corresponding to interesting biological hypothesis have been found by means of formal concepts holding in this data set [16,5].

Figure 4 shows the running time of formal concepts extractions in the biological data set with varying the minimal frequency threshold. The competitors in the experiment are three different algorithms used for computing frequent closed sets, namely ac-miner [17], closet [18] and CHARM [19].

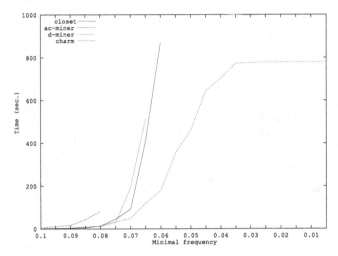

Fig. 4. An application [5]

3   Looking for Actionable Patterns

Let us now come back on our main goal that is actionable pattern discovery based on collections of formal concepts. The challenge is to mine relevant formal concepts that can indeed be interpreted by human experts to catalyze knowledge discovery. In case this is not possible, we may also look for application-independent and/or application-dependent post-processing techniques that can be applied on formal concepts to support the discovery of actionable patterns.

- When, among other things, a data set captures correctly a phenomenon of interest, collections of formal concepts can be huge and they may contain large collections of spurious patterns (false positives) and/or irrelevant patterns w.r.t. domain knowledge. This is somehow inherent to (unsupervized) local pattern discovery techniques. The two main directions of research to improve this situation concern (a) the use of randomization techniques for statistical validity assessment for the extracted patterns (see, e.g., [20]) and (b) the use of user-defined constraints to specify subjective interestingness issues based on domain knowledge (e.g., using $\mathcal{C}_{member}$ or $\mathcal{C}_{inter}$ primitive constraints).
- When considering the problem of erroneous true values (the property should not be satisfied but we record that it is satisfied), assuming that this is fundamentally rare, many "small" formal concepts may hold. Therefore, we need to avoid computing these too small patterns by using minimal size or minimal area constraints (See Section 2). In practice, using these constraints $\mathcal{C}^1_{size}$, $\mathcal{C}^2_{size}$ and/or $\mathcal{C}_{area}$ can be extremely efficient.
- When considering the problem of erroneous false values (the property is satisfied but we record that it is not satisfied), assuming again that this can not be too frequent, the number of extracted formal concepts will increase

fast w.r.t. what it should be if the data were correct. This is a clear need for fault-tolerance and we discuss several proposals hereafter (see Section 3.1).

- The number of patterns, even if only relevant ones have been collected, is indeed a problem. Knowledge discovery needs for human expert assessment of patterns and browsing or inspecting thousands of patterns is definitively not possible. The solution can come from (a) the design of pattern databases and advanced querying tools, and (b) summarization techniques based on, for instance, clustering methods. Considering problems and solutions for pattern database management is related to the emergent topic of inductive databases and will not be discussed further in this paper. Considering summarization techniques has been successfully applied (see Section 3.2).

3.1 Introducing Fault-Tolerance

We first revisit the definition of formal concepts for a fairly natural introduction of fault-tolerance. In the following, we say that a bi-set (X, Y) is included in a bi-set (X', Y') denoted $(X, Y) \subseteq (X', Y')$ iff $(X \subseteq X') \wedge (Y \subseteq Y')$.

Definition 3. *Assume $\mathcal{Z}_l(x, Y)$ denotes the number of false values of an object x on the attributes in Y, i.e., $|\{y \in Y | (x, y) \notin \mathbf{r}\}|$. Similarly, let $\mathcal{Z}_c(y, X) = |\{x \in X | (x, y) \notin \mathbf{r}\}|$ be the number of false values of an attribute y on the objects in X.*

Definition 4 (FC). *A bi-set $(X, Y) \in 2^{\mathcal{O}} \times 2^{\mathcal{P}}$ is a formal concept in $\mathbf{r}$ iff*
(2.1) $\forall x \in X, \mathcal{Z}_l(x, Y) = 0 \quad \wedge \quad \forall y \in Y, \mathcal{Z}_c(y, X) = 0$
(2.2) $\forall x \in \mathcal{O} \setminus X, \mathcal{Z}_l(x, Y) \geq 1 \quad \wedge \quad \forall y \in \mathcal{P} \setminus Y, \mathcal{Z}_c(y, X) \geq 1$

Sub-constraint 2.1 expresses that a formal concept contains only true values. Sub-constraint 2.2 denotes that formal concept relevancy is enhanced by a maximality property. It is now straightforward to introduce a declarative specification of fault-tolerance.

Definition 5 (DRBS [11]). *Given integer parameters δ and ϵ, a bi-set $(X, Y) \in 2^{\mathcal{O}} \times 2^{\mathcal{P}}$ is called a DRBS pattern (Dense and Relevant Bi-Set) in $\mathbf{r}$ iff*
(3.1) $\forall x \in X, \mathcal{Z}_l(x, Y) \leq \delta \quad \wedge \quad \forall y \in Y, \mathcal{Z}_c(y, X) \leq \delta$
(3.2) $\forall e \in \mathcal{O} \setminus X, \forall x \in X, \mathcal{Z}_l(e, Y) \geq \mathcal{Z}_l(x, Y) + \epsilon$
$\quad\quad \wedge \quad \forall e' \in \mathcal{P} \setminus Y, \forall y \in Y, \mathcal{Z}_c(e', X) \geq \mathcal{Z}_c(y, X) + \epsilon$
(3.3) It is maximal, i.e., $\nexists(X', Y') \in 2^{\mathcal{O}} \times 2^{\mathcal{P}}$ s.t. (X', Y') is a DRBS pattern and $(X, Y) \subseteq (X', Y')$.

DRBS patterns have at most δ false values per object and per attribute (Sub-constraint 3.1) and are such that each outside object (resp. attribute) has at least ϵ false values plus the maximal number of false values on the inside objects (resp. attributes) according to Sub-constraint 3.2. The size of a DRBS pattern increases with δ such that, when $\delta > 0$, it happens that several bi-sets are included in each other. Only maximal bi-sets are kept (Sub-constraint 3.3). Notice that δ and ϵ can be chosen differently on objects and on attributes. It is clear that when $\delta = 0$ and $\epsilon = 1$, DRBS $\equiv$ FC.

Table 4. A boolean context $\mathbf{r}_6$

	p_1	p_2	p_3	p_4	p_5	p_6	p_7
o_1	1	0	1	0	1	0	0
o_2	1	1	1	1	0	1	0
o_3	0	1	1	1	1	1	1
o_4	0	0	0	1	1	1	0
o_5	1	0	0	0	0	1	0
o_6	1	1	1	1	1	0	0
o_7	1	1	1	1	1	0	0

Table 5. Permutations on $\mathbf{r}_6$ to illustrate Example 1

	p_1	p_2	p_3	p_4	p_5	p_6	p_7
o_1	1	0	1	0	1	0	0
o_2	1	1	1	1	0	1	0
o_3	0	1	1	1	1	1	1
o_4	0	0	0	1	1	1	0
o_6	1	1	1	1	1	0	0
o_7	1	1	1	1	1	0	0
o_5	1	0	0	0	0	1	0

Example 1. *If $\delta = \epsilon = 1$, $(X, Y) = (\{o_1, o_2, o_3, o_4, o_6, o_7\}, \{p_3, p_4, p_5\})$ is a DRBS pattern in $\mathbf{r}_6$ (see Table 4). Columns p_1, p_2, p_6 and p_7 contain at least two false values on X, and o_5 contains three false values on Y (see Table 5).*

Collections of DRBS patterns can be computed in rather small data sets by using the correct and complete algorithm DR-MINER [11]. It is again based on the DUAL-MINER principle [15]. Notice that a preliminary approach for specifying "symmetrical" fault-tolerant formal concepts had been introduced in [9] (i.e., the so-called $\alpha\beta$-concepts) and that it has been compared with the DRBS pattern domain in [12].

Let us now consider a rather different (and say pragmatic) extension of formal concepts which is not symmetrical. It has been designed thanks to the previous work on one of the few approximate condensed representations of frequent sets, the so-called δ-free sets [17]. δ-free sets are some kind of generators whose counted frequencies enable to infer the frequency of many sets (sets included in their so-called δ-closures) without further counting but with a bounded error. The δ-freeness constraint on attributes sets has been formalized in terms of the size of the supported sets of objects (this so-called frequency has to be different from the frequency of all its subsets by at least δ). Notice also that the so-called generators in [21] or key patterns in [22] are special cases of δ-free sets ($\delta = 0$). The 0-closure is the classical closure operator and applying it on each 0-free set is one way to produce every closed set and thus every formal concept.

The idea for the so-called FBS patterns (Free set Based Bi-Set) is to consider bi-sets built on the δ-closure of δ-free attribute sets associated to their supporting sets of objects [23,10]. The intuition is that it will provide strong associations between sets of objects and sets of attributes. To avoid technical details, let us just comment an example.

Example 2. *If* $\delta = 1$, $\{p_2\}$ *is a 1-free set and* $(\{o_2, o_3, o_6, o_7\}, \{p_1, p_2, p_3, p_4, p_5\})$ *is a FBS pattern in* $\mathbf{r}_6$ *(see Table 6). The 1-closure of* $\{p_2\}$ *is* $\{p_1, p_2, p_3, p_4, p_5\}$ *because* p_1, p_3, p_4, *and* p_5 *are the attributes which are almost always true (1 exception is accepted) for the objects that have the property denoted by* p_2, *i.e., objects* o_2, o_3, o_6, *and* o_7.

When $\delta=0$, attribute 0-free sets are the minimal elements of the equivalence classes of the relation "has the same supporting set of objects". Then, the 0-closure provides a closed set of attributes that associated to its supporting set of object gives a formal concept. In other terms, when $\delta = 0$, FBS $\equiv$ FC.

Table 6. A permutation on $\mathbf{r}_6$ to illustrate Example 2

	p_1	p_2	p_3	p_4	p_5	p_6	p_7
o_2	1	1	1	1	0	1	0
o_3	0	1	1	1	1	1	1
o_6	1	1	1	1	1	0	0
o_7	1	1	1	1	1	0	0
o_1	1	0	1	0	1	0	0
o_4	0	0	0	1	1	1	0
o_5	1	0	0	0	0	1	0

The extraction of FBS patterns can be extremely efficient thanks to δ-freeness anti-monotonicity. Notice that FBS patterns are bi-sets with a bounded number of exception per column but every bi-set with a bounded number of exception per column is not necessarily a FBS pattern.

One crucial issue that can explain the added-value of formal concepts is the ease of interpretation thanks to the Galois connection. What happens with these DRBS and FBS extensions? For each bi-set (X, Y), do we have a function which associates X and Y? If a function exists which associates to each set X (resp. Y) at most a unique set Y (resp. X), the interpretation of each bi-set is much easier. Furthermore, if the two functions are monotonically decreasing, i.e. when the size of X (resp. Y) increases, the size of its associated set Y (resp. X) decreases. This property is meaningful since the more we have objects inside a bi-set, the less there are attributes that can be associated to describe them (or vice versa).

In a FBS pattern, we have no function from 2^P to 2^O but we have a function from 2^O to 2^P. The definition of this pattern is indeed not symmetrical. In many data sets, including huge and dense ones, complete collections of FBS can be extracted efficiently but we need for a better characterization of more relevant FBS patterns which might remain easy to extract from huge databases, e.g., what is the impact of different δ values for the δ-free-set part and the δ-closure computation? How can we avoid an unfortunate distribution of the false values among the same objects?

By construction, a DRBS has been defined such that we have a bounded number of exceptions per object and per attribute. Two interesting properties have been proven [11,16]. First, when $\epsilon > 0$, DRBS patterns are embedded by two functions ϕ (resp. ψ) which associate to X (resp. Y) a unique set Y (resp. X). Then, for a fixed δ, we have monotonicity properties of ϕ and ψ. Unfortunately, the functions loose this property on the whole DRBS collection. Furthermore, we have not identified yet an intentional definition of these functions.

Notice that [12] contains an empirical evaluation of these different pattern domains on both artificially noised data sets and a real-life medical data set. These extensions of formal concepts have been specified in a constraint-based mining framework, i.e., we have a declarative specification of the constraints on the patterns such that we can work on correct and complete implementations for computing them. Notice however that it is computationally challenging to work with the most elegant extension, i.e., DRBS. Other researchers have considered fault-tolerant pattern mining. To the best of our knowledge, most of the related work has concerned mono-dimensional patterns and/or the use of heuristic techniques [24,25]. [26] is one of the interesting proposal for geometrical tile mining (i.e., dense bi-sets which involve contiguous elements given orders on both dimensions). More recently, other attempts to relax closeness have been considered [27,28]. Fault-tolerance in general and its application to closed sets and formal concepts in particular definitively appears as an important topic for real-life data mining. Let us now consider another pragmatic approach for finding actionable patterns based on collections of formal concepts.

3.2   Post-processing Collections of Formal Concepts

The number of formal concepts which hold in a data set and which can be computed thanks to, for instance, user-defined constraints, can be huge. We already pointed out that many factors can have a dramatic impact on the number of formal concepts. For instance, when an error of measurement or an intrinsic variability of the observed phenomenon lead to a zero value whereas the value true should be obtained, we have to face with an explosion of the number of formal concepts. The fault-tolerant extensions discussed above are definitively not the ultimate solution: indeed, when tractable, the extractions still provide too many patterns. One simple idea is to post-process the formal concepts and more generally the extracted bi-sets to group the ones which are similar enough. It is rather straightforward to design similarity measures between bi-sets and one can perform, for instance, a hierarchical clustering method to group formal concepts. These groups can be interpreted by computing some kind of "quasi-formal concepts" that are in fact the bi-sets made of the union of the intents and the union of the extents of all the formal concepts that belong to a cluster [29]. This has been applied successfully to a real application in the domain of human gene expression data analysis [30]. In this application, mining a 90×5327 boolean gene expression matrix has given rise to 64 836 formal concepts. When considering size constraints that have enforced the formal concepts to imply at least five biological samples and five genes, only 1 669 patterns have been selected: this is however too much for a human interpretation for each of them. We applied the hierarchical clustering and then we have built about 50 quasi-synexpression groups, i.e., sets of genes strongly associated to sets of biological samples. One of them has been carefully interpreted and this has given rise to interesting biological new hypothesis [30].

Following the same ideas, we can also consider the possibility to build clusters or co-clusters by exploiting the formal concepts. A co-clustering (see, e.g., [31]) provides linked partitions on both dimensions (objects and attributes) and, in Boolean data, it tends to compute rectangles with mainly true (resp. false) values. Heuristic techniques (i.e., local optimization) enable to compute one bi-partition. In fact, a bi-clustering provides a global structure over the data while fault-tolerant extensions of formal concepts are typical local patterns which can lead to the discovery of unexpected but yet relevant local associations. In [32], co-clusters are computed from collections of bi-sets like formal concepts: it is a KMEANS like clustering that does not work on objects or attributes but on bi-sets. Notice also that once a bi-partition has been computed, by such a technique or with another co-clustering approach, we can again use the bi-sets for characterization purposes [10]. Notice that computing clusters or co-clusters based on formal concepts is different from selecting formal concepts that may constitute a collection of clusters or co-clusters (see, e.g., [33]).

In a conceptual clustering framework, Mineau et al. [34] present simple pre-pruning and post-pruning techniques that can be applied in reasonable time on large classification structures. The paper presents three such techniques: one is based on the definition of constraints over the generalization language, the other two are based on discrimination metrics applied on links between classes or on the classes themselves.

Another interesting relationship that may be studied further is the formal analogies between tiling as considered in [35] and co-clustering. Also, the quite active area of subspace clustering is clearly related to local pattern detection and constraint-based mining of bi-sets (see [36,37] for surveys).

4   Conclusion

We have discussed several aspects of actionable pattern discovery from collections of formal concepts. Thanks to constraint-based mining algorithms, computing complete collections of formal concepts that satisfy user-defined constraints is feasible for some useful constraints. This framework has been used to specify fault-tolerant extensions of formal concepts. This is definitively needed for mining large and noisy Boolean data sets. The DRBS pattern domain appears as a well-designed class but the price to pay is its computational complexity. The good news are that (a) DRBS pattern extractions may involve further user-defined constraints which can be used for efficient pruning, and (b) one can look for more efficient data structures and thus a more efficient DR-MINER implementation. A pragmatic usage could be to extract some bi-sets, e.g., formal concepts, and then select some of them (say $B = (X, Y)$) for further extensions towards fault-tolerant patterns: the computation of a DRBS patterns (say $B' = (X', Y')$) such that the constraint $B \subseteq B'$ is enforced. We also mentioned post-processing techniques that, for instance, cluster patterns like formal concepts not only to decrease the number of hypothesis that have to be interpreted by human experts but also to enhance their relevancy, e.g., achieving some kind of fault-tolerance.

Let us now conclude this paper by introducing a few open questions related to both Formal Concept Analysis and constraint-based mining of patterns.

Constraints and formal concept mining. One important direction of research remains constraint-based mining of formal concepts for hard constraints, that is constraints that are neither monotonic nor anti-monotonic and for which new enumeration strategies have to be designed. Constraints that refer to statistical measures (e.g., based on standard deviation), are typical examples of such hard constraints. Even though some types of hard constraints have been studied on simple pattern domains like itemsets (e.g., optimization constraints that look for the best k patterns w.r.t. an objective measure), so far, few researchers have tried to upgrade these algorithms to the 2-dimensional case (i.e., for bi-set mining). New constraint types must be designed as well. For instance, [38] has proposed to increase the relevancy of set patterns by means of constraints that exploit textual resources in the context of biological data analysis. This is a promising direction of research.

Extensions of formal concepts. The constraint-based mining framework supports the exploitation of domain knowledge to increase a priori relevancy of patterns. We notice however that in more and more applications, the data can hardly be presented as a single binary relation. Let us take an example: assume that we have genes with a lot of information about them (e.g., functions, associated transcription factors), biological experiments enabling to measure the expression of these genes and finally details about the experiments. Consider now that we want to extract the sets of house-keeping genes that are over-expressed in the same biological experiments that are related to muscle tissues. The extraction task sounds like a formal concept extraction, same words same ideas, but here the data are structured into multiple n-ary relations ($n > 2$). One hot topic is to revisit the principles of pattern mining in binary relation when considering n-ary relations. For instance, the counterpart of formal concepts in cubes (3-ary relations) has been recently investigated [39,40]. Also, extending the ideas of closed set and condensed representation mining in a multi-relational setting is a timely challenge. [41] presents a formal concept analysis-based class hierarchy design that can be viewed as normal forms for class hierarchies where each normal form addresses particular design goals. An overview of work in the area is presented by highlighting the formal concept analysis notions that are involved. [42] introduces a first formalization of a network of contexts, i.e., the data is represented by the means of several contexts. Descriptions of conceptual coherences within the formalized network of contexts are introduced.

Storing and querying collections of patterns. Following the inductive database perspective [43,2], patterns may be considered as first-class citizens and we have to design databases that not only contain data but also patterns like formal concepts or clustering results. Technically, many challenges concern the efficient management of large collections of set patterns because relational databases are not targeted towards this type of data. Also, the design of general-purpose query languages to support knowledge discovery by means of queries remains an open problem.

Local-to-Global. Many global patterns (e.g., classifiers, clusterings) can be considered as collections of local patterns that satisfy some kind of global constraints. These local patterns are themselves satisfying local constraints. The popular association-based classification approach [44] is an obvious example of a Local to Global (L2G) scheme: standard association rules are the local patterns (i.e., local constraints are the minimal frequency and minimal confidence constraints). The various proposals for building classifiers from them are then based on different global constraints on these collections of association rules. Clustering can be considered within a L2G framework as well (see, e.g;, [32]). A better understanding of cross-fertilization between local pattern detection and global pattern discovery is an extremely active research direction that may deliver interesting insights in the next few years. Among others, the exciting challenge of constrained clustering may benefit from such a Local to Global approach (see, e.g., [45] for a constrained co-clustering based on formal concepts).

Acknowledgements. The authors thank Céline Robardet and Ruggero G. Pensa for their participation to most of the research results discussed in this paper. This research is partially funded by EU contract IST-FET IQ FP6-516169.

References

1. Ganter, B., Stumme, G., Wille, R. (eds.): Formal Concept Analysis. LNCS (LNAI), vol. 3626. Springer, Heidelberg (2005)
2. Boulicaut, J.F.: Inductive databases and multiple uses of frequent itemsets: the cInQ approach. In: Meo, R., Lanzi, P.L., Klemettinen, M. (eds.) Database Support for Data Mining Applications. LNCS (LNAI), vol. 2682, pp. 1–23. Springer, Heidelberg (2004)
3. Calders, T., Rigotti, C., Boulicaut, J.F.: A survey on condensed representations for frequent sets. In: Boulicaut, J.-F., De Raedt, L., Mannila, H. (eds.) Constraint-Based Mining and Inductive Databases. LNCS (LNAI), vol. 3848, pp. 64–80. Springer, Heidelberg (2006)
4. Valtchev, P., Missaoui, R., Godin, R.: Formal concept analysis for knowledge discovery and data mining: The new challenges. In: Eklund, P.W. (ed.) ICFCA 2004. LNCS (LNAI), vol. 2961, pp. 352–371. Springer, Heidelberg (2004)
5. Besson, J., et al.: Constraint-based formal concept mining and its application to microarray data analysis. Intelligent Data Analysis 9(1), 59–82 (2005)
6. Kuznetsov, S.O., Obiedkov, S.A.: Comparing performance of algorithms for generating concept lattices. Experimental and Theoretical Artificial Intelligence 14(2-3), 189–216 (2002)

7. Stumme, G., et al.: Computing iceberg concept lattices with titanic. Data & Knowledge Engineering 42, 189–222 (2002)
8. Besson, J., Robardet, C., Boulicaut, J.F.: Constraint-based mining of formal concepts in transactional data. In: Dai, H., Srikant, R., Zhang, C. (eds.) PAKDD 2004. LNCS (LNAI), vol. 3056, pp. 615–624. Springer, Heidelberg (2004)
9. Besson, J., Robardet, C., Boulicaut, J.F.: Mining formal concepts with a bounded number of exceptions from transactional data. In: Goethals, B., Siebes, A. (eds.) KDID 2004. LNCS, vol. 3377, pp. 33–45. Springer, Heidelberg (2005)
10. Pensa, R., Boulicaut, J.F.: From local pattern mining to relevant bi-cluster characterization. In: Famili, A.F., et al. (eds.) IDA 2005. LNCS, vol. 3646, pp. 293–304. Springer, Heidelberg (2005)
11. Besson, J., Robardet, C., Boulicaut, J.F.: Mining a new fault-tolerant pattern type as an alternative to formal concept discovery. In: Schärfe, H., Hitzler, P., Øhrstrøm, P. (eds.) ICCS 2006. LNCS (LNAI), vol. 4068, pp. 144–157. Springer, Heidelberg (2006)
12. Besson, J., et al.: Constraint-based mining of fault-tolerant patterns from boolean data. In: Bonchi, F., Boulicaut, J.-F. (eds.) KDID 2005. LNCS, vol. 3933, pp. 55–71. Springer, Heidelberg (2006)
13. Goethals, B., Zaki, M.: Proceedings of the IEEE ICDM Workshop on Frequent Itemset Mining Implementations FIMI 2003, Melbourne, USA (2003)
14. Ganter, B.: Two basic algorithms in concept analysis. Technical report, Technische Hochschule Darmstadt, Germany, Preprint 831 (1984)
15. Bucila, C., et al.: DualMiner: A dual-pruning algorithm for itemsets with constraints. Data Mining and Knowledge Discovery 7(4), 241–272 (2003)
16. Besson, J.: Découvertes de motifs pertinents pour l'analyse du transcriptome: application à l'insulino-résistance. PhD thesis, INSA-Lyon, 69621 Villeurbanne cedex, France. (in French) (2005)
17. Boulicaut, J.F., Bykowski, A., Rigotti, C.: Free-sets: a condensed representation of boolean data for the approximation of frequency queries. Data Mining and Knowledge Discovery 7(1), 5–22 (2003)
18. Pei, J., Han, J., Mao, R.: CLOSET an efficient algorithm for mining frequent closed itemsets. In: Proceedings ACM SIGMOD Workshop DMKD 2000 (2000)
19. Zaki, M.J., Hsiao, C.J.: CHARM: An efficient algorithm for closed itemset mining. In: Proceedings SIAM DM 2002, Arlington, USA (2002)
20. Gionis, A., et al.: Assessing data mining results via swap randomization. In: Proceedings ACM SIGKDD 2006, Philadelphia, USA, pp. 167–176 (2006)
21. Pasquier, N., et al.: Efficient mining of association rules using closed itemset lattices. Information Systems 24, 25–46 (1999)
22. Bastide, Y., et al.: Mining frequent patterns with counting inference. SIGKDD Explorations 2, 66 (2000)
23. Pensa, R., Boulicaut, J.F.: Towards fault-tolerant formal concept analysis. In: Bandini, S., Manzoni, S. (eds.) AI*IA 2005. LNCS (LNAI), vol. 3673, pp. 212–223. Springer, Heidelberg (2005)
24. Yang, C., Fayyad, U., Bradley, P.S.: Efficient discovery of error-tolerant frequent itemsets in high dimensions. In: Proceedings ACM SIGKDD, pp. 194–203. ACM Press, New York (2001)
25. Seppänen, J.K., Mannila, H.: Dense itemsets. In: Proceedings ACM SIGKDD 2004, Seattle, USA, pp. 683–688. ACM Press, New York (2004)
26. Gionis, A., Mannila, H., Seppänen, J.K.: Geometric and combinatorial tiles in 0-1 data. In: Boulicaut, J.-F., et al. (eds.) PKDD 2004. LNCS (LNAI), vol. 3202, pp. 173–184. Springer, Heidelberg (2004)

27. Cheng, J., Ke, Y., Ng, W.: Delta-tolerance closed frequent itemsets. In: Proceedings IEEE ICDM 2006, Hong Kong, China, pp. 139–148 (2006)
28. Cheng, H., Yu, P.S., Han, J.: Ac-close: Efficiently mining approximate closed itemsets by core pattern recovery. In: Proceedings IEEE ICDM 2006, Hong Kong, China, pp. 839–844 (2006)
29. Robardet, C., et al.: Using classification and visualization on pattern databases for gene expression data analysis. In: Proceedings PaRMa'04 co-located with EDBT 2004, Heraclion-Crete, Greece. CEUR Proceedings, vol. 96, pp. 107–118 (2004)
30. Blachon, S., et al.: Clustering formal concepts to discover biologically relevant knowledge from gene expression data. Silico Biology 7(0033), 1–15 (2007)
31. Dhillon, I.S., Mallela, S., Modha, D.S.: Information-theoretic co-clustering. In: Proceedings ACM SIGKDD 2003, Washington, USA, pp. 89–98. ACM Press, New York (2003)
32. Pensa, R., Robardet, C., Boulicaut, J.F.: A bi-clustering framework for categorical data. In: Jorge, A.M., et al. (eds.) PKDD 2005. LNCS (LNAI), vol. 3721, pp. 643–650. Springer, Heidelberg (2005)
33. Durand, N., Crémilleux, B.: Ecclat: a new approach of clusters discovery in categorical data. In: Proceedings ES 2002, Cambridge, UK, pp. 177–190. Springer, Heidelberg (2002)
34. Mineau, G., Godin, A.B.,, R.: Simple pre- and post-pruning techniques for large conceptual clustering structures. In: Electronic Transactions on Artificial Intelligence (ETAI), pp. 1–20 (2000)
35. Geerts, F., Goethals, B., Mielikäinen, T.: Tiling databases. In: Suzuki, E., Arikawa, S. (eds.) DS 2004. LNCS (LNAI), vol. 3245, pp. 278–289. Springer, Heidelberg (2004)
36. Parsons, L., Haque, E., Liu, H.: Subspace clustering for high dimensional data: a review. ACM SIGKDD Exploration Newsletter 6, 90–105 (2004)
37. Madeira, S.C., Oliveira, A.L.: Biclustering algorithms for biological data analysis: A survey. IEEE/ACM Trans. Comput. Biol. Bioinf. 1, 24–45 (2004)
38. Kléma, J., et al.: Mining plausible patterns from genomic data. In: Proceedings IEEE CBMS 2006, Salt Lake City, USA, pp. 183–190 (2006)
39. Jaschke, R., et al.: TRIAS: An algorithm for mining iceberg tri-lattices. In: Proceedings IEEE ICDM 2006, Hong Kong, China, pp. 907–911 (2006)
40. Ji, L., Tan, K.L., Tung, A.K.H.: Mining frequent closed cubes in 3D datasets. In: Proceedings VLDB 2006, Seoul, Korea, pp. 811–822 (2006)
41. Godin, R., Valtchev, P.: Formal concept analysis-based class hierarchy design in object-oriented software development. In: Ganter, B., Stumme, G., Wille, R. (eds.) Formal Concept Analysis. LNCS (LNAI), vol. 3626, pp. 192–207. Springer, Heidelberg (2005)
42. Wille, R.: Conceptual structures of multicontexts. In: 4th Int. Conf. on Conceptual Structures ICCS 1996, pp. 23–39. Springer, Heidelberg (1996)
43. Imielinski, T., Mannila, H.: A database perspective on knowledge discovery. Communications of ACM 39, 58–64 (1996)
44. Liu, B., Hsu, W., Ma, Y.: Integrating classification and association rule mining. In: Proceedings KDD 1998, pp. 80–86. AAAI Press, Menlo Park (1998)
45. Pensa, R.G., Robardet, C., Boulicaut, J.F.: Constraint-driven Co-Clustering of 0/1 Data. In: S.B., et al. (eds.) Constrained Clustering: Advances in Algorithms, Theory and Applications. Data Mining and Knowledge Discovery Series, Chapman & Hall/CRC Press (to appear, 2008)

Acquiring Generalized Domain-Range Restrictions[*]

Sebastian Rudolph

Institut AIFB, Universität Karlsruhe, Germany
rudolph@aifb.uni-karlsruhe.de

Abstract. Proposing a certain notion of logical completeness as a novel quality criterion for ontologies, we identify and characterise a class of logical propositions which naturally extend domain and range restrictions commonly known from diverse ontology modelling approaches. We argue for the intuitivity of this kind of axioms and show that they fit equally well into formalisms based on rules as well as ones based on description logics. Extending the attribute exploration technique from formal concept analysis (FCA), we present an algorithm for the efficient interactive specification of all axioms of this form valid in a domain of interest. We compile some results that apply when role hierarchies and symmetric roles come into play and demonstrate the presented method in a small example.

1 Introduction

Semantic technologies have gained significant interest in recent years as indicated by prominent conferences and workshop as well as a plethora of research projects. *Ontologies* constitute the central means within this area by providing logical descriptions of a considered domain based on which knowledge about the domain can be deduced automatically (this task usually being referred to as *reasoning*). Yet, the practical deployment of semantic technologies in a wider range of applications clearly requires new technical methods as well as methodologies assisting the knowledge engineer in designing medium to large size ontologies containing formalized knowledge beyond the usual subclass-superclass (i.e., taxonomic) relationships.

Though reasoning methods provide some assistance in this regard (e.g., allowing to check for local and global consistency of the formalized knowledge as well as for an ontology's "capability" to logically entail wanted consequences), there are other quality criteria for ontologies that cannot be met by reasoning support alone. One of those central criteria – well-nigh currently neglected in knowledge representation research – is that of *completeness*. More precisely, a knowledge base KB can be said to be complete w.r.t. a certain logic(al fragment), if every statement expressible in that logic can be entailed from KB or declined by KB (e.g. by showing the validity of its negation). Remarkably, Formal Concept Analysis has provided powerful tools to achieve the mentioned kind of completeness for some logical fragments already more then twenty years ago and subsequently successfully applied in numerous domains.

---

[*] This work is supported by the Deutsche Forschungsgemeinschaft (DFG) under the ReaSem project. We furthermore thank Pascal Hitzler and Johanna Völker for their valuable comments.

R. Medina and S. Obiedkov (Eds.): ICFCA 2008, LNAI 4933, pp. 32–45, 2008.
© Springer-Verlag Berlin Heidelberg 2008

Clearly, completeness w.r.t. expressive formalisms (as, say, OWL1.1-completeness) is a goal which cannot be reasonably fulfilled for non-trivial ontologies. Hence (in analogy to identifying tractable fragments of DLs that allow relatively expressive modelling while still being of low reasoning complexity) we argue for identifying fragments being satisfactorily expressive and intuitive to the user as well as still computationally easy to handle, such that the completeness of a KB w.r.t. those fragments is both desirable and achievable.

Hence in our paper, we characterise a group of axioms which meet those requirements and canonically generalise both domain and range statements. Furthermore we provide a method for their interactive acquisition that in the end yields a knowledge base being complete w.r.t. the class of these axioms. In Section 2, after some initial motivation, we introduce and define this type of domain axioms expressible equivalently by DL (resp. OWL) statements or by rules. Section 3 presents *Role Exploration*, a method for – given a role (resp. binary predicate) and a set of "interesting" classes (resp. unary predicates) – interactively acquiring all axioms of this type valid in the described domain of interest.[1] This method is based on the aforementioned attribute exploration algorithm from formal concept analysis. Section 4 discusses how one could take advantage of additional knowledge about roles, namely role hierarchies and role symmetry, by modifications of the Role Exploration algorithm. In Section 5, we demonstrate Role Exploration by further elaborating an example for the setting brought up in Section 2. Finally, Section 6 concludes and gives an outlook to further research.

In the sequel, we assume the reader to be familiar basic notions from description logics (see [1] for a comprehensive and detailed overview) and rule-based languages [2].

2 Generalised Domain-Range Restrictions: Characterisation and Properties

Imagine the following situation: suppose, in a knowledge base describing persons and personal relationships, we have a role denoted with married which is to express whether a person is married to another person. So, clearly an ontology engineer would state that both domain and range of that role would have to be subclasses of Person, being expressed by the DL statements $\exists$married.$\top \sqsubseteq$ Person and $\forall$married.Person or by the rules married$(X, Y) \to$ Person(X) and married$(X, Y) \to$ Person(Y). In an OWL (Web Ontology Language, W3C recommendation [3]) ontology this could be expressed using the domain and range language constructs for object properties as follows:

```
<owl:ObjectProperty rdf:ID="Married">
  <rdfs:domain rdf:resource="#Person"/>
  <rdfs:range rdf:resource="#Person"/>
</owl:ObjectProperty>
```

Yet, what one would certainly like to additionally state is that males can marry only females and vice versa.[2] Obviously, this is not possible via the usual OWL domain and

---

[1] In order not to confuse the two meanings of the term "domain", we use *domain of interest* whenever referring to the meaning "universe of discourse" or "set of all entities".

[2] For the sake of the example we refer to a situation without same-sex marriages. However, this is not meant to reflect any personal attitude of the author towards this topic.

range constructs. However, the DL axioms Male ⊑ ∀married.Female and Female ⊑ ∀married.Male (as well as their OWL DL counterparts) or the rules married(X, Y) ∧ Male(X) → Female(Y) and married(X, Y) ∧ Female(X) → Male(Y) express exactly this relationship.

Staying with this kind of examples, note that there are countries (such as India), where the minimal age to get (and hence, to be) married is sex-dependant.[3] The corresponding regulation is no domain or range restriction in the classical sense either, yet can be stated by DL axioms like Male ⊓ ∃married.⊤ ⊑ Age21plus and Female ⊓ ∃married.⊤ ⊑ Age18plus or – in a rule language – by married(X, Y) ∧ Male(X) → Age21plus(X) and married(X, Y) ∧ Female(Z) → Age18plus(Y).

Having demonstrated the utility and intuitivity of this kind of modelling axioms, we introduce a type of statements capturing all of them while being still computationally easy to handle.

Definition 1. *Given a set C of named classes and a role* R, *a* GENERALIZED DOMAIN-RANGE RESTRICTION *(short: GDRR) is a rule having the following form*

$$R(X, Y) \land \bigwedge_{A \in \mathbf{A}} A(X) \land \bigwedge_{B \in \mathbf{B}} B(Y) \rightarrow \bigwedge_{C \in \mathbf{C}} C(X) \land \bigwedge_{D \in \mathbf{D}} D(Y)$$

where $\mathbf{A}, \mathbf{B}, \mathbf{C}, \mathbf{D} \subseteq C$ *and* R *is a role name. Note, that for* $\mathbf{C} \cup \mathbf{D} = \emptyset$, *the rule will have an empty head (also denoted by* □*) and, hence, will be interpreted as integrity constraint.*

Put into words, the GDRR presented in the above definition would mean the following: "For any two elements X and Y of the domain of interest that are connected by a role R and where X fulfills (all of) $\mathbf{A}$ as well as Y fulfills (all of) $\mathbf{B}$, we know that X additionally fulfills $\mathbf{C}$ and Y additionally fulfills $\mathbf{D}$."

The next theorem guarantees that for every GDRR, there is a semantically equivalent general concept inclusion axiom (GCI) in any sufficiently expressive DL (while these expressiveness requirements are very low).

Theorem 1. *The GDRR*

$$R(X, Y) \land \bigwedge_{A \in \mathbf{A}} A(X) \land \bigwedge_{B \in \mathbf{B}} B(Y) \rightarrow \bigwedge_{C \in \mathbf{C}} C(X) \land \bigwedge_{D \in \mathbf{D}} D(Y)$$

is equivalent to both of the following GCIs:[4]

$$\prod_{A \in \mathbf{A}} A \sqcap \exists R.\left(\prod_{B \in \mathbf{B}} B\right) \sqsubseteq \prod_{C \in \mathbf{C}} C \sqcap \forall R.\left(\left(\bigsqcup_{B \in \mathbf{B}} \neg B\right) \sqcup \left(\prod_{D \in \mathbf{D}} D\right)\right),$$

$$\prod_{B \in \mathbf{B}} B \sqcap \exists R^-.\left(\prod_{A \in \mathbf{A}} A\right) \sqsubseteq \prod_{D \in \mathbf{D}} D \sqcap \forall R^-.\left(\left(\bigsqcup_{A \in \mathbf{A}} \neg A\right) \sqcup \left(\prod_{C \in \mathbf{C}} C\right)\right).$$

---

[3] In the Indian *Child Marriage Restraint Act* of 1929, amended in 1978, child is defined as "[...] a person, who, if a male, has not completed twenty-one years of age, and if a female, has not completed eighteen years of age [...]" [4].

[4] Where we set $\prod_{E \in \mathbf{E}} E$ to be ⊤ whenever $\mathbf{E} = \emptyset$.

Although the GCI obtained by the uniform translation provided by Theorem 1 might look cumbersome and counterintuitive, note that obviously any GDRR having a conjunction of atoms in the head can be split into several GDRRs with single-atom heads. Each of those will be equivalent to a more intuitive GCI, as stated by the following corollary.

Corollary 1. *1. The GDRR of the shape*

$$R(X, Y) \wedge A_1(X), \ldots, A_n(X), B_1(Y), \ldots, B_k(Y) \rightarrow \square$$

is equivalent to each of the GCIs

$$A_1 \sqcap \ldots \sqcap A_n \sqcap \exists R.(B_1 \sqcap \ldots \sqcap B_k) \sqsubseteq \bot$$

$$B_1 \sqcap \ldots \sqcap B_k \sqcap \exists R^-.(A_1 \sqcap \ldots \sqcap A_n) \sqsubseteq \bot$$

2. The GDRR of the shape

$$R(X, Y) \wedge A_1(X), \ldots, A_n(X), B_1(Y), \ldots, B_k(Y) \rightarrow C(X)$$

is equivalent to each of the GCIs

$$A_1 \sqcap \ldots \sqcap A_n \sqcap \exists R.(B_1 \sqcap \ldots \sqcap B_k) \sqsubseteq C$$
$$B_1 \sqcap \ldots \sqcap B_k \sqsubseteq \forall R^-.(\neg A_1 \sqcup \ldots \sqcup \neg A_n \sqcup C)$$
$$B_1 \sqcap \ldots \sqcap B_k \sqcap \exists R^-.(A_1 \sqcap \ldots \sqcap A_n \sqcap \neg C) \sqsubseteq \bot$$

3. The GDRR of the shape

$$R(X, Y) \wedge A_1(X), \ldots, A_n(X), B_1(Y), \ldots, B_k(Y) \rightarrow C(Y)$$

is equivalent to each of the GCIs

$$A_1 \sqcap \ldots \sqcap A_n \sqsubseteq \forall R.(\neg B_1 \sqcup \ldots \sqcup \neg B_k \sqcup C)$$
$$A_1 \sqcap \ldots \sqcap A_n \sqcap \exists R.(B_1 \sqcap \ldots \sqcap B_k \sqcap \neg C) \sqsubseteq \bot$$
$$B_1 \sqcap \ldots \sqcap B_k \sqcap \exists R^-.(A_1 \sqcap \ldots \sqcap A_n) \sqsubseteq C$$

Note that therefore, each of the description logics $\mathcal{ALE}$ and $\mathcal{ELI}$ is sufficient to express GDRRs; for the first two types, even $\mathcal{EL}$ will do.

Considering the rule representation, note that we refrain from using negated atoms. Hence the proposed type of rules belongs to the fragment of Horn clauses. Following the general framework for defining Horn DLs from [5], the DL representation of GDRRs belongs to Horn-$\mathcal{ALE}$ (whereas $\mathcal{ELI}$ is already Horn anyway). Likewise, they also naturally fall in the DLP [6] fragment. Mark that, although no negated atoms are allowed, we can nevertheless express certain kinds of negative statements by using rules with empty heads (also called *integrity constraints*, as mentioned in Definition 1). For example, the statement "a child is not allowed to marry", normally modelled with a DL axiom like Child $\sqsubseteq \neg\exists$married.$\top$, can equivalently be expressed by the GDRR married(X, Y), Child$(X) \rightarrow \square$.

Hence, GDRRs identify a class of logical statements useful to characterise roles beyond the common domain-range restrictions still being both intuitive and computationally friendly (witnessed by their containment in the abovementioned fragments). Related to that, they also fulfill a certain computationally advantageous locality condition: given the set Δ of all entities of a domain of interest, checking whether a certain GDRR is satisfied therein can be done by separately checking all entity pairs connected by the role R. Mark that this is not the case for any "simple looking" GCI, take for example $\exists$has.Sorrow $\sqsubseteq$ $\exists$has.Liqueur – a proposition well-known from German poetry.[5]

3 Acquisition of GDRRs Via Role Exploration

In this section, we will propose a way to exhaustively determine all GDRRs of a certain shape (i.e., referring to a role R and a set of relevant atomic classes C) valid in a domain of interest, i.e., assuring "GDRR-completeness" of the resulting knowledge in the sense introduced in Section 1. This method is based on the attribute exploration algorithm well known from formal concept analysis. The algorithm we present will consequently ask an expert for the validity of GDRRs in the domain of interest and end up with a revised knowledge base and a complete (as defined later) set of GDRRs.

The attribute exploration algorithm our work is based on was introduced in [8]. Attribute exploration with partial or incomplete information has been dealt with in several variants e.g. in [9,10]. In [11], FCA and DL were combined for the first time by using complex concept descriptions to define new attributes in formal contexts. In [12], attribute exploration was used to determine the concept hierarchy of conjunctions on atomic concepts. The idea to use attribute exploration as a way to interactively refine an ontological knowledge base was brought up in [13] and thoroughly described in [14], where also an extension to the case with partial information was proposed. A concise algorithm for exploration with partly known objects has been provided in [15].

3.1 FCA and Attribute Exploration with Partial Information

We refrain from introducing the most basic FCA notions and instead refer the reader to [16].

For our considerations, we work with a generalised notion of this data structure, allowing for partial specification (i.e., it might be unknown, whether an object has an attribute or not). This is an important extension for a knowledge representation setting, since (due to the open world assumption), it is reasonable to assume that not all (even not all relevant) facts about a described entity are known.

Definition 2. *A* PARTIAL FORMAL CONTEXT $\mathbb{K}^?$ *is a quadruple* $(G, M, I^\square, I^\diamond)$ *where both* $(G, M, I^\square)$ *and* $(G, M, I^\diamond)$ *are formal contexts and* $I^\square \subseteq I^\diamond$.
A formal context $\mathbb{K} = (G, M, I)$ *will be called* COMPLETION *of* $\mathbb{K}^?$, *if* $I^\square \subseteq I \subseteq I^\diamond$.

The intuitive meaning of this definition is the following: $gI^\square m$ means, it is certain that object g has the attribute m, while $gI^\diamond m$ means, it is possible that object g has the

---

[5] "Es ist ein Brauch von alters her: *wer Sorgen hat, hat auch Likör!*" (emphasis by the author) to be found in Chapter 16 of [7].

attribute m or – in other words – it is *not* certain that object g does *not* have the attribute m. An intuitive visualization would be a table with rows corresponding to the objects and columns corresponding to the attributes, having crosses where $gI^\square m$, blanks where *not* $gI^\diamond m$ and question marks everywhere else.

Naturally, a completion of a partial formal context will be obtained by substituting each question mark by either a cross or a blank.

In FCA, *implications* constitute the central means of expressing knowledge. We formally specify this rather straightforward notion together with some further useful theory in the following definition.

Definition 3. *Let M be an arbitrary set. An* IMPLICATION *on M is a pair (A, B) with $A, B \subseteq M$. To support intuition, we write $A \to B$ instead of (A, B).*

$A \to B$ HOLDS *in a formal context $\mathbb{K} = (G, M, I)$, if for all $g \in G$, we have that $A \subseteq g^I$ implies $B \subseteq g^I$. We then write $\mathbb{K} \models A \to B$.*

We say, a partial formal context $\mathbb{K} = (G, M, I^\square, I^\diamond)$ ADMITS *an implication $A \to B$, if for all $g \in G$ we have that $A \subseteq g^{I^\square}$ implies $B \subseteq g^{I^\diamond}$. For $C \subseteq M$ and a set $\mathfrak{I}$ of implications on M, let $C^{\mathfrak{I}}$ denote the smallest set with $C \subseteq C^{\mathfrak{I}}$ that additionally fulfills*

$$A \subseteq C^I \quad implies \quad B \subseteq C^I$$

*for every implication $A \to B$ in $\mathfrak{I}$.[6] If $C = C^{\mathfrak{I}}$, we call C $\mathfrak{I}$-*CLOSED*. We say $\mathfrak{I}$* ENTAILS *$A \to B$ if $B \subseteq A^{\mathfrak{I}}$.[7] An implication set $\mathfrak{I}$ will be called* NON-REDUNDANT*, if for any $(A \to B) \in \mathfrak{I}$ we have that $B \not\subseteq A^{\mathfrak{I} \setminus \{A \to B\}}$. A set $\mathfrak{I}$ implications holding in a context $\mathbb{K}$ will be called* COMPLETE*, if every implication $A \to B$ holding in $\mathbb{K}$ is entailed by $\mathfrak{I}$. $\mathfrak{I}$ will be called an* IMPLICATION BASE *of a formal context $\mathbb{K}$ if it is non-redundant and complete.*

Note that implication entailment is decidable in linear time w.r.t. the size of $\mathfrak{I}$ [17,18]. Therefore, knowing the implication base in a logical setting allows fast handling of the whole corresponding implicational theory. Moreover, for every formal context, there exists a canonical implication base [19].

The method of attribute exploration allows to acquire the implication base of a domain of interest being just implicitly known by an expert in an interview-like process. Due to space reasons, we omit to display its technical details and refer the reader to the thorough presentation in [15].

Essentially, the following happens: the aspect of the domain of interest that shall be explored is formalized as a formal context $\mathbb{K} = (U, M, I)$. Usually, it is not known completely in advance. However, possibly, some entities of the domain of interest $g \in U$ are already known, as well as some attributes that g has or has not, constituting an initial partial formal context.

During runtime, the algorithm presents questions of the form

"Does the implication $A \to B$ hold in the context $\mathbb{K} = (U, M, I)$?"

to the human expert. The expert might confirm this. In this case, $A \to B$ is archived as part of $\mathbb{K}$'s implicational base $\mathfrak{IB}$. The other case would be that $A \to B$ does not hold

---

[6] Note, that this is well-defined, since the mentioned properties are closed wrt. intersection.

[7] Actually, this is a syntactic shortcut. Yet, it can be easily seen that this coincides with the usual entailment notion.

in (U, M, I). But then, there must exist a $g \in U$ with $A \in g^I$ and $B \notin g^I$. The expert is asked to input this g and – roughly speaking – enough evidence for qualifying g as a counterexample by augmenting the partial context such that $A \subseteq g^{I^\square}$ and $B \not\subseteq g^{I^\lozenge}$. The procedure terminates when the implicational knowledge of the $\mathbb{K}$ is completely acquired, i.e., the implications admitted by the partial formal context built from the entered counterexamples are the same as those entailed by $\mathfrak{KB}$.

In our approach, we will exploit the capability of attribute exploration to efficiently determine a propositional implicational theory. Notwithstanding, we extend the underlying language[8] from purely propositional to GDRRs.

3.2 Role Contexts

In this work, we employ attribute exploration in a way that is structurally very similar to the approach in [21], where this technique was used for specifying dynamic systems. In this setting, roles would be interpreted as actions that can be taken, classes are used to describe states and the models of a corresponding theory can be interpreted as state transition systems. Yet this technique easily carries over to the more general setting of knowledge specification as firstly sketched by the author in [14].

Definition 4. *Let* $\mathcal{KB}$ *be a DL knowledge base and, as usual, an interpretation* $\mathcal{I}$ *of* $\mathcal{KB}$ *be defined as* $(\Delta, \cdot^\mathcal{I})$, *where* Δ *is the individual set and* $\cdot^\mathcal{I}$ *a function mapping class names to subsets of* Δ *and role names to subsets of* $\Delta \times \Delta$.

For a given interpretation $\mathcal{I}$ *together with a set* C *of named classes and a role* R, *the* ROLE CONTEXT $\mathbb{K}_\mathsf{R}$ *is defined as formal context* (G, M, I) *with*

- $G := \mathsf{R}^\mathcal{I} = \{(\delta_1, \delta_2) \mid \delta_1, \delta_2 \in \Delta, (\delta_1, \delta_2) \in \mathsf{R}^\mathcal{I}\}$
  the objects of $\mathbb{K}_\mathsf{R}$ *are those individual pairs connected by the role* R,
- $M := \{\mathsf{C_d}, \mathsf{C_r} \mid \mathsf{C} \in C\}$
  the attribute set of $\mathbb{K}_\mathsf{R}$ *contains two "copies" of* C: *the* DOMAIN ATTRBUTES *indexed with* d *the* RANGE ATTRIBUTES *indexed with* r, *and*
- $I \subseteq G \times M$ *with* $(\delta_1, \delta_2)I\mathsf{C_d} \Longleftrightarrow \delta_1 \in \mathsf{C}^\mathcal{I}$ *and* $(\delta_1, \delta_2)I\mathsf{C_r} \Longleftrightarrow \delta_2 \in \mathsf{C}^\mathcal{I}$.
  the domain attributes indicate for an R-*connected pair of entities, whether the corresponding class contains the first entity of that pair, while the range attributes describe the second entity.*

The following theorem shows how the validity of a GDRR in an interpretation can be read from a corresponding role context.

Theorem 2. *An interpretation* $\mathcal{I}$ *satisfies a GDRR*

$$R(X, Y) \land \bigwedge_{A \in \mathbf{A}} A(X) \land \bigwedge_{B \in \mathbf{B}} B(Y) \to \bigwedge_{C \in \mathbf{C}} C(X) \land \bigwedge_{D \in \mathbf{D}} D(Y)$$

if and only if the corresponding role context $\mathbb{K}_\mathsf{R}$ *satisfies the implication*

$$\{\mathsf{A_d} \mid A \in \mathbf{A}\} \cup \{\mathsf{B_r} \mid B \in \mathbf{B}\} \to \bot \quad if\ \mathbf{C} \cup \mathbf{D} = \emptyset\ and$$

$$\{\mathsf{A_d} \mid A \in \mathbf{A}\} \cup \{\mathsf{B_r} \mid B \in \mathbf{B}\} \to \{\mathsf{C_d} \mid C \in \mathbf{C}\} \cup \{\mathsf{D_r} \mid D \in \mathbf{D}\} \quad otherwise.$$

---

[8] There exist already other language extensions, e.g. to Horn-logic with a bounded variable set, see [20].

This theorem enables us to "translate" any implication in a role context into an equivalent GDRR and via Theorem 1 further into a GCI. So, for a given implication i from $\mathbb{K}_R$, let $DL^+(i)$ denote an equivalent GCI with the pure role and $DL^-(i)$ an equivalent GCI with the inverse role.

Now, the basic idea for the knowledge acquisition method we are going to propose is to carry out attribute exploration (with uncertain knowledge) on the context $\mathbb{K}_R$. Thereby, our basic assumption is that there exists a distinguished interpretation I' entirely (but implicitly) known by the human expert that we want to specify in terms of GDRRs.

3.3   Reasoner-Aided Exploration

The general work flow of exploration based knowledge base refinement was first described by the author in [13] and has been subsequently applied in diverse approaches [22,14,15,23]. Basically, three entities are involved:

- the exploration algorithm consecutively asking questions,
- a reasoner trying to cope with those questions based on (terminological or grounded) information being present a priori (thereby minimising the expert's "workload"), and
- an (ideally omniscient) human expert dealing with those questions that cannot be answered by the reasoner.

For the sake of clarity, we will describe a rather concrete instantiation of this framework. Nevertheless, there are several degrees of freedom in certain parts of the algorithm in that certain additional computation steps could be carried out, which do not alter the outcome of the algorithm but might have significant influence on its performance. We indicate such optional steps in the algorithm leaving questions related to optimisation for future research.

So let $\mathcal{KB}$ be an OWL DL knowledge base and $\mathcal{R}$ be an OWL DL reasoner. Let furthermore C be a set of named classes and R a role[9] occurring in $\mathcal{KB}$.

Initialisation. We initialise a partial "working" context $\mathbb{K}_R^? = (G, M, I^\square, I^\diamond)$ by setting $G := \emptyset$, $M := \{C_d, C_r \mid C \in C\}$. It will be successively enriched during the exploration.

Scan for a-priori Data (optional). Although any exploration process can be carried out starting from scratch, i.e. without any objects known in advance, such information may be advantageous by making possible hypotheses obsolete. Besides the possibility of manually providing such information, there are two possible ways of extracting this kind of information from a given knowledge base, which we call the *lazy* and the *greedy* way, depending on whether reasoning is employed or not.

So, the lazy way of data search would, for all role statements $R(a, b) \in \mathcal{KB}$, add (a, b) to the object set G of $\mathbb{K}_R^?$ and set

$$I^\square := I^\square \cup \{((a,b), C_d) \mid C(a) \in \mathcal{KB}, C \in C\} \cup \{((a,b), C_r) \mid C(b) \in \mathcal{KB}, C \in C\} \text{ and}$$

$$I^\diamond := I^\diamond \cup \{((a,b), C_d) \mid \neg C(a) \notin \mathcal{KB}, C \in C\} \cup \{((a,b), C_r) \mid \neg C(b) \notin \mathcal{KB}, C \in C\}.$$

---

[9] The corresponding OWL DL term being *object property*.

Clearly, this would just add the relevant information explicitly present in $\mathcal{KB}$ to the working context.

Contrarily, the greedy way would employ reasoning to acquire more complete information to start with. In this case, for any role statement $R(a, b)$ that can be inferred from $\mathcal{KB}$ by $\mathcal{R}$, the pair (a, b) would be added to G. Employing $\mathcal{R}$ further, we then set

$$I^{\square} := I^{\square} \cup \{((a,b), C_d) \mid \mathcal{KB} \models C(a), C \in C\} \cup \{((a,b), C_r) \mid \mathcal{KB} \models C(b), C \in C\} \text{ and}$$

$$I^{\diamond} := I^{\diamond} \cup \{((a,b), C_d) \mid \mathcal{KB} \not\models \neg C(a), C \in C\} \cup \{((a,b), C_r) \mid \mathcal{KB} \not\models \neg C(b), C \in C\}.$$

Although the greedy way would deliver more starting information which might shorten the subsequent exploration process, this advantage might be vitiated by the large number of possibly time consuming reasoner calls.

Scan for a-priori GDRRs (optional). The exploration algorithms also allows for entering already known implications before starting the actual exploration process. Like in the case with a-priori data, this could accelerate the exploration process, since some hypotheses can be taken for granted.

In order to acquire this kind of information, we check for every GCI occurring in $\mathcal{KB}$, whether it syntactically entails[10] a GDRR (w.r.t. R and C) and if so, add the respective implication i to the set of implications known in advance. Note that also GCIs that represent just class hierarchies are interesting in this regard, since e.g. $C \sqsubseteq D$ would entail any GDRR $R(X, Y), C(X) \rightarrow D(X)$ as well as $R(X, Y), C(Y) \rightarrow D(Y)$.

Exploration. Now we start the exploration process on the partial working context. Every hypothetical implication i the algorithm comes up with is transformed into a subsumption statement $DL^+(i)$. The following two steps can be carried out in arbitrary order (or in parallel), whereas it is impossible that both succeed (which allows to refrain from either one if the other is known to have succeeded).

- Employ $\mathcal{R}$ to check whether $\mathcal{KB} \models DL^+(i)$. If so, silently confirm i to the exploration algorithm and continue the exploration.
- Employ $\mathcal{R}$ to check whether $\mathcal{KB} \cup \{DL^+(i)\}$ is unsatisfiable. If this is the case, this means that $\mathcal{KB}$ forces any model to contain a pair of individuals (i_1, i_2) serving as a counterexample for i.

If none of the above cases applies, the human expert has to decide whether the proposed GDRR is valid in the described domain of interest, i.e., whether $I' \models DL^+(i)$. If the expert agrees, i will be confirmed to the exploration algorithm and additionally – since the expert has revealed genuinely new information – $DL^+(i)$ will be added to $\mathcal{KB}$. After that, the exploration continues with a new hypothesis.

In case the GDRR is denied (either by $\mathcal{R}$ or by the expert), a counterexample must be provided. If $\mathcal{R}$ was able to show the unsatisfiability of $\mathcal{KB} \cup \{C \sqsubseteq D\}$, it might even be able to automatically provide a counterexample in the following way. Let $A \rightarrow B$ be the implication in question, and set $\mathcal{G}^+ := \{DL^+(A \rightarrow \{b\}) \mid b \in B\}$ and $\mathcal{G}^- := \{DL^-(A \rightarrow \{b\}) \mid b \in B\}$. Now, for every GCI $C \sqsubseteq D$ contained in $\mathcal{G}^+ \cup \mathcal{G}^-$, we use $\mathcal{R}$ to retrieve

---

[10] Hereby we mean entailment that can be detected by easy (i.e. tractable) syntactic transformations. Due to lack of space, we postpone an elaboration of this part to future work.

instances of $C \sqcap \neg D$. If one such instance, is found, we add a new pair (e_1, e_2) to G and set $I^{\square} := I^{\square} \cup \{(e_1, e_2)\} \times A$ as well as $I^{\diamond} := I^{\diamond} \cup \{(e_1, e_2)\} \times (M \setminus \{b, \bot\})$. In this case, the exploration process can be continued without consulting the expert.

However, even if the unsatisfiability of $\mathcal{KB} \cup \{C \sqsubseteq D\}$ can be shown, there might be no named individual in the $\mathcal{KB}$ witnessing this in the sense just described. Then – as well as in the case when the expert had to deny the hypothetical GDRR himself – he has to manually add information to the knowledge base in a way that a counterexample can be retrieved by the method described above. Obviously, this can be achieved in any case by entering an R-connected individual pair i_1 and i_2 with appropriate class assertions, but there are other ways (as adding instances for one concept description from $\mathcal{G}^+$ or $\mathcal{G}^-$). Then a (partial) counterexample description can be generated automatically in the above described way.

Termination. After the exploration finishes, we have obtained a twofold result:

- A refined version of $\mathcal{KB}$ which is "GDRR-complete" w.r.t C and R meaning the following: Every GDRR involving the role R and concepts from C is either entailed by $\mathcal{KB}$ or adding it to $\mathcal{KB}$ leads to unsatisfiability. Hence, $\mathcal{KB}$ completely characterises I' in terms of this class of GDRRs.
- An implication base $\mathfrak{IB}$, accumulated by the exploration process. $\mathfrak{IB}$ allows to check *in linear time* for *every* GDRR on R and C whether it is valid in I' or not.

4 Interplay with Other Role Properties

Considering OWL DL, there are lots of other features which can be used to characterise roles. In the sequel we will briefly review how some of this information can be taken advantage of during the role exploration process.

Symmetric Roles. Quite frequently, roles are known to be symmetric. This might be expressed by the DL statement $R \equiv R^-$ or the rule $R(X, Y) \rightarrow R(Y, X)$; OWL even provides a dedicated language construct for this. In this case, the symmetry carries over to $\mathbb{K}_R$ in the following sense: for every implication $A \rightarrow B$ holding in $\mathbb{K}_R$, the implication $\psi(A) \rightarrow \psi(B)$ with

$$\psi : \begin{cases} C_d \mapsto C_r \\ C_r \mapsto C_d \\ \bot \mapsto \bot \end{cases} \text{ for all } C \in C$$

holds in $\mathbb{K}_R$ as well. In [24], attribute exploration has been extended in order to take this kind of symmetries into account, allowing the acquisition of implicational knowledge "modulo permutations" on the attribute set.

Role Hierarchies. A standard feature in expressive description logics (and as well contained in OWL DL) is the definition of role hierarchies. For two given roles R_1, R_2, the role R_1 is subsumed by the role R_2, (DL notation: $R_1 \sqsubseteq R_2$) if $R_1^{I} \subseteq R_2^{I}$. It takes just little consideration that in this case, every implication valid in $\mathbb{K}_{R_2}$ is also valid in $\mathbb{K}_{R_1}$. This can be exploited for the exploration in the following way: Assume for both R_1 and R_2, all valid GDRRs w.r.t. C have to be determined. The most efficient way to do

Person $\sqsubseteq$ Male $\sqcup$ Female	Person $\sqsubseteq$ Child $\sqcup$ Adult	Catholic $\sqcap$ Priest $\sqsubseteq$ Male
Male $\sqcap$ Female $\sqsubseteq \bot$	Child $\sqcap$ Adult $\sqsubseteq \bot$	Catholic $\sqcap$ Protestant $\sqsubseteq \bot$
married $\equiv$ married$^-$	$\exists$married.$\top \sqsubseteq$ Person	$\top \sqsubseteq \forall$married.Person

Fig. 1. Example knowledge base $\mathcal{KB}$ about marriages

so would then be to first carry out the procedure for R_2 and use the acquired implication base as a-priori knowledge for the next procedure, thereby reducing the amount of hypothetical GDRRs brought up by the algorithm.

5 An Example: So, Who Marries Whom?

For a small demonstration how the presented technique would be applied in practice, let us stay with the example from Section 2. Let $\mathcal{KB}$ be the knowledge base given in Fig. 1. Now imagine, this knowledge base is to be refined with respect to the role married. Let

$$C := \{\text{Person, Male, Female, Child, Adult, Catholic, Protestant, Priest}\}$$

be the set of interesting class names. So the set of attributes of the role context would be

$$M := \{ \text{Person}_d, \text{Person}_r, \text{Male}_d, \text{Male}_r, \text{Female}_d, \text{Female}_r, \text{Child}_d, \text{Child}_r, \text{Adult}_d,$$
$$\text{Adult}_r, \text{Catholic}_d, \text{Catholic}_r, \text{Protestant}_d, \text{Protestant}_r, \text{Priest}_d, \text{Priest}_r, \bot\}$$

Note that the role married is defined to be symmetric; therefore, the respective additional considerations from the previous section apply. Assume the following married couples already to be known: Andreas & Christiane, Anupriya & Kedar, as well as Astrid & Thomas. So, after initialisation, the starting context would have a shape as depicted in Fig. 2.

In the sequel, we review the hypothetical implications the exploration algorithm comes up with and explain how they are handled by the reasoner and (resp. or) the human expert.

1. Question: $\emptyset \to \{\text{Person}_d, \text{Adult}_d, \text{Person}_r, \text{Adult}_r\}$ (In words – mark that the empty premise requires the conclusion to be universally true): "If two entities marry, are they both persons and adults?")

	Person$_d$	Person$_r$	Male$_d$	Male$_r$	Female$_d$	Female$_r$	Child$_d$	Child$_r$	Adult$_d$	Adult$_r$	Catholic$_d$	Catholic$_r$	Protestant$_d$	Protestant$_r$	Priest$_d$	Priest$_r$	$\bot$
Andreas & Christiane	×	×	×			×			×	×			×	×			
Anupriya & Kedar	×	×		×	×				×	×			×	×			
Astrid & Thomas	×	×		×	×				×	×			×	×	×		

Fig. 2. Starting context for the GDRR-exploration of the role married

Passing the corresponding GCI (which would be $\exists$married.$\top \sqsubseteq$ Person $\sqcap$ Adult $\sqcap$ $\forall$married.(Person $\sqcap$ Adult)) to the OWL DL reasoner does not yield an answer, since it cannot be derived from the given knowledge base. Hence, the human expert has to be asked and would confirm this implication – since we assume a legal system where child marriages are prohibited. So the GCI is added to $\mathcal{KB}$ as a new axiom.

2. Question: {Male$_d$} $\rightarrow$ {Female$_r$} (In words: "If a male is married, is he necessarily married to a female?")
   This axiom which we already encountered in Section 2 is obviously true but cannot be derived from $\mathcal{KB}$. Therefore, it is passed to the human expert, who again would confirm it which leads to another update of $\mathcal{KB}$

3. Question: {Female$_d$} $\rightarrow$ {Male$_r$} (In words: "If a female is married, is she necessarily married to a male?")
   Mark that this axiom is not redundant, since all information specified so far does not exclude the possibility of female-female marriages. Again, the human expert would be asked, confirm the validity and update $\mathcal{KB}$ anew.

4. Question: {Female$_d$, Male$_d$} $\rightarrow$ {$\bot$} (In words: "Is it impossible that somebody married is male and female at the same time?")
   Obviously, the validity of this statement follows from the axiom Male$\sqcap$Female $\sqsubseteq$ $\bot$ contained in the original knowledge base and is therefore silently answered by the reasoner without bothering the expert.

5. Question: {Child$_d$} $\rightarrow$ {$\bot$} (In words: "Is it impossible for a child to be married?")
   It takes little consideration that this axiom can be derived from the updated knowledge base containing Child $\sqcap$ Adult $\sqsubseteq$ $\bot$ as well as the axiom that was added to the $\mathcal{KB}$ as a result of the first question. Thus it is tacitly confirmed by the reasoner as well.

6. Question: {Catholic$_d$} $\rightarrow$ {$\bot$} (In words: "Is it impossible that a Catholic marries?")
   In fact, since none of the marrying individuals entered so far is Catholic, this is a reasonable hypothesis. Of course it cannot be proved from the current KB, but it cannot be rejected either. Again the expert would have to decide on this. This time, he would decline the hypothesis and enter information witnessing this – possibly a married couple of whom at least one is a Catholic.

In this fashion, the exploration proceeds until it terminates. Only one of the hypotheses presented in the sequel has to be confirmed by the human expert (and consequently added to the knowledge base), namely {Catholic$_d$, Priest$_d$} $\rightarrow$ {$\bot$} – an axiom, the validity of which might become subject to change in the centuries to come.

6 Conclusion and Future Work

We have motivated and identified a class of OWL axioms that generalise the well-known domain and range restrictions in an intuitive way and can be expressed both in DL-based as well as rule-based formalisms. Moreover, we have proposed an interactive method for refining a knowledge base with respect to a given role (binary predicate) by acquiring all GDRRs valid in a certain domain of interest. As indicated by the given example, we are sure that the proposed technique will be of great help to domain experts and ontology engineers in specifying their domain since it ensures both consistency of the result and completeness in the above described sense.

There are several directions into which we will proceed with our work. An interesting question directly related to the logical fragment of GDRRs is to what extent role involving OWL axioms present in current ontologies can be expressed in the rather restricted form of GDRRs. This would yield an empirical justification for our claim that the identified fragment is of practical interest.

As to the theoretical foundations, an integration of the presented exploration technique with Relational Exploration [14] seems to be promising. Together with the observation, that in recent years, there have been several similar approaches yet differing in the explored logical fragments as well as the additionally used exploration features, the quest for a unifying general theoretical framework would be beneficial since it could both grant theoretical insights as well as spawn versatile joint work towards an integrated implementation which will proof very useful in the context of knowledge specification for the semantic web.

From the perspective of algorithm implementation and optimization, one question longing for empirical clarification is that for the optimal choice of the optional parts of the algorithm, especially, whether "greedy" or "lazy" scan for a-priori information should be applied (this amounts to the question: reasoning whenever possible vs. reasoning only if necessary). Of course, the optimal choice depends on the performance of the reasoner employed w.r.t. the several mentioned reasoning tasks. Since different reasoners might perform differently well in subsumption checking opposed to instance retrieval, it might even be advisable to use several different reasoners.

Finally, the method presented here fits perfectly into recently started work towards a synergetic integration of exploration techniques with complementary approaches from lexical ontology learning aiming at systems that can be beneficially applied in practical situations, as sketched in [23].

In the end, we are very confident, that "completeness-eligible" fragments of common knowledge representation languages in combination with exploration-based techniques will help to establish unprecedented quality standards for ontologies.

References

1. Baader, F., et al. (eds.): The Description Logic Handbook: Theory, Implementation, and Applications. Cambridge University Press, Cambridge (2003)
2. Horrocks, I., Patel-Schneider, P.F.: A proposal for an OWL rules language. In: Proc. of the Thirteenth International World Wide Web Conference (WWW 2004), pp. 723–731. ACM, New York (2004)
3. McGuinness, D., v. Harmelen, F.: OWL Web Ontology Language Overview. W3C Recommendation (February 10, 2004), http://www.w3.org/TR/owl-features/
4. Ministry of Law and Justice, Government of India: The Child Marriage Restraint Act. Act No. 19 of 1929 (1929), http://indiacode.nic.in
5. Krötzsch, M., Rudolph, S., Hitzler, P.: Complexity boundaries for horn description logics. In: N.N. (ed.) Proceedings of AAAI 2007. LNCS (LNAI), Springer, Heidelberg (2007)

6. Grosof, B., et al.: Description logic programs: Combining logic programs with description logics. In: Proc. of the International World Wide Web Conference (WWW 2003), pp. 48–57. ACM, New York (2003)
7. Busch, W.: Die fromme Helene. Diogenes (2003)
8. Ganter, B.: Two basic algorithms in concept analysis. Technical Report 831, FB4, TH Darmstadt (1984)
9. Burmeister, P.: Merkmalimplikationen bei unvollständigem Wissen. In: Lex, W. (ed.) Arbeitstagung Begriffsanalyse und Künstliche Intelligenz, TU Clausthal, pp. 15–46 (1991)
10. Ganter, B.: Attribute exploration with background knowledge. Theoretical Computer Science 217, 215–233 (1999)
11. Prediger, S.: Terminologische Merkmalslogik in der formalen Begriffsanalyse. In: Stumme, G., Ganter, B. (eds.) Begriffliche Wissensverarbeitung: Methoden und Anwendungen, pp. 99–124. Springer, Heidelberg (2000)
12. Baader, F.: Computing a minimal representation of the subsumption lattice of all conjunctions of concepts defined in a terminology. In: Proceedings of KRUSE 1995, Santa Cruz, USA, pp. 168–178 (1995)
13. Rudolph, S.: An FCA method for the extensional exploration of relational data. In: Ganter, B., de Moor, A. (eds.) Using Conceptual Structures, Contributions to ICCS 2003, Dresden, Germany, Shaker, Aachen, pp. 197–210 (2003)
14. Rudolph, S.: Relational Exploration - Combining Description Logics and Formal Concept Analysis for Knowledge Specification. Dissertation, Universitätsverlag Karlsruhe (2006)
15. Baader, F., et al.: Completing description logic knowledge bases using formal concept analysis. In: Veloso, M.M. (ed.) IJCAI, pp. 230–235 (2007)
16. Ganter, B., Wille, R.: Formal Concept Analysis: Mathematical Foundations. Springer, New York (1997)
17. Dowling, W.F., Gallier, J.H.: Linear-time algorithms for testing the satisfiability of propositional Horn formulae. J. Log. Program. 1, 267–284 (1984)
18. Maier, D.: The Theory of Relational Databases. Computer Science Press (1983)
19. Guigues, J.L., Duquenne, V.: Familles minimales d'implications informatives resultant d'un tableau de données binaires. Math. Sci Humaines 95, 5–18 (1986)
20. Zickwolff, M.: Rule Exploration: First Order Logic in Formal Concept Analysis. PhD thesis, FB4, TH Darmstadt (1991)
21. Ganter, B., Rudolph, S.: Formal concept analysis methods for dynamic conceptual graphs. In: Delugach, H.S., Stumme, G. (eds.) ICCS 2001. LNCS (LNAI), vol. 2120, pp. 143–156. Springer, Heidelberg (2001)
22. Rudolph, S.: Exploring relational structures via FLE. In: Wolff, K.E., Pfeiffer, H.D., Delugach, H.S. (eds.) ICCS 2004. LNCS (LNAI), vol. 3127, pp. 196–212. Springer, Heidelberg (2004)
23. Rudolph, S., Völker, J., Hitzler, P.: Supporting lexical ontology learning by relational exploration. In: Hill, R., Polovina, S., Priss, U. (eds.) Proceedings of ICCS 2007. LNCS (LNAI), Springer, Heidelberg (2007)
24. Ganter, B.: Finding closed sets under symmetry. Technical Report 1307, FB4, TH Darmstadt (1990)

A Finite Basis for the Set of $\mathcal{EL}$-Implications Holding in a Finite Model

Franz Baader* and Felix Distel**

Theoretical Computer Science, TU Dresden, Germany
{baader,felix}@tcs.inf.tu-dresden.de

Abstract. Formal Concept Analysis (FCA) can be used to analyze data given in the form of a formal context. In particular, FCA provides efficient algorithms for computing a minimal basis of the implications holding in the context. In this paper, we extend classical FCA by considering data that are represented by relational structures rather than formal contexts, and by replacing atomic attributes by complex formulae defined in some logic. After generalizing some of the FCA theory to this more general form of contexts, we instantiate the general framework with attributes defined in the Description Logic (DL) $\mathcal{EL}$, and with relational structures over a signature of unary and binary predicates, i.e., models for $\mathcal{EL}$. In this setting, an implication corresponds to a so-called general concept inclusion axiom (GCI) in $\mathcal{EL}$. The main technical result of this paper is that, in $\mathcal{EL}$, for any finite model there is a *finite* set of implications (GCIs) holding in this model from which all implications (GCIs) holding in the model follow.

1 Introduction

Classical Formal Concept Analysis [12] assumes that data from an application are given by a formal context, i.e., by a set of objects G, a set of attributes M, and an incidence relation I that states whether or not an object satisfies a certain attribute. To analyze the data given by such a context, FCA provides tools for computing a minimal basis for the implications between sets of attributes holding in the context [11,8]. An implication $A \rightarrow B$ between sets of attributes A, B holds in a given context if all objects satisfying every attribute in A also satisfy every attribute in B. A classical result by Duquenne and Guigues [13] says that such a unique minimal basis always exists. If the set of attributes is finite, which is usually assumed, this basis is trivially finite as well.

From a model-theoretic or (first-order predicate) logical point of view, a formal context is a very simple relational structure where all predicates (the attributes) are unary. In many applications, however, data are given by more complex relational structures where objects can be linked by relations of arities greater than 1. In order to take these more complex relationships between objects into account

---

* Partially supported by NICTA, Canberra Research Lab.

** Supported by the Cusanuswerk.

when analyzing the data, we consider concepts defined in a certain logic rather than simply sets of atomic attributes (i.e., conjunctions of unary predicates). Intuitively, a concept is a formula with one free variable, and thus determines a subset of the domain (the extension of the concept) for any model of the logic used to construct these formulae. We show that, under certain conditions on this logic, many of the basic results from FCA can be extended to this more general framework. Basically, this requirement is that a finite set of objects (i.e., elements of the domain of a given model) always has a most specific concept describing these objects. The operator that goes from a finite set of objects to its most specific concept corresponds to the prime operator in classical FCA, which goes from a set of objects A to the set of attributes A' that all objects from the set have in common. The classical prime operator in the other direction, which goes from a set of attributes B to the set of objects B' satisfying all these attributes, has as its corresponding operator the one that goes from a concept to its extension.

We instantiate this general framework with concepts defined in the Description Logic $\mathcal{EL}$ [2,3], i.e., formal contexts are replaced by finite models of this DL and attributes are $\mathcal{EL}$-concepts. Though being quite inexpressive, $\mathcal{EL}$ has turned out to be very useful for representing biomedical ontologies such as SNOMED [22] and the Gene Ontology [23]. A major advantage of using an inexpressive DL like $\mathcal{EL}$ is that it allows for efficient reasoning procedures [3,5]. Actually, it turns out that $\mathcal{EL}$ itself does not satisfy the requirements on the logic needed to transfer results from FCA since objects need not have a most specific concept. However, if we extend $\mathcal{EL}$ to $\mathcal{EL}_{\text{gfp}}$ by allowing for cyclic concept definitions interpreted with greatest fixpoint semantics, then the resulting logic satisfies all the necessary requirements. Implications in this setting correspond to so-called general concept inclusion axioms (GCIs), which are available in modern ontology languages such as OWL [14] and are supported by most DL systems [15].

The main technical result of this paper is that, in $\mathcal{EL}$ and in $\mathcal{EL}_{\text{gfp}}$, the set of GCIs holding in a finite model always has a finite basis, i.e., although there are in general infinitely many such GCIs, we can always find a finite subset from which the rest follows. We construct such a finite basis first for $\mathcal{EL}_{\text{gfp}}$, and then show how this basis can be modified to yield one for $\mathcal{EL}$. Due to the space limitation, we cannot give complete proofs of these results. They can be found in [4].

Related work. There have been previous approaches for dealing with more complex contexts involving relations between objects. So-called power context families [24] allow for the representation of relational structures by using a separate (classical) context for each arity, where the objects of the context for arity n are n-tuples. As such, power context families are just an FCA-style way of representing relational structures. In order to make use of the more complex relational structure given by power context families, Prediger [16,18,17] and Priss [19] allow the knowledge engineer to define new attributes, and provide means for handling the dependencies between the newly defined attributes and existing attributes by means of formal concept analysis. However, rather than considering all complex attributes definable by the logical language, as our approach does, they restrict the attention to finitely many attributes explicitly defined by the knowledge engineer.

Similar to our general framework, Ferré [6,7] considers complex attributes definable by some logical language. The equivalent of a formal context, called logical context in [6,7], associates a formula (i.e., a complex attribute) with each object. Since the authors assume that formulae form a join-semilattice, the formula associated with a set of objects is obtained as the join of the formulae associated with the elements of the set. Our general framework can be seen as an instance of the one defined in [6,7], where the association of formulae to (sets of) objects is defined using the semantics of the logic in question. However, Ferré's work does not consider implications, which is the main focus of the present paper (see [4] for a more detailed comparison of our approach with the one in [6,7]). An approach similar to the one of [6,7] was developed in [10] motivated by an application in biochemistry.

The work whose objectives are closest to ours is the one by Rudolph [20,21], who considers attributes defined in the DL $\mathcal{FLE}$, which is more expressive than $\mathcal{EL}$. However, instead of using one generalized context with infinitely many complex attributes, he considers an infinite family of contexts, each with finitely many attributes, obtained by restricting the so-called role depth of the concepts. He then applies attribute exploration [9] to the classical contexts obtained this way, in each step increasing the role depths until a certain termination condition applies. Rudolph shows that, for a finite model, this condition will always be satisfied eventually, and that the implication bases of the contexts considered up to that step contain enough information to decide, for any GCI between $\mathcal{FLE}$-concepts, whether this GCI holds in the given model or not. However, these implication bases do not appear to yield a basis for all the GCIs holding in the given finite model, though it might be possible to modify Rudolph's approach such that it produces a basis in our sense. The main problem with this approach is, however, that the number of attributes grows very fast when the role depth grows (this number increases at least by one exponential in each step).

2   The General Framework

In classical FCA, a formal context (G, M, I) consists of a set of objects G, a set of attributes M, and an incidence relation $I \subseteq G \times M$. Such a formal context induces two operators (both usually denoted by $\cdot'$), one mapping each set of objects A to the set of attributes A' these objects have in common, and the other mapping each set of attributes B to the set of objects satisfying these attributes. A formal concept is a pair (A, B) such that $G \supseteq A = B'$ and $M \supseteq B = A'$. The set A is the extensional description of the concept whereas B is its intensional description. The two $\cdot'$ operators form a Galois connection, and if applied twice yield closure operators $\cdot''$ on the set of objects and the set of attributes, respectively.

In our general framework, we assume that intensional descriptions of sets of objects are given by concept descriptions. A *concept description language* is a pair $(\mathcal{L}, \mathcal{I})$, where $\mathcal{L}$ is a set, whose elements are called *concept descriptions*, and

$\mathcal{I}$ is a set of tuples $i = (\Delta_i, \cdot^i)$, called *models*, consisting of a non-empty set Δ_i (of objects) and a mapping $\cdot^i : \mathcal{L} \to \mathfrak{P}(\Delta_i) : f \mapsto f^i$ that assigns an *extension* $f^i \subseteq \Delta_i$ to each concept description $f \in \mathcal{L}$.

Intuitively, models correspond to formal contexts, and the operator $\cdot^i$ corresponds to the $\cdot'$ operator that assigns an extension B' to each set of attributes B. In order to define an analogon to the $\cdot'$ operator in the other direction, we introduce the subsumption preorder on concept descriptions: $f_1 \in \mathcal{L}$ *is subsumed by* $f_2 \in \mathcal{L}$ (written $f_1 \sqsubseteq f_2$) if $f_1^i \subseteq f_2^i$ for all models $i \in \mathcal{I}$. If $f_1 \sqsubseteq f_2$ and $f_2 \sqsubseteq f_1$, then we say that f_1 and f_2 are *equivalent* ($f_1 \equiv f_2$).

Given a set of objects A in a formal context, its intensional description A' is the largest set of attributes B such that $A \subseteq B'$. Since $B_1' \subseteq B_2'$ if $B_1 \supseteq B_2$, such a largest set should correspond to the least one w.r.t. subsumption. This motivates the following definition.

Definition 1 (Most specific concept). *Let $(\mathcal{L}, \mathcal{I})$ be a concept description language, $i \in \mathcal{I}$ be a model, and $X \subseteq \Delta_i$. Then $f \in \mathcal{L}$ is a* most specific concept *for X in i if*

$$X \subseteq f^i, \tag{1}$$

and f is a least concept description with this property, i.e., for all $g \in \mathcal{L}$ with $X \subseteq g^i$ we have $f \sqsubseteq g$.

The most specific concept of a set $X \subseteq \Delta_i$ need not exist, but if it exists then it is unique up to equivalence. In case X has a most specific concept in i, we denote it (or, more precisely, an arbitrary element of its equivalence class) by X^i. The concept description X^i is called the *intensional description* of the set of objects X. An example of a concept description language for which X^i always exists is $\mathcal{EL}_{\mathrm{gfp}}$, which will be introduced in Section 3 below.

The following lemma shows that the mappings

$$\cdot^i : \mathfrak{P}(\Delta_i) \to \mathcal{L} \quad \text{and} \quad \cdot^i : \mathcal{L} \to \mathfrak{P}(\Delta_i)$$

do indeed form a Galois-connection with properties similar to the $\cdot'$ operators in classical FCA. Because of these similarities to FCA we will sometimes use the term *description context* for a model $i \in \mathcal{I}$.

Lemma 2. *Let $(\mathcal{L}, \mathcal{I})$ be a concept description language such that X^i exists for every $i \in \mathcal{I}$ and every $X \subseteq \Delta_i$. Let $i \in \mathcal{I}$ be a model, $X, X_1, X_2 \subseteq \Delta_i$ sets of objects, and $f, f_1, f_2 \in \mathcal{L}$ concept descriptions. Then the following holds:*

(a) $X_1 \subseteq X_2 \Rightarrow X_1^i \sqsubseteq X_2^i$,
(b) $f_1 \sqsubseteq f_2 \Rightarrow f_1^i \subseteq f_2^i$,
(c) $X \subseteq X^{ii}$,
(d) $f^{ii} \sqsubseteq f$,

(e) $X^i \equiv X^{iii}$,
(f) $f^i = f^{iii}$,
(g) $X \subseteq f^i \Leftrightarrow X^i \sqsubseteq f$.

Proofs of these facts can be obtained by adapting the proofs from classical FCA. They can be found in [4], but also in [6,7] since the framework introduced above can be seen as an instance of the framework defined in [6,7].

In the remainder of this section, we assume that $(\mathcal{L}, \mathcal{I})$ is an arbitrary, but fixed, concept description language. All definitions given below are implicitly parameterized with this language. Our goal is to characterize the subsumption relations that are valid in a given description context of this language by determining a minimal basis of implications comparable to the Duquenne-Guigues basis in classical FCA. We start by defining the notion of an implication and by showing some general results that hold for arbitrary concept description languages. Later on, we will look at the concept description language $\mathcal{EL}_{\mathrm{gfp}}$ in more detail.

Definition 3 (Implication). *An implication is a pair (f_1, f_2) of concept descriptions $(f_1, f_2) \in \mathcal{L} \times \mathcal{L}$, which we will usually denote as $f_1 \to f_2$. We say that the implication $f_1 \to f_2$ holds in the description context $\iota = (\Delta_\iota, \iota)$ if $f_1^\iota \subseteq f_2^\iota$.*

Obviously, we have $f_1 \sqsubseteq f_2$ iff $f_1 \to f_2$ holds in every description context $\iota \in \mathcal{I}$. However, as said above, we are now interested in the implications that hold in a fixed description context rather than in all of them.

In order to define the notion of a basis of the implications holding in a description context, we must first define a consequence operator on implications. Let $\mathcal{B} \subseteq \mathcal{L} \times \mathcal{L}$ be a set of implications and $f_1 \to f_2$ an implication. If $f_1 \to f_2$ holds in all description contexts $i \in \mathcal{I}$ in which all implications from $\mathcal{B}$ hold, then we say that $f_1 \to f_2$ *follows* from $\mathcal{B}$. It is not hard to see that the relation *follows* is

- *reflexive*, i.e. every implication $f_1 \to f_2 \in \mathcal{B}$ follows from $\mathcal{B}$, and
- *transitive*, i.e. if $f_1 \to f_2$ follows from $\mathcal{B}_2$, and every implication in $\mathcal{B}_2$ follows from $\mathcal{B}_1$, then $f_1 \to f_2$ follows from $\mathcal{B}_1$.

Definition 4 (Basis). *For a given description context ι we say that $\mathcal{B} \subseteq \mathcal{L} \times \mathcal{L}$ is a basis for the implications holding in ι if $\mathcal{B}$ is*

- sound *for ι, i.e., it contains only implications holding in ι;*
- complete *for ι, i.e., any implication that holds in ι follows from $\mathcal{B}$; and*
- minimal *for ι, i.e., no strict subset of $\mathcal{B}$ is complete for ι.*

Since the above definitions use only the $\cdot^\iota$ operator that assigns an extension to every concept description, but not the one in the other direction, they also make sense for concept description languages where the most specific concept of a set of objects need not always exist. An example of such a language is $\mathcal{EL}$, i.e., the sublanguage of $\mathcal{EL}_{\mathrm{gfp}}$ that does not allow for cyclic concept definitions (see Section 3 below).

The description language $(\mathcal{L}', \mathcal{I}')$ is a *sublanguage* of the description language $(\mathcal{L}, \mathcal{I})$ if $\mathcal{L}' \subseteq \mathcal{L}$ and $\mathcal{I}' = \{i|_{\mathcal{L}'} \mid i \in \mathcal{I}\}$, where $i|_{\mathcal{L}'}$ is the restriction of i to $\mathcal{L}'$, i.e., $\Delta_i = \Delta_{i|_{\mathcal{L}'}}$ and $\cdot^{i}|_{\mathcal{L}'}$ is the restriction of the mapping $\cdot^i$ to $\mathcal{L}'$.

Proposition 5. *Assume that $(\mathcal{L}', \mathcal{I}')$ is a sublanguage of $(\mathcal{L}, \mathcal{I})$, that $f_1 \to f_2 \in \mathcal{L}' \times \mathcal{L}'$, and that $\mathcal{B} \subseteq \mathcal{L}' \times \mathcal{L}'$. Then $f_1 \to f_2$ follows from $\mathcal{B}$ in $(\mathcal{L}, \mathcal{I})$ iff $f_1 \to f_2$ follows from $\mathcal{B}$ in $(\mathcal{L}', \mathcal{I}')$.*

This proposition will be used later on to transfer results from $\mathcal{EL}_{\mathrm{gfp}}$ to $\mathcal{EL}$.

In the remainder of this section, we will characterize complete subsets of the set of all implications holding in a given description context ι. Whenever we use the $\cdot^{\iota}$ operator from sets of objects to concept descriptions, we implicitly assume that it is defined.

Analogously to the situation in classical FCA, we can restrict the attention to implications whose right-hand sides are closed under the operator $\cdot^{\iota\iota}$.

Lemma 6. *If the implication $f_1 \to f_2$ holds in ι, then it follows from $\{f_1 \to f_1^{\iota\iota}\}$, and the set $\{f_1 \to f_1^{\iota\iota}\}$ is sound for ι.*

Proof. By Lemma 2(f), all implications of the form $f \to f^{\iota\iota}$ hold in ι, which yields soundness of $\{f_1 \to f_1^{\iota\iota}\}$.

Let $f_1 \to f_2$ be any implication that holds in ι, i.e., $f_1^{\iota} \subseteq f_2^{\iota}$. By Lemma 2(g), this is equivalent to

$$f_1^{\iota\iota} \sqsubseteq f_2. \tag{2}$$

Let i be some model in which $f_1 \to f_1^{\iota\iota}$ holds. By definition this means that $f_1^i \subseteq (f_1^{\iota\iota})^i$. Using Lemma 2(g) again we obtain $f_1^{ii} \sqsubseteq f_1^{\iota\iota}$. Together with (2) and transitivity of $\sqsubseteq$, this yields $f_1^{ii} \sqsubseteq f_2$, and hence $f_1^i \subseteq f_2^i$. Thus, we have shown that $f_1 \to f_2$ holds in any model i in which $f_1 \to f_1^{\iota\iota}$ holds. $\qquad\square$

Corollary 7. *The set of implications $\{f \to f^{\iota\iota} \mid f \in \mathcal{L}\}$ is complete for ι.*

Having reduced the number of right-hand sides that need to be considered, our goal is now to restrict the left-hand sides. This is possible if we can find a so-called dominating set of concept descriptions.

Definition 8 (Dominating set). *The set $\mathcal{D} \subseteq \mathcal{L}$ dominates the description context ι if, for every $f \in \mathcal{L}$, there is some $g \in \mathcal{D}$ such that $f \sqsubseteq g$ and $f^{\iota} = g^{\iota}$.*

It is sufficient to consider implications whose left-hand sides belong to a dominating set.

Lemma 9. *If $\mathcal{D} \subseteq \mathcal{L}$ dominates ι, then $\mathcal{B} := \{f \to f^{\iota\iota} \mid f \in \mathcal{D}\}$ is sound and complete for ι.*

Proof. Soundness has already been shown. To show completeness, let $f_1 \to f_2$ be an implication that holds in ι. By Lemma 6, $f_1 \to f_2$ follows from $f_1 \to f_1^{\iota\iota}$. Hence it is sufficient to show that $f_1 \to f_1^{\iota\iota}$ follows from $\mathcal{B}$. Since $\mathcal{D}$ dominates ι, there exists $g \in \mathcal{D}$ such that $f_1 \sqsubseteq g$ and $g^{\iota} = f_1^{\iota}$.

Let i be a model in which all implications of $\mathcal{B}$ hold. From $f_1 \sqsubseteq g$ and Lemma 2(b) it follows that $f_1^i \subseteq g^i$. Since $g \to g^{\iota\iota} \in \mathcal{B}$ holds in i, we also have $g^i \subseteq (g^{\iota\iota})^i$, and thus $f_1^i \subseteq (g^{\iota\iota})^i$. In addition, $g^{\iota} = f_1^{\iota}$ yields $g^{\iota\iota} = f_1^{\iota\iota}$. Thus, $f_1^i \subseteq (f_1^{\iota\iota})^i$, which shows that $f_1 \to f_1^{\iota\iota}$ holds in i. $\qquad\square$

The sound and complete set of implications $\mathcal{B}$ induced by a dominating set $\mathcal{D}$ need not be a basis since it need not be minimal. However, if $\mathcal{D}$ is finite, then $\mathcal{B}$ is finite as well. Thus, a subset of $\mathcal{B}$ that is a basis can be obtained by removing redundant elements.

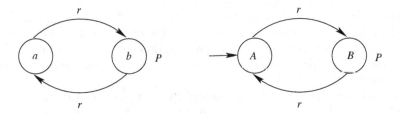

Fig. 1. A model (left) and a description graph (right)

3 $\mathcal{EL}_{\text{gfp}}$ as an Instance of the General Framework

We start by defining $\mathcal{EL}$, and then show how it can be extended to $\mathcal{EL}_{\text{gfp}}$. Concept descriptions of $\mathcal{EL}$ are built from a set $\mathcal{N}_c$ of concept names and a set $\mathcal{N}_r$ of role names, using the constructors top concept, conjunction, and existential restriction:

- concept names and the top concept $\top$ are $\mathcal{EL}$-concept descriptions;
- if C, D are $\mathcal{EL}$-concept descriptions and r is a role name, then $C \sqcap D$ and $\exists r.C$ are $\mathcal{EL}$-concept descriptions.

In the following, we assume that the sets $\mathcal{N}_c$ and $\mathcal{N}_r$ of concept and role names are finite. This assumption is reasonable since in practice data are usually represented over a finite signature.

Models of this language are pairs $(\Delta_I, \cdot^I)$ where Δ_I is a finite,[1] non-empty set, and $\cdot^I$ maps role names r to binary relations $r^I \subseteq \Delta_I \times \Delta_I$ and $\mathcal{EL}$-concept descriptions to subsets of Δ_I such that

$$\top^I = \Delta_I, \qquad\qquad (C \sqcap D)^I = C^I \cap D^I, \quad \text{and}$$

$$(\exists r.C)^I \;\; = \{d \in \Delta_i \mid \exists e \in C^I \text{ such that } (d, e) \in r^I\}.$$

Subsumption and equivalence between $\mathcal{EL}$-concept descriptions is defined as in our general framework, i.e., $C \sqsubseteq D$ iff $C^I \sqsubseteq D^I$ for all models I, and $C \equiv D$ iff $C \sqsubseteq D$ and $D \sqsubseteq C$.

Unfortunately, $\mathcal{EL}$ itself cannot be used to instantiate our framework since in general a set of objects need not have a most specific concept in $\mathcal{EL}$. This is illustrated by the following simple example. Assume that $\mathcal{N}_c = \{P\}$, $\mathcal{N}_r = \{r\}$, and consider the model I with $\Delta_I = \{a, b\}$, $r^I = \{(a, b), (b, a)\}$, and $P^I = \{b\}$ (see the left-hand side of Fig. 1 for a graphical representation of this model). To see that the set $\{a\}$ does not have a most specific concept, consider the $\mathcal{EL}$-concept descriptions

$$C_k := \underbrace{\exists r. \exists r \ldots \exists r.}_{k \text{ times}} \top.$$

---

[1] Usually, the semantics given for description logics allows for models of arbitrary cardinality. However, in the case of $\mathcal{EL}$ the restriction to finite models is without loss of generality since it has the finite model property, i.e., a subsumption relationship holds w.r.t. all models iff it holds w.r.t. all finite models.

We have $\{a\} \subseteq C_k^I = \{a, b\}$ for all k, and thus a most specific concept C for $\{a\}$ would need to satisfy $C \sqsubseteq C_k$ for all $k \geq 0$. However, it is easy to see that $C \sqsubseteq C_k$ can only be true if the role depth of C, i.e., the maximal nesting of existential restrictions, is at least k. Since any $\mathcal{EL}$-concept description has a finite role depth, this shows that such a most specific concept C cannot exist.

However, most specific concepts always exist in $\mathcal{EL}_{\mathrm{gfp}}$, the extension of $\mathcal{EL}$ by cyclic concept definitions interpreted with greatest fixpoint (gfp) semantics.[2] In $\mathcal{EL}_{\mathrm{gfp}}$, we assume that the set of concept names is partitioned into the set $\mathcal{N}_{\mathrm{prim}}$ of primitive concepts and the set $\mathcal{N}_{\mathrm{def}}$ of defined concept. A *concept definition* is of the form

$$B_0 \equiv P_1 \sqcap \ldots \sqcap P_m \sqcap \exists r_1.B_1 \sqcap \ldots \sqcap \exists r_n.B_n$$

where $B_0, B_1, \ldots, B_n \in \mathcal{N}_{\mathrm{def}}$, $P_1, \ldots, P_m \in \mathcal{N}_{\mathrm{prim}}$, and $r_1, \ldots, r_n \in \mathcal{N}_r$. The empty conjunction (i.e., $m = 0 = n$) stands for $\top$. A *TBox* is a finite set of concept definitions such that every defined concept occurs at most once as a left-hand side of a concept definition.

Definition 10 ($\mathcal{EL}_{\mathrm{gfp}}$-concept description). *An $\mathcal{EL}_{\mathrm{gfp}}$-concept description is a tuple $(A, \mathcal{T})$ where $\mathcal{T}$ is a TBox and A is a defined concept occurring on the left-hand side of a definition in $\mathcal{T}$.*

For example, $(A, \mathcal{T})$ with $\mathcal{T} := \{A \equiv \exists r.B, B \equiv P \sqcap \exists r.A\}$ is an $\mathcal{EL}_{\mathrm{gfp}}$-concept description. Any $\mathcal{EL}_{\mathrm{gfp}}$-concept description $(A, \mathcal{T})$ can be represented by a directed, rooted, edge- and node-labeled graph: the nodes of this graph are the defined concepts in $\mathcal{T}$, with A being the root; the edge label of node B_0 is the set of primitive concepts occurring in the definition of B_0; and every conjunct $\exists r_i.B_i$ in the definition of B_0 gives rise to an edge from B_0 to B_i with label r_i. In the following, we call such graphs *description graphs*. The description graph associated with the $\mathcal{EL}_{\mathrm{gfp}}$-concept description from our example is shown on the right-hand side of Fig. 1, where A is the root.

Models of $\mathcal{EL}_{\mathrm{gfp}}$ are of the form $I = (\Delta_I, \cdot^I)$ where Δ_I is a finite, non-empty set, and $\cdot^I$ maps role names r to binary relations $r^I \subseteq \Delta_I \times \Delta_I$ and primitive concepts to subsets of Δ_I. The mapping $\cdot^I$ is extended to $\mathcal{EL}_{\mathrm{gfp}}$-concept descriptions $(A, \mathcal{T})$ by interpreting the TBox $\mathcal{T}$ with gfp-semantics: consider all extensions of I to the defined concepts that satisfy the concept definitions in $\mathcal{T}$, i.e., assign the same extension to the left-hand side and the right-hand side of each definition. Among these extensions of I, the *gfp-model of $\mathcal{T}$ based on I* is the one that assigns the largest sets to the defined concepts (see [1] for a more detailed definition of gfp-semantics). The *extension* $(A, \mathcal{T})^I$ of $(A, \mathcal{T})$ in I is the set assigned to A by the gfp-model of $\mathcal{T}$ based on I.

Again, subsumption and equivalence of $\mathcal{EL}_{\mathrm{gfp}}$-concept descriptions is defined as in the general framework. The following theorem shows that the description language $\mathcal{EL}_{\mathrm{gfp}}$ we have just defined is indeed an instance of the framework introduced in Section 2.

---

[2] Because of the space restriction, we can only give a very compact introduction of this DL. See [1,4] for more details.

Theorem 11. *In $\mathcal{EL}_{\mathrm{gfp}}$, the most specific concept of a set of objects always exists.*

The proof of this theorem given in [4] is based on the methods and results from [2]. It proceeds in two steps. First, it is shown how to construct the most specific concept of a singleton set $\{a\}$. The main idea is that the graph representing the model can also be viewed as the description graph of an $\mathcal{EL}_{\mathrm{gfp}}$-concept description, where the root is the node corresponding to a. In the example (see Fig. 1), we have simply renamed the lower case individual names into upper case concept names. The $\mathcal{EL}_{\mathrm{gfp}}$-concept description $(A, \mathcal{T})$ represented by the description graph on the right-hand side of Fig. 1 is the most specific concept of $\{a\}$ in the model represented by the graph on the left-hand side of Fig. 1. The most specific concept of a set of objects $\{a_1, \ldots, a_n\}$ is the least common subsumer (lcs) of the most specific concepts of the singleton sets $\{a_i\}$. In [2] it is shown that the lcs in $\mathcal{EL}_{\mathrm{gfp}}$ always exists and how to compute it.

4   A Finite Basis for Implications in $\mathcal{EL}_{\mathrm{gfp}}$

We show that the set of implications holding in a given model always has a finite basis in $\mathcal{EL}_{\mathrm{gfp}}$. A first step in this direction is to show that it is enough to restrict the attention to implications with acyclic $\mathcal{EL}_{\mathrm{gfp}}$-concept descriptions as left-hand sides. The $\mathcal{EL}_{\mathrm{gfp}}$-concept description $(A, \mathcal{T})$ is *acyclic* if the graph associated with it is acyclic. It is easy to see that there is a 1–1-relationship between $\mathcal{EL}$-concept descriptions and acyclic $\mathcal{EL}_{\mathrm{gfp}}$-concept descriptions. For example, $(A, \{A \equiv B \sqcap \exists r.B, B \equiv P\})$ corresponds to $P \sqcap \exists r.P$, and $\exists r.P$ corresponds to $(A, \{A \equiv \exists r.B, B \equiv P\})$. This shows that $\mathcal{EL}$ can indeed be seen as a sublanguage of $\mathcal{EL}_{\mathrm{gfp}}$. In the following, we will not distinguish an acyclic $\mathcal{EL}_{\mathrm{gfp}}$-concept description from its equivalent $\mathcal{EL}$-concept description.

Given an $\mathcal{EL}_{\mathrm{gfp}}$-concept description, its *node size* is the number of nodes in the description graph corresponding to it. For an acyclic $\mathcal{EL}_{\mathrm{gfp}}$-concept description, we define its *depth* to be the maximal length of a path starting at the root in the description graph corresponding to it. Any $\mathcal{EL}_{\mathrm{gfp}}$-concept description $(A, \mathcal{T})$ can be approximated by acyclic $\mathcal{EL}_{\mathrm{gfp}}$-concept descriptions $(A, \mathcal{T})_d$ of increasing depth d. To obtain $(A, \mathcal{T})_d$, the description graph associated with $(A, \mathcal{T})$ is unraveled into a (possibly infinite) tree, and then all branches are cut at depth d. It is easy to see that $(A, \mathcal{T}) \sqsubseteq (A, \mathcal{T})_d$ holds for all $d \geq 0$.

Lemma 12. *Let $\mathcal{U}$ be an $\mathcal{EL}_{\mathrm{gfp}}$-concept description of node size m, I a model of cardinality n, and $d = m \cdot n + 1$. Then $a \in (\mathcal{U}_d)^I$ implies $a \in \mathcal{U}^I$.*

A detailed proof of this lemma can be found in [4].

Theorem 13. *In $\mathcal{EL}_{\mathrm{gfp}}$, the set of acyclic concept descriptions dominates every description context I.*

Proof. Let $\mathcal{U}$ be an $\mathcal{EL}_{\mathrm{gfp}}$-concept description and I a description context. We must find an acyclic $\mathcal{EL}_{\mathrm{gfp}}$-concept description $\mathcal{V}$ such that $\mathcal{U} \sqsubseteq \mathcal{V}$ and $\mathcal{U}^I = \mathcal{V}^I$.

Let m be the node size of $\mathcal{U}$, n the cardinality of I, and $d = m \cdot n + 1$. We know that $\mathcal{U} \sqsubseteq \mathcal{U}_d$, and thus also $\mathcal{U}^I \subseteq (\mathcal{U}_d)^I$. Lemma 12 shows that the inclusion in the other direction holds as well. Thus, $\mathcal{V} := \mathcal{U}_d$ does the job. $\square$

By Lemma 9, this theorem immediately implies the following corollary.

Corollary 14. *For any description context I of $\mathcal{EL}_{\mathrm{gfp}}$, the set*

$$\{\mathcal{U} \to \mathcal{U}^{II} \mid \mathcal{U} \text{ is an acyclic } \mathcal{EL}_{\mathrm{gfp}}\text{-concept description}\}$$

is sound and complete for I.

The complete set of implications given in the corollary is, of course, infinite. Also note that, though the left-hand sides $\mathcal{U}$ of implications in this set are acyclic, the right-hand sides $\mathcal{U}^{II}$ need not be acyclic. We show next that there is also a *finite* sound and complete set of implications. As mentioned before, a finite basis can then be obtained by removing redundant elements.

Theorem 15. *In $\mathcal{EL}_{\mathrm{gfp}}$, for any description context I, there exists a* finite *set $\mathcal{B}$ of implications that is sound and complete for I.*

Proof. By Corollary 14 it suffices to find a finite and sound set of implications from which all implications of the form $\mathcal{U} \to \mathcal{U}^{II}$, where $\mathcal{U}$ is an acyclic $\mathcal{EL}_{\mathrm{gfp}}$-concept description, follow. To this purpose, consider the set $\mathcal{E} := \{\mathcal{U}^I \mid \mathcal{U} \text{ is an } \mathcal{EL}_{\mathrm{gfp}}\text{-concept description}\}$, and let $\mathcal{C}$ be a set of $\mathcal{EL}_{\mathrm{gfp}}$-concept descriptions that contains, for each set $X \in \mathcal{E}$, exactly one element $\mathcal{V}$ with $\mathcal{V}^I = X$. Because of Theorem 13, we can assume without loss of generality that $\mathcal{C}$ contains only acyclic descriptions. Since Δ_I is finite, the sets $\mathcal{E}$ and $\mathcal{C}$ are also finite.

Consider the following finite set of implications, which is obviously sound:

$$\mathcal{B} := \{P \to P^{II} \mid P \in \mathcal{N}_{\mathrm{prim}} \cup \{\top\}\}$$
$$\cup \{\exists r.C \to (\exists r.C)^{II} \mid r \in \mathcal{N}_r, C \in \mathcal{C}\}$$
$$\cup \{C_1 \sqcap C_2 \to (C_1 \sqcap C_2)^{II} \mid C_1, C_2 \in \mathcal{C}\}.$$

We show that, for any acyclic $\mathcal{EL}_{\mathrm{gfp}}$-concept description $\mathcal{U}$, the implication $\mathcal{U} \to \mathcal{U}^{II}$ follows from $\mathcal{B}$. Since $\mathcal{U}$ is acyclic, we can view it as an $\mathcal{EL}$-concept description. The proof is by induction on the structure of this description.

Base case: $\mathcal{U} = P \in \mathcal{N}_{\mathrm{prim}} \cup \{\top\}$. Then $P \to P^{II}$ is in $\mathcal{B}$ by definition. Thus, it also follows from $\mathcal{B}$.

Step case 1: $\mathcal{U} = \exists r.\mathcal{V}$ for some $r \in \mathcal{N}_r$ and some $\mathcal{EL}$-concept description $\mathcal{V}$. Let J be a description context in which all implications from $\mathcal{B}$ hold. The semantics of existential restrictions yields

$$\mathcal{U}^J = (\exists r.\mathcal{V})^J = \{x \in \Delta_J \mid \exists y \in \mathcal{V}^J : (x, y) \in r^J\}.$$

By the induction hypothesis, $\mathcal{V} \to \mathcal{V}^{II}$ follows from $\mathcal{B}$, and thus holds in J. Therefore $\mathcal{V}^J \subseteq (\mathcal{V}^{II})^J$, which yields

$$\mathcal{U}^J \subseteq \{x \in \Delta_J \mid \exists y \in (\mathcal{V}^{II})^J : (x, y) \in r^J\}.$$

Now, choose $C \in \mathcal{C}$ such that $C^I = \mathcal{V}^I$. Lemma 2(g) yields $\mathcal{V}^{II} \sqsubseteq C$, and thus

$$\mathcal{U}^J \subseteq \{x \in \Delta_J \mid \exists y \in C^J : (x,y) \in r^J\} = (\exists r.C)^J.$$

Since $\exists r.C \to (\exists r.C)^{II} \in \mathcal{B}$ holds in J by assumption, we get

$$\mathcal{U}^J \subseteq ((\exists r.C)^{II})^J = (\{x \in \Delta_I \mid \exists y \in C^I : (x,y) \in r^I\}^I)^J =$$
$$= (\{x \in \Delta_I \mid \exists y \in \mathcal{V}^I : (x,y) \in r^I\}^I)^J = ((\exists r.\mathcal{V})^{II})^J = (\mathcal{U}^{II})^J.$$

Thus, we have shown that $\mathcal{U} \to \mathcal{U}^{II}$ holds in every context J in which all implications from $\mathcal{B}$ hold.

Step case 2: $\mathcal{U} = \mathcal{U}_1 \sqcap \mathcal{U}_2$ for $\mathcal{EL}$-concept descriptions $\mathcal{U}_1, \mathcal{U}_2$. Let J be a description context in which all implications from $\mathcal{B}$ hold. By the induction hypothesis, $\mathcal{U}_1^J \subseteq (\mathcal{U}_1^{II})^J$ and $\mathcal{U}_2^J \subseteq (\mathcal{U}_2^{II})^J$. Therefore

$$\mathcal{U}^J = (\mathcal{U}_1 \sqcap \mathcal{U}_2)^J = \mathcal{U}_1^J \cap \mathcal{U}_2^J \subseteq (\mathcal{U}_1^{II})^J \cap (\mathcal{U}_2^{II})^J.$$

We choose $C_1, C_2 \in \mathcal{C}$ such that $C_1^I = \mathcal{U}_1^I$ and $C_2^I = \mathcal{U}_2^I$. Then

$$\mathcal{U}^J \subseteq (C_1^{II})^J \cap (C_2^{II})^J \subseteq C_1^J \cap C_2^J = (C_1 \sqcap C_2)^J,$$

where the second inclusion holds due to Lemma 2(d). Since the implication $C_1 \sqcap C_2 \to (C_1 \sqcap C_2)^{II} \in \mathcal{B}$ holds in J, we get

$$\mathcal{U}^J \subseteq ((C_1 \sqcap C_2)^{II})^J = ((C_1^I \cap C_2^I)^I)^J = ((\mathcal{U}_1^I \cap \mathcal{U}_2^I)^I)^J =$$
$$= ((\mathcal{U}_1 \sqcap \mathcal{U}_2)^{II})^J = (\mathcal{U}^{II})^J.$$

This shows that $\mathcal{U} \to \mathcal{U}^{II}$ follows from $\mathcal{B}$.   □

Corollary 16. *In $\mathcal{EL}_{\mathrm{gfp}}$, for any description context I there exists a finite basis for the implications holding in I.*

Proof. Starting with $\mathcal{B}^* := \mathcal{B}$, where in the beginning all implications are unmarked, take an unmarked implication $\mathcal{U} \to \mathcal{V} \in \mathcal{B}^*$. If this implication follows from $\mathcal{B}^*$, then remove it, i.e., $\mathcal{B}^* := \mathcal{B}^* \setminus \{\mathcal{U} \to \mathcal{V}\}$; otherwise, mark $\mathcal{U} \to \mathcal{V}$. Continue with this until all implications in $\mathcal{B}^*$ are marked. The final set $\mathcal{B}^*$ is the desired basis.   □

5   A Finite Basis for Implications in $\mathcal{EL}$

Although the sublanguage $\mathcal{EL}$ of $\mathcal{EL}_{\mathrm{gfp}}$ is not an instance of our general framework, we can nevertheless show the above corollary also for this language. Because of Proposition 5, it is sufficient to show that in $\mathcal{EL}_{\mathrm{gfp}}$ any description context I has a finite basis consisting of implications where both the left-hand and the right-hand sides are acyclic.

The following proposition will allow us to construct a finite set of implications *with acyclic right-hand sides* from which a given implication $\mathcal{U} \to \mathcal{U}^{II}$ (with potentially cyclic right-hand side) follows. Recall that, for any $\mathcal{EL}_{\mathrm{gfp}}$-concept description $\mathcal{U}$, we obtain the acyclic description $\mathcal{U}_d$ by unraveling the description graph and then cutting all branches at depth d.

Proposition 17. *Let k_0 be a non-negative integer, I a description context, and $\mathcal{U}$ be an $\mathcal{EL}_{\mathrm{gfp}}$-concept description. Then the implication $\mathcal{U} \to \mathcal{U}^{II}$ follows from*

$$\mathcal{B} := \{(X^I)_{k_0} \to (X^I)_{k_0+1} \mid X \subseteq \Delta_I\} \cup \{\mathcal{U} \to (\mathcal{U}^{II})_{k_0}\}.$$

Proof. The proof depends on the following technical result, whose proof can be found in [4].

(∗) For any set $X \subseteq \Delta_I$, there exist sets $\mathcal{P} \subseteq \mathcal{N}_{\mathrm{prim}}$ and $\mathcal{Y} \subseteq \mathcal{N}_r \times \mathfrak{P}(\Delta_I)$ such that

$$X^I \equiv \prod_{P \in \mathcal{P}} P \sqcap \prod_{(r,Y) \in \mathcal{Y}} \exists r.Y^I.$$

The above equivalence is actually an abbreviation for saying that X^I is of the form $(A, \mathcal{T})$ where $\mathcal{T}$ consists of the following concept definitions:

- $A \equiv \prod_{P \in \mathcal{P}} P \sqcap \prod_{(r,Y) \in \mathcal{Y}} \exists r.B_{r,Y}$;
- the definitions in the TBoxes $\mathcal{T}_{r,Y}$ for $(r, Y) \in \mathcal{Y}$ where $Y^I = (B_{r,Y}, \mathcal{T}_{r,Y})$.

Note that the sets of defined concepts in the TBoxes $\mathcal{T}_{r,Y}$ can be assumed to be pairwise disjoint and not to contain A.

To prove the proposition, we first show, by induction on ℓ, that the implications $(X^I)_\ell \to (X^I)_{\ell+1}$ follow from $\mathcal{B}$ for all $\ell \geq k_0$. For $\ell = k_0$ this is trivial because $(X^I)_{k_0} \to (X^I)_{k_0+1} \in \mathcal{B}$.

Now, assume that $(Y^I)_k \to (Y^I)_{k+1}$ follows from $\mathcal{B}$ for every $Y \subseteq \Delta_I$ and every $k, k_0 \leq k < \ell$. Let J be a model in which all implications from $\mathcal{B}$ hold. Then, by the induction hypothesis, we get

$$((Y^I)_k)^J \subseteq ((Y^I)_{k+1})^J \tag{3}$$

for all $k_0 \leq k < \ell$ and all $Y \subseteq \Delta_I$. By (∗), for any set $X \subseteq \Delta_I$, there exist sets $\mathcal{P} \subseteq \mathcal{N}_{\mathrm{prim}}$ and $\mathcal{Y} \subseteq \mathcal{N}_r \times \mathfrak{P}(\Delta_I)$ such that

$$X^I \equiv \prod_{P \in \mathcal{P}} P \sqcap \prod_{(r,Y) \in \mathcal{Y}} \exists r.Y^I.$$

It is easy to see that this implies

$$(X^I)_\ell \equiv \prod_{P \in \mathcal{P}} P \sqcap \prod_{(r,Y) \in \mathcal{Y}} \exists r.(Y^I)_{\ell-1} \tag{4}$$

and

$$(X^I)_{\ell+1} \equiv \prod_{P \in \mathcal{P}} P \sqcap \prod_{(r,Y) \in \mathcal{Y}} \exists r.(Y^I)_\ell. \tag{5}$$

Thus, we have

$$\left((X^I)_\ell\right)^J \overset{(4)}{=} \left(\prod_{P \in \mathcal{P}} P \sqcap \prod_{(r,Y) \in \mathcal{Y}} \exists r.(Y^I)_{\ell-1}\right)^J$$

$$= \prod_{P \in \mathcal{P}} P^J \sqcap \prod_{(r,Y) \in \mathcal{Y}} \{x \in \Delta_J \mid \exists y \in ((Y^I)_{\ell-1})^J : (x,y) \in r^J\}.$$

From (3) we obtain $((Y^I)_{\ell-1})^J \subseteq ((Y^I)_\ell)^J$, and thus

$$\left((X^I)_\ell\right)^J \subseteq \prod_{P \in \mathcal{P}} P^J \sqcap \prod_{(r,Y) \in \mathcal{Y}} \{x \in \Delta_J \mid \exists y \in ((Y^I)_\ell)^J : (x,y) \in r^J\}$$

$$= \left(\prod_{P \in \mathcal{P}} P \sqcap \prod_{(r,Y) \in \mathcal{Y}} \exists r.(Y^I)_\ell\right)^J$$

$$\overset{(5)}{=} \left((X^I)_{\ell+1}\right)^J.$$

Hence we have shown that $(X^I)_\ell \to (X^I)_{\ell+1}$ follows from $\mathcal{B}$, which concludes the induction proof.

Now, let J again be a model in which all implications from $\mathcal{B}$ hold, and let $x \in \mathcal{U}^J$. We must show that this implies $x \in (\mathcal{U}^{II})^J$. We have $x \in ((\mathcal{U}^{II})_{k_0})^J$ because $\mathcal{U} \to (\mathcal{U}^{II})_{k_0} \in \mathcal{B}$. Hence $x \in ((\mathcal{U}^{II})_k)^J$ for all $k \leq k_0$ since $(\mathcal{U}^{II})_{k_0} \sqsubseteq (\mathcal{U}^{II})_k$ for all $k \leq k_0$. From what we have shown above, we know that

$$(\mathcal{U}^{II})_k \to (\mathcal{U}^{II})_{k+1}$$

follows from $\mathcal{B}$ for all $k \geq k_0$. Thus $((\mathcal{U}^{II})_k)^J \subseteq ((\mathcal{U}^{II})_{k+1})^J$ holds in J for all $k \geq k_0$, which yields $x \in ((\mathcal{U}^{II})_k)^J$ also in this case.

Therefore $x \in ((\mathcal{U}^{II})_k)^J$ for $k = |\mathcal{G}_\mathcal{U}| \cdot |\Delta_J| + 1$, independently of whether this number is smaller or larger than k_0. It follows directly from Lemma 12 that $x \in (\mathcal{U}^{II})^J$. Thus, we have shown that

$$\mathcal{U}^J \subseteq (\mathcal{U}^{II})^J$$

if all implications from $\mathcal{B}$ hold in J. This means that $\mathcal{U} \to \mathcal{U}^{II}$ follows from $\mathcal{B}$. $\qquad\square$

Having proved Proposition 17, we are almost finished with constructing a finite, sound and complete set of acyclic implications for the implications holding in a description context I. The idea is to replace any implication $\mathcal{U} \to \mathcal{U}^{II}$ in the finite, sound and complete set of implications constructed in the proof of Theorem 15 by the corresponding implications from Proposition 17.

The remaining problems is, however, that the set of implications obtained this way need not be sound for I. Indeed, if k_0 is too small, then the implications in $\{(X^I)_{k_0} \to (X^I)_{k_0+1} \mid X \subseteq \Delta_I\}$ need not hold in I. Therefore, we define for every $X \subseteq \Delta_I$

$$d_X := m_X \cdot n + 1,$$

where m_X is the node size of X^I and n is the cardinality of the model I. The number k_0 is the maximum of these numbers, i.e.,

$$k_0 := \max_{X \subseteq \Delta_I} d_X. \tag{6}$$

Then, because $d_X \leq k_0$ for every $X \subseteq \Delta_I$, we have

$$X^I \sqsubseteq (X^I)_{k_0+1} \sqsubseteq (X^I)_{k_0} \sqsubseteq (X^I)_{d_X}.$$

By Lemma 2(b), this implies

$$X^{II} \subseteq ((X^I)_{k_0+1})^I \subseteq ((X^I)_{k_0})^I \subseteq ((X^I)_{d_X})^I.$$

From Lemma 12 we obtain $X^{II} \supseteq ((X^I)_{d_X})^I$, and thus

$$X^{II} = ((X^I)_{k_0+1})^I = ((X^I)_{k_0})^I = ((X^I)_{d_X})^I.$$

In particular, this shows

$$((X^I)_{k_0})^I \subseteq ((X^I)_{k_0+1})^I.$$

Hence, all implications in $\{(X^I)_{k_0} \to (X^I)_{k_0+1} \mid X \subseteq \Delta_I\}$ hold in I.

Theorem 18. *In $\mathcal{EL}_{\mathrm{gfp}}$, for any description context I, there exists a* finite *set $\mathcal{B}$ of implications that is sound and complete for I, and such that all concept descriptions occurring in $\mathcal{B}$ are acyclic.*

Proof. Let $\mathcal{C}$ be the set of acyclic $\mathcal{EL}_{\mathrm{gfp}}$-concept descriptions defined in the proof of Theorem 15. We have shown in that proof that the set

$$
\begin{aligned}
\mathcal{B}_\star := \ & \{P \to P^{II} \mid P \in \mathcal{N}_{\mathrm{prim}} \cup \{\top\}\} \\
& \cup \{\exists r.C \to (\exists r.C)^{II} \mid r \in \mathcal{N}_r, C \in \mathcal{C}\} \\
& \cup \{C_1 \sqcap C_2 \to (C_1 \sqcap C_2)^{II} \mid C_1, C_2 \in \mathcal{C}\}
\end{aligned}
$$

is complete for I.

Let k_0 be defined as in (6). Then, by Proposition 17, the fact that $\mathcal{B}_\star$ is complete also implies that the following set of implications is complete for I:

$$
\begin{aligned}
\mathcal{B} := \ & \{(X^I)_{k_0} \to (X^I)_{k_0+1} \mid X \subseteq \Delta_I\} \\
& \cup \{P \to (P^{II})_{k_0} \mid P \in \mathcal{N}_{\mathrm{prim}} \cup \{\top\}\} \\
& \cup \{\exists r.C \to ((\exists r.C)^{II})_{k_0} \mid r \in \mathcal{N}_r, C \in \mathcal{C}\} \\
& \cup \{C_1 \sqcap C_2 \to ((C_1 \sqcap C_2)^{II})_{k_0} \mid C_1, C_2 \in \mathcal{C}\}.
\end{aligned}
$$

Regarding soundness, we have shown above that, due to the fact that k_0 was chosen large enough, all implications of the form $(X^I)_{k_0} \to (X^I)_{k_0+1}$ hold I. The implications $P \to (P^{II})_{k_0}$ hold because $P \to P^{II}$ holds in I, and $P^{II} \sqsubseteq (P^{II})_{k_0}$. The same arguments can be used to show that the implications of the forms $\exists r.C \to ((\exists r.C)^{II})_{k_0}$ and $C_1 \sqcap C_2 \to ((C_1 \sqcap C_2)^{II})_{k_0}$ hold in I.

The left-hand sides of implications in $\mathcal{B}$ are acyclic since the elements of $\mathcal{C}$ are acyclic, primitive concepts and $\top$ are acyclic, and any concept description of the form $\mathcal{U}_k$ is acyclic. This last argument also shows that the right-hand sides of implications in $\mathcal{B}$ are acyclic. □

Since $\mathcal{B}$ contains only acyclic $\mathcal{EL}_{\mathrm{gfp}}$-concept descriptions, it can also be viewed as a set of implications in $\mathcal{EL}$. Proposition 5, together with Theorem 18, shows that $\mathcal{B}$ is also complete for the $\mathcal{EL}$-implications holding in I. As argued before, the existence of a finite, sound and complete set also implies the existence of a basis.

Corollary 19. *In $\mathcal{EL}$, for any description context I, there exists a finite basis for the implications holding in I.*

6   Conclusion

We have shown that any description context I (i.e., any finite relational structure over a finite signature of unary and binary predicate symbols) has a finite basis for the $\mathcal{EL}$- and $\mathcal{EL}_{\mathrm{gfp}}$-implications holding in I. Such a basis provides the knowledge engineer with interesting information on the application domain described by the context. The knowledge engineer can, for example, use these implications as starting point for building an ontology describing this domain.

In this paper, we have concentrated on showing the existence of a finite basis. Of course, if this approach is to be used in practice, we also need to find efficient algorithms for computing the basis. After that, the next step will be to generalize attribute exploration [9] to our more general setting. This would allow us to consider also relational structures that are not explicitly given, but rather "known" by a domain expert.

Finally, we will also try to show similar results for other DLs. For the DL $\mathcal{FL}_0$, which differs from $\mathcal{EL}$ in that existential restrictions are replaced by value restrictions, we are quite confident that this is possible. For more expressive DLs, like $\mathcal{ALC}$, this is less clear.

References

1. Baader, F.: Least common subsumers and most specific concepts in a description logic with existential restrictions and terminological cycles. In: Gottlob, G., Walsh, T. (eds.) Proc. of the 18th Int. Joint Conf. on Artificial Intelligence (IJCAI 2003), Acapulco, Mexico, pp. 319–324 (2003)
2. Baader, F.: Terminological cycles in a description logic with existential restrictions. In: Gottlob, G., Walsh, T. (eds.) Proc. of the 18th Int. Joint Conf. on Artificial Intelligence (IJCAI 2003), Acapulco, Mexico, pp. 325–330 (2003)
3. Baader, F., Brandt, S., Lutz, C.: Pushing the $\mathcal{EL}$ envelope. In: Kaelbling, L.P., Saffiotti, A. (eds.) Proc. of the 19th Int. Joint Conf. on Artificial Intelligence (IJCAI 2005), Edinburgh (UK), pp. 364–369 (2005)
4. Baader, F., Distel, F.: A finite basis for the set of $\mathcal{EL}$-implications holding in a finite model. LTCS-Report 07-02, Theoretical Computer Science, TU Dresden, Germany (2007), http://lat.inf.tu-dresden.de/research/reports.html
5. Baader, F., Lutz, C., Suntisrivaraporn, B.: CEL—a polynomial-time reasoner for life science ontologies. In: Furbach, U., Shankar, N. (eds.) IJCAR 2006. LNCS (LNAI), vol. 4130, pp. 287–291. Springer, Heidelberg (2006)
6. Ferré, S.: Systèmes d'information logiques: un paradigme logico-contextuel pour interroger, naviguer et apprendre. PhD thesis, IRISA, France (2002)
7. Ferré, S., Ridoux, O.: Introduction to logical information systems. Information Processing & Management 40(3), 383–419 (2004)
8. Ganter, B.: Algorithmen zur Formalen Begriffsanalyse. In: Ganter, B., Wille, R., Wolff, K.E. (eds.) Beiträge zur Begriffsanalyse, pp. 241–254. B.I. Wissenschaftsverlag (1987)
9. Ganter, B.: Attribute exploration with background knowledge. Theoretical Computer Science 217(2), 215–233 (1999)
10. Ganter, B., Kuznetsov, S.O.: Pattern structures and their projections. In: Delugach, H.S., Stumme, G. (eds.) ICCS 2001. LNCS (LNAI), vol. 2120, pp. 129–144. Springer, Heidelberg (2001)

11. Ganter, B., Wille, R.: Implikationen und Abhängigkeiten zwischen Merkmalen. In: Degens, P.O., Hermes, H.-J., Opitz, O. (eds.) Die Klassifikation und ihr Umfeld, Indeks-Verlag, Frankfurt (1986)
12. Ganter, B., Wille, R.: Formal Concept Analysis: Mathematical Foundations. Springer, New York (1997)
13. Guigues, J.-L., Duquenne, V.: Familles minimales d'implications informatives résultant d'un tableau de données binaires. Math. Sci. Humaines 95, 5–18 (1986)
14. Horrocks, I., Patel-Schneider, P.F., van Harmelen, F.: From SHIQ and RDF to OWL: The making of a web ontology language. Journal of Web Semantics 1(1), 7–26 (2003)
15. Möller, R., Haarslev, V.: Description logic systems. In: Baader, F., et al. (eds.) The Description Logic Handbook: Theory, Implementation, and Applications, pp. 282–305. Cambridge University Press, Cambridge (2003)
16. Prediger, S.: Logical scaling in formal concept analysis. In: Delugach, H.S., et al. (eds.) ICCS 1997. LNCS, vol. 1257, pp. 332–341. Springer, Heidelberg (1997)
17. Prediger, S.: Terminologische Merkmalslogik in der Formalen Begriffsanalyse. In: Stumme, G., Wille, R. (eds.) Begriffliche Wissensverarbeitung: Methoden und Anwendungen, pp. 99–124. Springer, Heidelberg (1999)
18. Prediger, S., Wille, R.: The lattice of concept graphs of a relationally scaled context. In: Tepfenhart, W.M. (ed.) ICCS 1999. LNCS, vol. 1640, pp. 401–414. Springer, Heidelberg (1999)
19. Priss, U.: The formalization of WordNet by methods of relational concept analysis. In: Fellbaum, C. (ed.) WordNet: An Electronic Lexical Database and some of its applications, MIT Press, Cambridge (1998)
20. Rudolph, S.: Exploring relational structures via FLE. In: Wolff, K.E., Pfeiffer, H.D., Delugach, H.S. (eds.) ICCS 2004. LNCS (LNAI), vol. 3127, pp. 196–212. Springer, Heidelberg (2004)
21. Rudolph, S.: Relational Exploration: Combining Description Logics and Formal Concept Analysis for Knowledge Specification. PhD thesis, Technische Universität Dresden (2006)
22. Spackman, K.A., Campbell, K.E., Cote, R.A.: SNOMED RT: A reference terminology for health care. J. of the American Medical Informatics Association, Fall Symposium Supplement, 640–644 (1997)
23. The Gene Ontology Consortium. Gene Ontology: Tool for the unification of biology. Nature Genetics 25, 25–29 (2000)
24. Wille, R.: Conceptual graphs and formal concept analysis. In: Delugach, H.S., et al. (eds.) ICCS 1997. LNCS, vol. 1257, pp. 290–303. Springer, Heidelberg (1997)

Lexico-Logical Acquisition of OWL DL Axioms*
An Integrated Approach to Ontology Refinement

Johanna Völker and Sebastian Rudolph

Institut AIFB, Universität Karlsruhe, Germany
{voelker,rudolph}@aifb.uni-karlsruhe.de

Abstract. In order to overcome human and time resource problems in the task of ontology design, we propose to combine the LExO approach to learning expressive ontology axioms from textual definitions with Relational Exploration – a technique based on the well-known attribute exploration algorithm from FCA which is used to interactively clarify underspecified logical dependencies. By forcing particular modeling decisions the exploration of classes and class extension relationships guarantees completeness with respect to a certain logical fragment and increases the overall quality of the ontology. Providing an implementation as well as an example, we demonstrate how ontology learning and exploration complement each other in a synergetic way.

1 Introduction

In the prospering Semantic Web research field, ontologies – logical domain specifications useful for automatically drawing conclusions about the described domain – have taken a central role. Yet, building ontologies is a difficult and time-consuming task, requiring to combine the knowledge of domain experts with the skill and experience of ontology engineers resulting in a high demand on scarce expert resources. Moreover, the size of knowledge bases needed in real world applications easily exceeds the modeling capabilities of any human expert. On the other hand, both quality and expressivity of the ontologies generated automatically by the state-of-the-art ontology learning systems fail to meet the expectations of people who argue in favor of powerful, knowledge-intensive applications based on ontological reasoning.

In order to overcome this bottleneck, it is necessary to thoroughly assist the modeling process by providing hybrid semi-automatic methods which (i) intelligently suggest potentially relevant knowledge elements (complex domain axioms or facts) extracted from resources such as domain relevant text corpora and (ii) provide guidance during the knowledge specification process by asking decisive questions in order to clarify still undefined parts of the knowledge base.

Obviously, those two requirements complement each other. The first one clearly falls into the area of natural language processing. By using existing methods for knowledge extraction from texts, passages can be identified which indicate the validity of certain

---

* This work has been supported by the European Commission under contract IST-2006-027595 NeOn, and by the Deutsche Forschungsgemeinschaft (DFG) under the ReaSem project.

pieces of knowledge. For the second requirement, strictly logic-based exploration techniques such as the the well-known and well-established attribute exploration from formal concept analysis (and its variants and extensions) are needed in order to obtain logically crisp propositions. We believe that integrating these two directions of knowledge acquisition in one scenario will help overcoming disadvantages of either approach. The framework proposed in this paper realizes this integration and shows its potential for practical applications.

In Section 2, we briefly introduce the description logic $\mathcal{SHOIN}$. Section 3, sketches the field of ontology learning before presenting LExO as one method for acquiring DL axioms from texts. Section 4 gives the necessary background for Relational Exploration (RE), a technique used for interactive knowledge specification based on FCA. In Section 5, we describe in detail how LExO and RE (possibly assisted by other ontology learning components) can be synergetically combined in the process of ontology engineering and evaluation. Implementation details as well as an example are given in Section 6. Finally, Section 7 concludes and gives an outlook to future research.

2 Preliminaries

Here, we will very briefly introduce the description logic $\mathcal{SHOIN}$. A $\mathcal{SHOIN}$ knowledge base (KB, also: ontology) is based on sets N_R (*role names*) $\mathbf{C}$ (*atomic concepts*) and $\mathbf{I}$ (*individuals*). The set of $\mathcal{SHOIN}$ *roles* is $\mathbf{R} = N_R \cup \{R^- \mid R \in N_R\}$. In the following, we leave this vocabulary implicit and assume that A, B are atomic concepts, a, b, i are individuals, and R, S are roles. Those can be used to define concept descriptions employing the constructors from the upper part of Table 1. We use C, D to denote concept descriptions. Moreover, a $\mathcal{SHOIN}$ KB consists of two finite sets of axioms that are referred to as *TBox* and *ABox*. The possible axiom types for each are displayed in the lower part of Table 1.

Note that we do not explicitly consider concept or role equivalence $\equiv$, since it can be modeled via mutual concept or role inclusions. We adhere to the common model-theoretic semantics for $\mathcal{SHOIN}$ with general concept inclusion axioms (GCIs): an interpretation $\mathcal{I}$ consists of a set Δ called *domain* together with a function $\cdot^{\mathcal{I}}$ mapping individual names to elements of Δ, class names to subsets of Δ, and role names to subsets of $\Delta \times \Delta$. This function is inductively extended to roles and concept descriptions and finally used to decide whether the interpretation satisfies given axioms (see Table 1).

$\mathcal{SHOIN}$ serves as the theoretical basis for the web ontology language OWL DL as defined in [1]. OWL DL constitutes a standardized knowledge representation language well established in the Semantic Web domain. It is a fragment of first order predicate logic with the advantage of being decidable and even the availability of optimized reasoners for it.

3 Lexical and Logical Knowledge Acquisition

Ontology generation from natural language text, or lexical resources – most commonly referred to as "ontology learning" – is a relatively new field of research which aims to support the tedious task of knowledge acquisition by automatic means.

Table 1. Role/concept constructors and axiom types in $\mathcal{SHOIN}$. Semantics refers to an interpretation $\mathcal{I}$ with domain Δ. As usual, we require to restrict number restrictions to simple roles, i.e. (roughly speaking and omitting further technical details) roles that do not include roles which are declared to be transitive.

Name	Syntax	Semantics	
inverse role	R^-	$\{(x,y) \mid (y,x) \in R^{\mathcal{I}}\}$	
top	$\top$	Δ	
bottom	$\bot$	$\emptyset$	
nominal	$\{i\}$	$\{i^{\mathcal{I}}\}$	
negation	$\neg C$	$\Delta \setminus C^{\mathcal{I}}$	
conjunction	$C \sqcap D$	$C^{\mathcal{I}} \cap D^{\mathcal{I}}$	
disjunction	$C \sqcup D$	$C^{\mathcal{I}} \cup D^{\mathcal{I}}$	
universal restriction	$\forall R.C$	$\{x \mid (x,y) \in R^{\mathcal{I}} \text{ implies } y \in C^{\mathcal{I}}\}$	
existential restriction	$\exists R.C$	$\{x \mid \text{for some } y \in \Delta, \ (x,y) \in R^{\mathcal{I}}, y \in C^{\mathcal{I}}\}$	
(unqualified) number	$\leq nR$	$\{x \mid \#\{y \in \Delta \mid (x,y) \in R^{\mathcal{I}}\} \leq n\}$	
restriction	$\geq nR$	$\{x \mid \#\{y \in \Delta \mid (x,y) \in R^{\mathcal{I}}\} \geq n\}$	
role inclusion	$S \sqsubseteq R$	$S^{\mathcal{I}} \subseteq R^{\mathcal{I}}$	TBox
transitivity	$\mathsf{Trans}(S)$	$S^{\mathcal{I}}$ is transitive	TBox
general concept inclusion	$C \sqsubseteq D$	$C^{\mathcal{I}} \subseteq D^{\mathcal{I}}$	TBox
concept assertion	$C(a)$	$a^{\mathcal{I}} \in C^{\mathcal{I}}$	ABox
role assertion	$R(a,b)$	$(a^{\mathcal{I}}, b^{\mathcal{I}}) \in R^{\mathcal{I}}$	ABox

However, many of today's ontology learning approaches build upon methods and ideas which were developed by (computational) linguists long before ontologies became a popular means of knowledge representation. Ontology learning techniques based, e.g., on lexico-syntactic patterns [2], or Harris' distributional hypothesis [3] draw from previous advances in lexical acquisition, and terminology research which have been to a major extent focusing on the extraction of lexical relations. However, there is a tacit agreement in the ontology learning community that there exists a certain correspondence between lexical relations (e.g. hyponymy, synonymy), and ontological axioms (e.g. subsumption, equivalence). This assumption which is not only prevalent in ontology learning, but also influences manual ontology engineering[1] led to a kind of "lexical", i.e., lexically inspired ontology generation implemented in frameworks such as OntoLearn [4], OntoLT [5] or Text2Onto [6].

One may argue that due to the differences between lexical semantics, and the model-theoretic semantics of description logics (see also [7]), this type of approach will always yield at best light-weight, semi-formal ontologies without precisely defined semantics, being grounded in natural language more than in logics. On the other hand, lexical approaches to ontology generation offer a lot of advantages: They can benefit from large amounts of lexical resources such as machine-readable dictionaries, encyclopedias, and all kinds of web documents that are available in abundance on the web. The resulting ontologies are usually close to the human way of modeling, since they provide lexicalizations of classes, individuals and properties, thus being easily comprehensible and

---

[1] In fact, if one tries to explain the semantics of subsumption to a non-logician, one often resorts to "clue phrases" similar to lexico-syntactic patterns which themselves reflect lexical relations.

reusable. Finally, most of these approaches are very flexible with respect to the degree of user interaction, and relatively easy to combine with other, complementary or supporting ontology learning methods.

Besides those lexical methods, a second direction of ontology learning has received more and more attention during the last couple of years. Approaches based on Inductive Logic Programming (ILP) [8,9] and Formal Concept Analysis (FCA) [10] have been developed in the logics community, for some reason widely unappreciated by lexical ontology learning research. Although there are a few approaches aiming to reconcile the two worlds by using either FCA [11,12] or ILP [13] for lexical ontology acquisition, none of them has been designed specifically for the refinement of OWL DL ontologies or knowledge bases. Common to all those approaches is their idea to acquire knowledge based on presented domain entities and their properties. However, this type of logical ontology generation is often less efficient than lexical approaches, and requires a relatively large amount of manually acquired knowledge (e.g. ABox statements for taxonomy induction). The resulting ontologies lack the traceability of a natural language grounding, and meaningful labels for complex class descriptions. Their expressivity is typically restricted to some variant of $\mathcal{ALC}$. On the other hand, those approaches have several advantages. Since they are based on already structured, formal data, they naturally come with a precisely defined, formal set-theoretic semantics. Thus being on "safe logical grounds", it is guaranteed that the acquired knowledge is also logically consistent.

Despite their respective advantages, both lexical and logical approaches to automatic (or semi-automatic) ontology engineering have failed to meet all the expectations of people arguing in favor of knowledge-intensive, reasoning-based applications, e.g., in domains such as bio-informatics or medicine. In particular, expressivity and quality of the resulting axiomatizations are often insufficient for practical use. In order to meet these fundamental requirements, a few lexical approaches towards learning more expressive ontologies, i.e. ontologies featuring the expressiveness of OWL DL, have been proposed recently [7,14]. But these approaches have to face a lot of challenges which need to be overcome in order to make them useable in practice. Obviously, the more expressive learned (or manually engineered) ontologies become, the more important it will be to provide automatic support for quality assurance, since the difficulty of a purely manual revision rises with the growing complexity of the ontology. On the other hand, applications relying on reasoning over complex ontologies make it necessary to consider a larger variety of qualitative aspects which must be taken into account as an ontology is being learned or constructed, including logical consistency, and completeness. Notwithstanding, there exist only very few frameworks aiming at a tight integration of methods for ontology learning and evaluation. Although, e.g., Haase et al. [15] propose a way to deal with logical inconsistencies in lexically generated ontologies the problem of modeling completeness has been largely neglected up to now.

In this paper, we therefore present an approach to ontology acquisition which effectively combines the strengths of the two complementary directions of research while at the same time compensating for many of their respective disadvantages. It relies upon Relational Exploration, an FCA-based approach to systematic, logical refinement (cf.

Section 4), and the automatic generation of formal class descriptions by means of natural language processing techniques which is described in the remainder of this Section.

LExO[2] (Learning EXpressive Ontologies) [7] is an approach towards the automatic generation of ontologies featuring the expressiveness of OWL DL. The core of LExO is a syntactic transformation of definitory natural language sentences into description logic axioms. Given a natural language definition of a class, LExO starts by analyzing the syntactic structure of the input sentence. The resulting dependency tree is then transformed into a set of OWL axioms by means of manually engineered transformation rules. Possible input resources for LExO include all kinds of definitory sentences, i.e. universal statements about concepts, that can be found in online glossaries such as Wikipedia[3], comments in the ontology, or simply given by a domain expert.

In order to exemplify the approach, we assume that we would like to refine the description of the class *Reviewer* the semantics of which could be informally described as follows: *A reviewer is a person who reviews a paper that has been submitted to a conference or workshop.*[4] We will come back to this example in Section 6.

A minimum set of rules for translating this sentence into a DL class description is given by Table 2 (for a more complete listing of possible transformation rules and further explanations see [7]).

Table 2. Transformation Rules for *Reviewer*

Rule	Natural Language Syntax	OWL Axioms
Disjunction	X: NP_0 or NP_1	$X \equiv (NP_0 \sqcup NP_1)$
Copula	X: NP_0 VBE NP_1	$NP_0 \equiv NP_1$
Relative Clause	X: NP_0 C(*rel*) VP_0	$X \equiv (NP_0 \sqcap VP_0)$
Verb with Prep. Compl.	X: V_0 $Prep_0$ NP(*pcomp-n*)$_0$	$X \equiv \exists V_0\_Prep_0.NP_0$

Depending on the concrete set of translation rules and modeling preferences of the user, a translation of this sentence into OWL DL could then yield the following axioms:

reviewer
$\equiv$ *a_person_who_reviews_a_paper_that_has_been_submitted_to_a_conference_or_workshop*
a_person_who_reviews_a_paper_that_has_been_submitted_to_a_conference_or_workshop
$\equiv$ *a_person* $\sqcap$ *reviews_a_paper_that_has_been_submitted_to_a_conference_or_workshop*
reviews_a_paper_that_has_been_submitted_to_a_conference_or_workshop
$\equiv$ $\exists$*reviews.a_paper_that_has_been_submitted_to_a_conference_or_workshop*
a_paper_that_has_been_submitted_to_a_conference_or_workshop
$\equiv$ *a_paper* $\sqcap$ *has_been_submitted_to_a_conference_or_workshop*
has_been_submitted_to_a_conference_or_workshop
$\equiv$ $\exists$*has_been_submitted_to.a_conference_or_workshop*
a_conference_or_workshop $\equiv$ (*a_conference* $\sqcup$ *workshop*)

---

[2] http://ontoware.org/projects/lexo/

[3] http://en.wikipedia.org

[4] Depending on the intended meaning of *Reviewer* other, broader definitions (e.g. covering reviews of journal articles, or research projects) might be more adequate, but we wanted to keep the example as simple as possible.

Obviously, the above set of axioms can be normalized, and turned into a semantically equivalent, unfolded, representation:

Reviewer ≡ *Person* ⊓ ∃*review.*(*Paper* ⊓ ∃*submitted_to.*(*Conference* ⊔ *Workshop*))

While such a compact class description might be easier to grasp at first glance (at least for ontology engineers being familiar with logics), the first axiomatization obviously conveys a lot of additional information to the human reader. The fact that each part of the overall class description (e.g. *Conference* ⊔ *Workshop*) is associated with an equivalent atomic class (e.g. *a_conference_or_workshop*) makes completely transparent how this axiomatization was constructed, and at the same time provides the user with an intuitive explanation of the semantics of each class description. Further advantages of the extended axiomatization are discussed in Section 5.

4 Relational Exploration

In order to sketch relational exploration (RE, introduced in [16] and thoroughly treated in [10]), we first need to briefly recall some basic notions from FCA (see [17] for further reference).

A *(formal) context* $\mathbb{K}$ is a triple (G, M, I) with an arbitrary set G (called *objects*), an arbitrary set M (called *attributes*), and a relation $I \subseteq G \times M$ (called *incidence relation*). We read gIm as: "object g has attribute m." Furthermore, let $g^I := \{m \mid gIm\}$. An *implication* on an arbitrary set M is written $A \to B$ with $A, B \subseteq M$. It *holds* in a formal context $\mathbb{K} = (G, M, I)$, if for all $g \in G$ we have that $A \subseteq g^I$ implies $B \subseteq g^I$. We then write $\mathbb{K} \models A \to B$. A set $\mathfrak{I}$ of implications *entails* $A \to B$ if $A \to B$ holds in all contexts wherein all implications from $\mathfrak{I}$ hold.

An implication set $\mathfrak{I}$ will be called *non-redundant*, if for any $(A \to B) \in \mathfrak{I}$ we have that $\mathfrak{I} \setminus \{A \to B\}$ does not entail $A \to B$. $\mathfrak{I}$ will be called *complete* w.r.t. a context $\mathbb{K}$, if every implication $A \to B$ holding in $\mathbb{K}$ is entailed by $\mathfrak{I}$. $\mathfrak{I}$ will be called an *implication base* of $\mathbb{K}$ if it is non-redundant and complete. Since implication entailment is known to be decidable in linear time [18], the implication base allows fast handling of an implicational theory. The classical attribute exploration algorithm [19,20] provides a method for efficiently determining an implicational base of a formal context that is only implicitly known by an expert.

The technique of RE extends this algorithm to a DL setting: Given an interpretation $\mathcal{I}$ on a domain Δ and a set M of $\mathcal{SHOIN}$ concept descriptions, the corresponding $\mathcal{I}$-*context* is defined by $\mathbb{K}_\mathcal{I}(M) := (\Delta, M, I)$ with $\delta I C :\Leftrightarrow \delta \in C^\mathcal{I}$. Then it can be easily shown, that implications in $\mathbb{K}_\mathcal{I}$ coincide with certain axioms w.r.t. their validity in $\mathcal{I}$: for $C, \mathcal{D} \subseteq M$, the implication $C \to \mathcal{D}$ holds in $\mathbb{K}_\mathcal{I}$ if and only if $\mathcal{I}$ satisfies the DL axiom $\sqcap C \sqsubseteq \sqcap \mathcal{D}$. Hence it is possible to explore DL axioms (more precisely: general concept inclusion axioms, short: GCIs) with this technique. In an interview-like process, a domain expert has to judge whether a proposed GCI is valid in the domain (formally: the interpretation $\mathcal{I}$) he is describing and in the negative case provide a counterexample.[5] Since OWL DL [1] – the standard language for representing ontologies – is based on description logics, the RE method easily carries over to any kind of ontologies specified in that language.

---

[5] This will be further elaborated and demonstrated in the subsequent sections.

Especially when working in an OWL or DL setting, the open world assumption is omnipresent; most of the known objects will not be completely specified, i.e., for certain classes it might be unknown whether the considered individual is an instance. Hence, it is essential for exploration methods to be capable of dealing with this kind of information. Lately, there has been significant work on applying FCA results on partial information (e.g. described in [21,20]) to the ontology refinement setting. An according approach (briefly sketched in [10]) has been fully theoretically elaborated and implemented as described in [22]. It allows to use partly specified objects as counterexamples for hypothetical implications. We decided to follow this approach, hence the implementation presented in the remainder of this paper allows for handling partial contexts.

The advantage of RE is that the obtained results are logically crisp and naturally consistent. Moreover, the acquired information is complete with respect to certain well-defined logic fragments of OWL DL.[6] Yet, one major shortcoming of RE is the following: due to the aimed-at completeness, the number of asked questions (and therefore, the runtime and the workload for the expert) grows rapidly with the number of involved concepts and roles which threatens to exceed the ontology designers resources.

In order to counter this we propose a combination of two strategies: firstly, we use an OWL DL reasoner to determine whether the answer to a question posed by the exploration algorithm can be deduced from a previously given background knowledge ontology. Secondly we use lexical ontology learning to determine a relatively small number of relevant classes to focus on. Both points will be elaborated in the next section.

5   An Integrated Approach to Ontology Refinement

In the sequel, we will describe how LExO and RE can be synergetically combined by giving a comprehensive description of the integrated algorithm. En route, we will briefly mention how other lexical ontology learning techniques could be beneficially used within that process. In addition to the LExO and RE component, an OWL DL reasoner will be applied in order to draw conclusions that are already implicitly present, i.e. entailed by the actual knowledge base making an intervention of the user obsolete.

Creation of new Definitions and Mappings. We start with an OWL DL ontology $\mathcal{KB}$ to be refined with respect to a (new or already contained) class C, for which a natural language definition is provided by some textual resource. This textual definition is then analyzed by LExO yielding a set $\mathcal{KB}'$ of OWL DL axioms as described in Section 3. Most likely, some (or even most) of the named classes those axioms refer to will not be present in $\mathcal{KB}$. Therefore, at least the primitive classes amongst those – i.e. those classes not stated to be equivalent to a complex class description[7] – should be linked to $\mathcal{KB}$. There are several ways for doing that. If textual definitions are available, LExO could be employed "recursively", i.e., it might be applied to the definitions of the classes in question in order to obtain other classes that can be linked to $\mathcal{KB}$ more easily. In any case, ontology *mappings* between $\mathcal{KB}$ and $\mathcal{KB}'$ could be either added manually or

---

[6] Which fragment precisely depends on which variant of RE is used.

[7] These are the classes occurring explicitly in the normal form (cf. Section 3).

established by one or several of the well-known mapping tools like FOAM[8] [23]. So let *Map* be a (possibly empty) set of respective mapping axioms.

Selection of Relevant Classes. In the next step, we stipulate the focus of the subsequent exploration, by selecting the named classes from $\mathcal{KB} \cup \mathcal{KB}'$ whose logical dependencies shall be further clarified. A natural default choice for this would be the set of all named classes from $\mathcal{KB}'$, as we might suppose the (remaining) classes from $\mathcal{KB}$ to be modeled in a sufficiently precise way – an assumption that might be disproved later on. However, it might be reasonable to include some of the classes from $\mathcal{KB}$ as well. Knowledge extraction methods that determine the relevance of terms (like those offered by Text2Onto [6]) could be employed for an automatic selection or to generate reasonable suggestions. In any case, let **C** denote the set of selected attributes.

After this selection of relevant named classes, a basic fact from FCA allows to further restrict **C**: put into DL notation, it assures the dispensability of a class $C \in \mathbf{C}$ whenever there is a set $\mathbf{D} = \{D_1, \ldots, D_n\} \subseteq \mathbf{C} \setminus \{C\}$ such that $C \equiv D_1 \sqcap \ldots \sqcap D_n$ follows from all knowledge $\mathcal{KB}_\Sigma := \mathcal{KB} \cup \mathcal{KB}' \cup Map$ stated so far.[9] It takes just a little consideration that this is the case iff

$$\mathcal{KB}_\Sigma \models \bigsqcap \Big\{ D \ \Big| \ D \in \mathbf{C} \setminus \{C\}, \ \mathcal{KB}_\Sigma \models C \sqsubseteq D \Big\} \sqsubseteq C,$$

such that the elimination of redundant classes from **C** requires just $O(|\mathbf{C}|^2)$ reasoner calls in the worst case. Let **C'** denote the result of this reduction process.

Exploration. Now we start RE as described in Section 4 on the concept set **C'**. A work flow diagram of the procedure is displayed in Figure 1. For every hypothetical DL axiom $C_1 \sqcap \ldots \sqcap C_n \sqsubseteq D_1 \sqcap \ldots \sqcap D_m$ brought up by the exploration algorithm:

- Employ the reasoner to check whether this GCI is a consequence of $\mathcal{KB}_\Sigma$. If so, confirm the implication and continue the exploration with the next hypothesis.
- Employ the reasoner to query for all individuals γ with $C_1 \sqcap \ldots \sqcap C_n \sqcap \neg D_i(\gamma)$ for an i from $1, \ldots, m$, i.e., for instances of the class which characterizes the property for being a material counterexample[10] for the hypothetical GCI. Let Γ be the set of individuals retrieved this way. If $\Gamma \neq \emptyset$, select one $\gamma \in \Gamma$ and check for every $C \in \mathbf{C}$ whether $C(\gamma)$ or $\neg C(\gamma)$. Then the counterexample together with the information about the attributes it provably has or has not is passed to the exploration algorithm. Optionally the human expert – possibly assisted by lexical knowledge retrieval tools – might be asked to complete the assertions for γ in order to get a more specific description for it. In any case, after providing γ, the exploration will proceed with the next hypothesis.
- If the DL axiom in question can be neither automatically proved nor declined (the latter meaning $\Gamma = \emptyset$), the human will be asked for the ultimate decision whether

---

[8] http://www.aifb.uni-karlsruhe.de/WBS/meh/foam/

[9] In FCA terms this can be conceived as a kind of a-priori attribute reduction. Note that this process is nondeterministic. In case two classes happen to be equiextensional, we nondeterministically remove one of them.

[10] Material counterexamples are objects for which is known which part of the conclusion they violate. The exploration algorithm (even the one dealing with partial knowledge) can only make use of this kind of counterexamples.

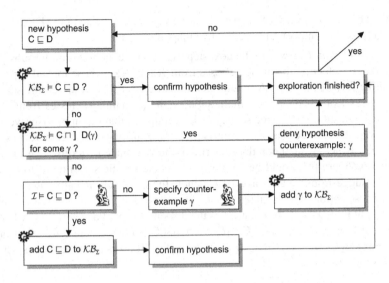

Fig. 1. Relational Exploration process (the gear wheels indicate ontology management activities including reasoning and updates, whereas the thinker icon marks user involvement)

the axiom is satisfied in the described domain $\mathcal{I}$ or not. Again, ontology learning tools could support him by suggesting answers endowed with a probability, or simply scanning a corpus for potential hints and presenting selected passages.

The exploration terminates after finitely many steps, yet it may also be stopped by the user beforehand. In the latter case, the internal order of the classes from the set $\mathbf{C}'$ is relevant since it determines the order of the posed questions. Hence, it is beneficial to sort those classes w.r.t. their relevance, possibly based on textual information. After the exploration cycle being finished, we have obtained a refined knowledge base $\mathcal{KB}_{\Sigma}$ containing the (possibly new) class C endowed with its definition (as extracted from the textual definition) and its interrelationships with concepts from the original knowledge base. Additionally, the "semantic neighborhood" of C has been made logically explicit by interactive exploration. In fact, any subsumption between conjunctions of classes from $\mathbf{C}$ can be decided (i.e. proven or disproven) based on the refined knowledge base. This also shows the advantage of introducing atomic classes for the complex concept descriptions occurring in the LExO output as demonstrated in Section 3: although RE as applied in this case[11] deals only with conjunctions on atomic classes, we introduce more expressivity "through the back-door" by having complex definitions for those named classes in our ontological background ready to be exploited by the reasoner.

The synergies provided by the presented combination are manifold: Firstly, the classes contained in the definitions provided by LExO provide a reasonable small to medium size "exploration scope" being crucial for a reasonable application of the RE technique. Secondly, we can use textual information for generating ontological informa-

---

[11] Actually, RE provides means for exploring GCIs in whole $\mathcal{ALE}$ with bounded role depth, however we restrict to conjunctions on atomic classes in this example.

tion (a source not accessible to purely logical approaches) yet being able to interactively clarify logical dependencies that have been left open by the text. The latter is done in a guided way ensuring completeness.

Overall, the proposed framework provides means for interactively integrating learned or manually acquired axiomatizations into an existing ontology, while at the same time facilitating their evaluation and refinement.

6 Implementation and Example

In order to prove the feasibility of a synthesis of ontology learning and RE as described in Section 5, we implemented a prototypical application named RELExO. Both sources and binaries of RELExO are available for public use and can be downloaded from its homepage[12] which has been set up to provide further information with respect to our experiments on ontology learning and relational exploration. RELExO relies upon KAON2[13] as an ontology management back-end and features a simple graphical user interface. Its architecture is depicted by Figure 2.

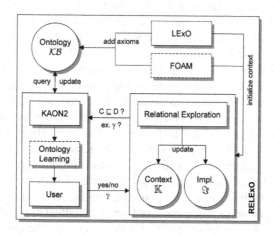

Fig. 2. RELExO Architecture

LExO, possibly complemented by other ontology learning components, generates or extends the initial set of axioms $\mathcal{KB}$ (mappings can be added by FOAM, if necessary), and initializes the partial context $\mathbb{K}$ by suggesting a set of attributes **C** to the user. The actual refinement process is handled by a RE component which manages the partial context $\mathbb{K}$ and the implication set $\mathfrak{I}$. Both are updated based on answers obtained from the "expert team" constituted by the KAON2 reasoner, an optional ontology learning component as well as the human knowledge engineer.

We now illustrate the integrated ontology refinement process which has been elaborated on in Section 5 by means of a real-world example. The complete material

---

[12] http://relexo.ontoware.org
[13] http://kaon2.semanticweb.org

necessary for reproducing this example, i.e. ontologies and screenshots, is contained in the RELExO distribution.

The **SWRC** (Semantic Web for Research Communities)[14] [24] ontology is a well-known ontology modeling the domain of Semantic Web research. Version 0.7 contains 71 classes, e.g., for different types of persons, publication, and events, 48 object properties, 46 datatype properties, and an overall number of 672 axioms. Its expressiveness is slightly beyond OWL DLP featuring subsumption, properties, and a few disjointness axioms. The ontology serves as a basis for semantic annotation in the AIFB web portal[15] which manages information about more than 2,000 persons, projects, and publications. For the purpose of our experiment, we exported all instance data stored in the AIFB portal into one single OWL file (more than 3 Megabytes in RDF syntax), and merged it with the corresponding TBox, i.e. the latest version of SWRC. After minor syntactic corrections (removing non XML-compliant characters), we obtained a considerably large ontology. Debugging with RaDON[16] revealed two inconsistencies caused by conflicting range specifications of data properties which could be fixed without difficulty.

Subsequently (in order to keep the example simple and rule out a few trivial questions that would otherwise come up in the exploration phase), we added axioms stating the disjointness of the SWRC top-level concepts *Person*, *Event*, and *Publication* – obviously true axioms yet not present in the current version of this ontology. Those axioms could also have been generated automatically by techniques for learning disjointness from [14]. However, adding these axioms turned the ontology inconsistent again as some individuals were inferred to instantiate both *Person* and *Publication*. The reason for this inconsistency was an incorrect use of the *editor* relationship in SWRC. Although its domain was restricted to *Person* (*"editor_of"*), the property was apparently conceived to have *"has_editor"* semantics by most of the annotators. We fixed this inconsistency by changing the definition of *editor* accordingly. Another problem became apparent after we had already started the exploration of the resulting ontology with RELExO. An individual (in our opinion) belonging to the class *ResearchPaper* was proposed as a counterexample, but could not be classified as a such. A closer look at both individual and ontology showed that it was assigned to the class *InProceedings* which was declared disjoint from *ResearchPaper*, the latter actually being empty. Since we found that this modeling decision is not justified by the associated comments in the ontology, we simply removed the disjointness axiom.

To demonstrate the RELExO approach, we assume that we would like to add a new class *Reviewer* to the SWRC ontology. Part of a change request could be a natural language description of this class such as *"a reviewer is a person who reviews a paper that has been submitted to a conference or workshop"* (cf. Section 3). Given this definitory sentence, LExO automatically suggests an axiomatization of *Reviewer* to the user who can correct or remove some of the generated axioms before they are added to the ontology. Applying FOAM for suggesting mappings between the newly introduced class

---

[14] http://ontoware.org/projects/swrc/
[15] http://www.aifb.uni-karlsruhe.de
[16] http://radon.ontoware.org

names and those already present in SWRC, we find *Paper* to be equivalent to *Research-Paper* and add a corresponding equivalence axiom to the extended ontology. Likewise, we find *Person*, *Conference* and *Workshop* already present in the original ontology.

In the next step, the set of "relevant" classes has to be selected. As mentioned in Section 5, it is reasonable to choose those atomic classes present in the definition of *Reviewer*. We decided to add two more classes denoting undergraduate and PhD students and (introducing abbreviations for overly long concept names from $\mathcal{KB}'$) we set:

$$\mathbf{C}' := \{\bot, CoW, Conference, SubCoW, Person, PhDStudent,$$
$$ResearchPaper, RevPSubCoW, Undergraduate, Workshop\}.$$

Based on this set of classes, the RE algorithm is started. The first hypothetical DL axiom, the exploration comes up with is $\top \sqsubseteq \bot$. Naturally, this hypothesis cannot be deduced from the ontology. Hence, following the description in Section 5, KAON2 will query the knowledge base for instances of $\top \sqcap \neg\bot$ which is equivalent to $\top$. Hence *all* ABox individuals are retrieved. Choosing one of the retrieved individuals, in our case *id1289instance*, we find it to be an instance of *ResearchPaper* and (since in our example, we chose the option to give the expert the opportunity to enhance the counterexample specification) add the information that it is an instance of *SubCoW*.

In a similar way, the next hypothesis posed – $\top \sqsubseteq ResearchPaper \sqcap SubCoW$ – is handled. Clearly, not every ABox individual is a research paper witnessed by the counterexample *id1303instance* being a journal article and hence neither a research paper (according to the underlying ontology) nor submitted to a conference or workshop.

However, the subsequent hypothesis $CoW \sqsubseteq \bot$ can neither be proved nor disproved by KAON2 using the information actually present in the ontology – since it does not contain any individuals being a conference or workshop. Therefore, the human expert will be asked for the final decision. Obviously, this hypothesis has to be denied and

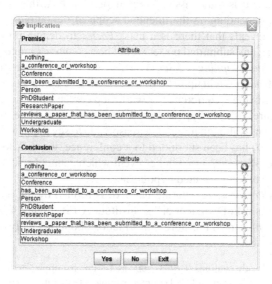

Fig. 3. Dialog for displaying the hypothetical axiom $CoW \sqcap SubCoW \sqsubseteq \bot$

Fig. 4. Specifying a counterexample. Every (non-)class-membership deducible from the knowledge base is automatically entered leaving just the open questions to the expert. RELExO can be configured to automatically display the web page associated with an individual's URI.

a counterexample for it is just any conference, so we enter *ICFCA_2008* and specify it as instance of *Conference*.[17] Note that due to the capability of dealing with partial information, the expert may leave open whether this individual belongs to the other considered classes. However, we employ the reasoner in order to determine all class-memberships deducible from the present information. In our case, it can be inferred that *ICFCA_2008* is also an instance of *CoW* and definitely no instance of *ResearchPaper*.

Consequently, the next question $CoW \sqsubseteq Conference$ comes up and has to be denied as well by entering the workshop instance *OntoLex_2007*.

Equally, the hypothesis $SubCoW \sqsubseteq ResearchPaper$ cannot be decided based on the present knowledge and is thus passed to the expert. In fact, this is the first "design decision" to make depending on the intended scope of the ontology. A look into the SWRC taxonomy reveals that there is a class *Poster* to denote posters presented at conferences. Indeed, any submitted poster would be a counterexample for the presented hypothesis, so we add *iMapping_Poster_SWUI_2006* to the knowledge base.

The next hypothesis brought up is $CoW \sqcap SubCoW \sqsubseteq \bot$ being an integrity constraint saying that nothing being a conference or workshop can be submitted (to a conference or workshop). Figure 3 shows how it is presented to the user. Here, we encounter another design decision. Although it might be reasonable to say that a workshop (actually: a workshop proposal) has been submitted to a conference, we stick to the intended semantics of the term *Workshop* as a kind of event which cannot be submitted and hence confirm the validity of the presented hypothesis.

The hypothesis $Person \sqsubseteq \bot$, coming up next, is refuted by the reasoner retrieving an individual who is a PhD student at the institute AIFB. Figure 4 shows the dialog wherein the user is presented the stored information about this individual and is asked to add the missing facts.

---

[17] This information already qualifies *ICFCA_2008* as a counterexample for the presented hypothesis. RELExO checks for every alleged counterexample whether it is indeed a such and rejects the input otherwise.

In this way, the exploration continues. During the process, some individuals are added and the following new axioms are confirmed:

- $SubCoW \sqcap Person \sqsubseteq \bot$ (a person cannot be submitted)
- $ResearchPaper \sqsubseteq SubCoW$ (every research paper has been submitted to a conference or workshop)[18]
- $RevPSubCoW \sqsubseteq Person$ (everybody reviewing a submitted paper is a person)
- $Person \sqcap PhDStudent \sqcap Undergraduate \sqsubseteq \bot$ (PhD students are disjoint with undergraduates)[19]
- $RevPSubCoW \sqcap Undergraduate \sqcap Person \sqsubseteq \bot$ (actually a "policy decision": undergraduates are not allowed to review papers)

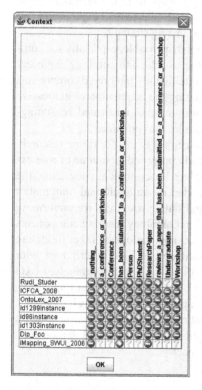

Fig. 5. Partial formal context resulting from the exploration

The formal context with the examples acquired during the exploration is displayed by Figure 5. It is automatically exported to the native ConExp[20] format and stored as a CEX file.

We end up with a refined SWRC ontology containing the new class *Reviewer* fully integrated into the existing ontology. Any subsumption between conjunctions of the specified interesting classes can be directly decided based on this refined SWRC ontology. This can be nicely demonstrated by starting RELExO again with the refined ontology: it terminates without ever asking the human expert for a decision, showing that all upcoming questions can be answered by the reasoner alone.

7 Conclusion and Outlook

In this paper, we have sketched a way to combine complementary approaches to ontological knowledge acquisition: the more intensional approach of distilling conceptual information from textual resources, and the extensional method of extracting hypothetical domain axioms based on given entities. We have instantiated this approach by designing and implementing a framework that integrates the LExO ontology learning application, a Relational Exploration component, and the KAON2 reasoner. To the best of our knowledge, RELExO is the first publicly available implementation of an exploration-based ontology refinement approach. It is open source and supports the standard ontology language OWL. In an example using the well-known SWRC

---

[18] We regard this justified by the existence of a class *Unpublished* disjoint to *ResearchPaper*.

[19] Another modeling flaw: this axiom should have been present in SWRC.

[20] http://conexp.sourceforge.net

ontology we have demonstrated the feasibility of our approach, and its applicability to real-world ontology engineering tasks.

Altogether, we are confident that the proposed framework will considerably alleviate the task of designing comprehensive and complex, yet logically consistent ontologies for knowledge-intensive applications. The number of design decisions to be made by the human user is minimized by the usage of textual resources and the employment of a reasoning back-end. Relational exploration provides guidance, ensuring that neither redundant information will be asked for nor important information is simply forgotten in the modeling process, and supports on-the-fly ontology evaluation: as in our example (where we were well-nigh inevitably confronted with design flaws in the used ontology), present modeling errors in the ontology are often indicated by "surprising" or counterintuitive questions asked by the algorithm. Hence from the methodological point of view, a cyclic ontology engineering process with intertwined exploration and manual refinement (or debugging) phases seems a promising strategy.

After all, human intervention will always remain indispensable, especially for complex knowledge modeling tasks. Notwithstanding, the workload to ontology engineers and domain experts can be drastically decreased by intelligently integrated components for semi-automatic ontology engineering. By facilitating the acquisition of expressive OWL ontologies, we hope to foster the development of more sophisticated, reasoning-based applications, and help to put semantic technologies into practice.

Pursuing this promising goal, we identify several central issues for future research. Firstly, we are planning to incorporate the just recently proposed technique of role exploration from [25]. In order to achieve an even tighter lexico-logical integration, the implementation of RELExO could be further extended by an additional (automatic) expert which uses ontology learning techniques, and online resources for confirming hypotheses, or suggesting counterexamples. Additional ontology learning components could be used to complement the LExO-generated axiomatizations by other modeling primitives (e.g. disjointness axioms), or to sort the attributes, i.e. class descriptions, with respect to the current domain or the user's interests. Finally, we will integrate RELExO into an ontology engineering environment such as the NeOn Toolkit[21], and improve its usability by adding a natural language generation component for translating hypotheses, i.e. logical implications, into natural language questions. In the end, we are confident that further extensive evaluations in real world application scenarios will demonstrate the advantages of our combined, lexico-logical approach.

References

1. McGuinness, D.L., van Harmelen, F.: OWL Web Ontology Language Overview (2004), CTAN: http://www.w3.org/TR/2004/REC-owl-features-20040210/
2. Hearst, M.: Automatic acquisition of hyponyms from large text corpora. In: Proceedings of the 14th International Conference on Computational Linguistics, pp. 539–545 (1992)
3. Harris, Z.S.: Word. Distributional Structure 10, 146–162 (1954)
4. Velardi, P., et al.: Evaluation of ontolearn, a methodology for automatic population of domain ontologies. In: Ontology Learning from Text: Methods, Applications and Evaluation, IOS Press, Amsterdam (2005)

---

[21] http://www.neon-toolkit.org

5. Buitelaar, P., Olejnik, D., Sintek, M.: OntoLT: A Protégé plug-in for ontology extraction from text. In: Fensel, D., Sycara, K.P., Mylopoulos, J. (eds.) ISWC 2003. LNCS, vol. 2870, Springer, Heidelberg (2003)
6. Cimiano, P., Völker, J.: Text2Onto - a framework for ontology learning and data-driven change discovery. In: Proc. of the 10th International Conference on Applications of Natural Language to Information Systems, pp. 227–238. Springer, Heidelberg (2005)
7. Völker, J., Hitzler, P., Cimiano, P.: Acquisition of OWL DL axioms from lexical resources. In: Franconi, E., Kifer, M., May, W. (eds.) ESWC 2007. LNCS, vol. 4519, Springer, Heidelberg (2007)
8. Fanizzi, N., et al.: Concept formation in expressive description logics. In: Boulicaut, J.-F., et al. (eds.) ECML 2004. LNCS (LNAI), vol. 3201, Springer, Heidelberg (2004)
9. Cohen, W.W., Hirsh, H.: Learning the classic description logic: Theoretical and experimental results. In: Doyle, J., Sandewall, E., Torasso, P. (eds.) Proceedings of the 4th International Conference on Principles of Knowledge Representation and Reasoning (KR 1994), Bonn, Germany, May 24-27, 1994, pp. 121–133. Morgan Kaufmann, San Francisco (1994)
10. Rudolph, S.: Relational Exploration - Combining Description Logics and Formal Concept Analysis for Knowledge Specification. Universitätsverlag Karlsruhe, Dissertation (2006)
11. Stumme, G., Maedche, A.: FCA-merge: Bottom-up merging of ontologies. In: Proc. 17th International Conference on Artificial Intelligence (IJCAI 2001), pp. 225–230 (2001)
12. Cimiano, P., Hotho, A., Staab, S.: Learning concept hierarchies from text corpora using formal concept analysis. Journal of Artificial Intelligence Research 24, 305–339 (2005)
13. Nedellec, C.: Corpus-based learning of semantic relations by the ILP system, asium. In: Cussens, J., Džeroski, S. (eds.) LLL 1999. LNCS (LNAI), vol. 1925, pp. 259–278. Springer, Heidelberg (2000)
14. Völker, J., et al.: Learning disjointness. In: Franconi, E., Kifer, M., May, W. (eds.) ESWC 2007. LNCS, vol. 4519, Springer, Heidelberg (2007)
15. Haase, P., Völker, J.: Ontology learning and reasoning - dealing with uncertainty and inconsistency. In: da Costa, P.C.G., et al. (eds.) Proc. of the Workshop on Uncertainty Reasoning for the Semantic Web (URSW), pp. 45–55 (2005)
16. Rudolph, S.: Exploring relational structures via FLE. In: Wolff, K.E., Pfeiffer, H.D., Delugach, H.S. (eds.) ICCS 2004. LNCS (LNAI), vol. 3127, pp. 196–212. Springer, Heidelberg (2004)
17. Ganter, B., Wille, R.: Formal Concept Analysis: Mathematical Foundations. Springer, New York (1997) Translator-C. Franzke
18. Maier, D.: The Theory of Relational Databases. Computer Science Press (1983)
19. Ganter, B.: Two basic algorithms in concept analysis. Technical Report 831, FB4, TH Darmstadt (1984)
20. Ganter, B.: Attribute exploration with background knowledge. Theoretical Computer Science 217, 215–233 (1999)
21. Burmeister, P.: Merkmalimplikationen bei unvollständigem Wissen. In: Lex, W. (ed.) Arbeitstagung Begriffsanalyse und Künstliche Intelligenz, TU Clausthal, pp. 15–46 (1991)
22. Baader, F., et al.: Completing description logic knowledge bases using formal concept analysis. In: Veloso, M.M. (ed.) IJCAI, pp. 230–235 (2007)
23. Ehrig, M., Sure, Y.: FOAM - framework for ontology alignment and mapping. results of the ontology alignment initiative. In: Ashpole, B., et al. (eds.) Proc. of the Workshop on Integrating Ontologies, vol. 156, pp. 72–76 (2005)
24. Sure, Y., Bloehdorn, S., Haase, P., Hartmann, J., Oberle, D.: The SWRC ontology - semantic web for research communities. In: Bento, C., Cardoso, A., Dias, G. (eds.) EPIA 2005. LNCS (LNAI), vol. 3808, pp. 218–231. Springer, Heidelberg (2005)
25. Rudolph, S.: Acquiring generalized domain-range restrictions. In: Medina, R., Obiedkov, S. (eds.) ICFCA 2008. LNCS (LNAI), vol. 4933, pp. 32–45. Springer, Heidelberg (2008)

From Concepts to Concept Lattice:
A Border Algorithm for Making Covers Explicit

Ben Martin and Peter Eklund

School of Information Systems and Technology
The University of Wollongong
Northfields Avenue, Wollongong, NSW 2522, Australia
monkeyiq@users.sourceforge.net,
peklund@uow.edu.au

Abstract. The paper presents a new border algorithm for making the covering relation of concepts explicit for iceberg concept lattices. The border algorithm requires no information from the formal context relying only on the formal concept set in order to explicitly state the covering relation between formal concepts. Empirical testing is performed to compare the border algorithm with a traditional algorithm based on the Covering Edges algorithm from Concept Data Analysis [4].

1 Introduction

The process of obtaining a concept lattice from a formal context can be seen as two distinct sub-tasks: finding the set of all formal concepts and making the covering relation of the formal concepts explicit. From a strictly mathematical perspective the second step may seem superfluous as the set of all formal concepts uniquely defines the covering relation as well [5]. However, from a performance perspective explicitly recording the covering relation given only the set of all formal concepts is a computationally intensive task [4].

An iceberg concept lattice is a concept lattice where concepts which have fewer formal objects than a given desired minimum threshold are omitted. The omitted formal concepts are usually from the lower portion of the concept lattice.

Where an iceberg concept lattice is acceptable while performing Formal Concept Analysis there are many algorithms from Data Mining which can relatively efficiently find the set of all formal concepts. This leaves the step of explicitly recording the covering relation for the formal concepts to be computed.

This paper presents a Border algorithm to perform this latter task. The border algorithm does not require any information from the formal context, relying solely on the set of formal concepts to make the covering relation explicit. Thus, the efficiency of the border algorithm is independent of the number of formal objects.

The research was motivated by applying Formal Concept Analysis to formal contexts which contain millions of formal objects where an iceberg concept lattice with up to a few thousand formal concepts might be desired.

For brevity, this paper will also refer to formal concepts, formal attribute and formal objects simply as concepts, attributes and objects respectively.

R. Medina and S. Obiedkov (Eds.): ICFCA 2008, LNAI 4933, pp. 78–89, 2008.

2   Data Mining and Formal Concept Analysis

Stumme et al [16] presented both the Titanic algorithm and highlighted the link
between finding the lattice closure in FCA and the problem of locating Closed
Frequent Itemsets (CFI) in Data Mining [8]. The Titanic algorithm is based on
the *Apriori* Data Mining algorithm [2].

This paragraph establishes the link between CFI in Data Mining and finding
the set of concepts (intents) from a formal context in FCA. The set of concept
intents uniquely determines the set of concepts. An itemset in Data Mining is
a set of formal attributes in the formal context. The support of an itemset is
the number of objects that have at least that itemset in their row in the formal
context.

In Formal Concept Analysis a formal context is a triple (O, A, I) where
O is a set of objects, A is a set of attributes, and I is a binary relation
between the objects and the attributes, i.e. $I \subseteq O \times A$. A formal concept of a formal context (O, A, I) is a pair (X, Y) where $X \subseteq O$, $Y \subseteq A$,
$X = \{o \in O \mid \forall m \in Y : (o, a) \in I\}$ and $Y = \{a \in A \mid \forall o \in X : (o, a) \in I\}$.

Expressing Data Mining in Formal Concept Analysis terms, an Itemset would
be a set $Y \subseteq A$ where Y'' might not equal Y.

The "Frequent" part of the Closed Frequent Itemset relates to a cut off thresh-
old in the Data Mining process. The support of an Itemset Y is $|Y'|$. For a given
minimum support value z, any itemset with a higher support than the minimum
is considered frequent, ie. $|Y'| > z$.

A closed itemset is an itemset $Y \subseteq A$ for which there exists no attribute
$y \in A\backslash Y$ such that the itemset $Y \cup \{y\}$ has the same support as Y. That is, any
itemset is closed if it can not be expanded to contain any other attribute and
not have a lesser support.

Thus, the itemset $Y \subseteq A$ can be considered closed if and only if $Y = Y''$.
Note that if $Y \neq Y''$ then there must be some $y \in A\backslash Y$ which is in Y''.

Thus, a CFI has a support above the minimal threshold and cannot be expanded to contain any other attribute without modifying its support. That is, a
Closed Frequent Itemset (CFI) $Y \subseteq A$ has $Y = Y''$ and $|Y'| > z$.

Much computational complexity can be avoided if the minimum support is set
slightly higher than zero [6]. For example, by using a minimum support in the
range of 2% to 5% the algorithm might perform 100 to 1,000 faster than when
using a minimum support of zero. At some point before reaching a minimum
support of zero many of the algorithms will become intractable [6].

Finding the set of all intents in FCA is equivalent to finding the set of all
Closed Frequent Itemsets (CFI) in Data Mining using a minimum support of
zero. When the minimum support is above zero the side effect is that some
concepts at the bottom of the concept lattice are not discovered. Such partial
concept lattices are called iceberg lattices [16].

Algorithms for finding closed frequent itemsets in the Data Mining community
are the subject of much research and recently an annual workshop to benchmark
both algorithms and their implementations was held [6].

The set of CFI for a given dataset implicitly contains the (iceberg) concept lattice for that dataset. In order to display such a concept lattice one must make explicit the covering relation of the CFI. Titanic explicitly records this covering relation while discovering the CFI whereas many Data Mining algorithms normally stop after finding just the CFI themselves.

The process of obtaining a concept lattice using Data Mining algorithms can be seen as two distinct subtasks: finding the CFI and finding the covering relation of the CFI. For further details of the first step, Data Mining CFI, see [6,12,13,17,9,14,7]. The focus of this paper is on making the covering relation explicit after the CFI have been found. This is discussed in Section 4 with empirical testing in Section 6.

3   A Baseline Algorithm

The algorithm "CoveringEdges" from Concept Data Analysis [4] will be used as a baseline implementation for empirical testing.

The Covering Edges [4] algorithm was implemented with a minor change to work against iceberg lattices as shown in Fig.1. The algorithm from Fig.1 will be referred to as the "covering edges" in the following discussion.

1. IceBergCoveringEdges(C, (G, M, I))
2. for each(X, Y) $\in$ C
3. Set count of any concept in C to 0
4. for each m $\in$ M \ Y
   (a) inters := X $\cap$ {m}'
   (b) Find (X_1, Y_1) $\in$ C such that X_1 = inters
   (c) **if((X_1, Y_1) exists) then**
       i. count(X_1, Y_1) := count(X_1, Y_1) + 1
       ii. if(|Y_1| - |Y|) = count(X_1, Y_1) then
           A. Add edge (X_1, Y_1) $\rightarrow$ (X, Y) to E

Fig. 1. Modified CoveringEdges using the same syntax as in Concept Data Analysis [4]. As the iceberg lattice does not contain all concepts, the modified version must check that the concept (X_1, Y_1) exists before proceeding.

4   A Border Algorithm

In the following discussion C will represent the set of all concepts. A naive algorithm for finding the covering relation among concepts would inspect each concept as a potential parent for every concept. The time complexity to find the parent-child relations between the concepts for this algorithm is proportional to $|C|^2$.

Various order theoretic properties of a concept lattice can be used to reduce the search space for covers and thus handle a larger $|C|$. The following border algorithm maintains a border set B as the concept lattice is inspected in a top down manner. If the mean size of B is y the complexity becomes proportional to $y \times |C|$.

The border algorithm is shown in Fig.2. The algorithm only requires the set of all intents F as input. The algorithm requires that the intent set be a partial order of least intent attribute cardinality to maximum intent attribute cardinality. For example, all intents with only one attribute will be sorted before all intents with two or more attributes. As each intent will uniquely define a concept the discussion will also simply mention "the concept" for $a \in F$ instead of the concept with the intent $\{a\}$.

There may be many concepts $D \subseteq F$ sharing the smallest intent cardinality. Any intent of D may be the intent of the top concept of the lattice. As such a new intent is created t which will act as the intent of the top concept of the lattice. After the algorithm has completed, if t has only a single child then it can be removed from the lattice. This allows the algorithm to quickly know what concept is the top concept of the concept lattice. The top concept is used to initialize the border set with so it must be known.

The algorithm works by starting with a border of the top lattice node and sequentially working through concepts in the ordered intent set $a \in F$ and forming the intersection with each intent in the border set to find parents of a. After each concept is checked against the border set any new parent-child relations are explicitly linked and the border set is updated.

The algorithm in Fig.2 will be referred to as the "intents only" algorithm in the following.

5   Application Example

The concept lattice shown in Figure 5 is used to present an example of the application of the border algorithm.

The steps that the border algorithm performs to find the concept lattice are shown in Figure 4. The current concept that is being worked on, the current border set as well as the Intents set from line 12 in Figure 2 before and after Maxima is called on it. All edge additions are shown at the time when they are performed on line 14 of Figure 2.

6   Performance Analysis

The hardware and software setup is described to a level of detail that should allow third parties to obtain similar empirical test results. In particular the way in which bitsets are implemented can have a huge impact on performance so certain low level details must be presented in the interests of reproducibility. The benchmark system is a dual core AMD X2 running at 2.2GHz with 2Gb of RAM at DDR400. The implementation is single threaded and single processed so only takes advantage of a single CPU core. The implementations use the PostgreSQL database to read the CFI. Testing was performed on a synthetic formal context generated with the IBM synthetic data generator [15] databases from the UCI dataset [3] and two filesystem examples. The filesystem examples include a formal context derived from the metadata of 67,000 document files [1] and 2,000

1. IntentsOnly(F)
2.
3. set $t := \{\}$
4. $F := F \cup \{t\}$
5. Border $:= \{t\}$
6.
7. for each $a \in F$
8.   Intents $:= \{\}$
9.   for each $b \in$ Border
10. $y := b \cap a$
11.   Intents $:=$ Intents $\cup \{y\}$
12.   Intents $:= Maxima($ Intents $)$
13.   for each $y \in$ Intents
14.   Add edge $y \to a$ to E
15.   Border $:=$ Border $\setminus \{y\}$
16.   Border $:=$ Border $\cup \{a\}$
17.
18. if $|children(t)| = 1$
19.   for each $c \in children(t)$
20.   Remove edge $t \to c$ from E
21.   F $:=$ F$\setminus\{t\}$

Fig. 2. Algorithm to make the order relation between concepts explicit. Input: F the set of concept intents partially ordered on the cardinality of the Intent size from smallest intent size to largest. Output: E an edge mapping from parent concept intent to child concept intent forming the covers for the concept lattice of F. The Intents set introduced on line 8 is also partially ordered from intents with the smallest cardinality to intents with the largest cardinality.

1. Maxima(Intents)
2.
3. Ret $:= \{\}$
4. for each $y \in$ reverse(Intents)
5.   ismin $:= 1$
6.   for each $r \in$ Ret
7.   if $\{y\} \cap \{r\} = \{r\}$
8.   ismin $:= 0$
9.   if ismin
10.   Ret $:=$ Ret$\cup\{y\}$
11.
12. return Ret

Fig. 3. The Maxima function returns the set of intents which are maximal from the given set of intents. The Intents set used as input is ordered from smallest intent cardinality to largest intent cardinality. Line 4 indicates that the ordered Intents poset is to be inspected in reverse order, from largest intent cardinality to smallest.

1. Current Concept = $\{\{a\}\}$
   Border = $\{\{\}\}$
   Intents = $\{\{\}\}$
   Maxima(Intents) = $\{\{\}\}$
   Add edge $\{\} \rightarrow \{a\}$

2. Current Concept = $\{\{c\}\}$
   Border = $\{\{a\}\}$
   Intents = $\{\{\}\}$
   Maxima(Intents) = $\{\{\}\}$
   Add edge $\{\} \rightarrow \{c\}$

3. Current Concept = $\{\{d\}\}$
   Border = $\{\{c\}, \{a\}\}$
   Intents = $\{\{\}, \{\}\}$
   Maxima(Intents) = $\{\{\}\}$
   Add edge $\{\} \rightarrow \{d\}$

4. Current Concept = $\{\{b\}\}$
   Border = $\{\{d\}, \{c\}, \{a\}\}$
   Intents = $\{\{\}, \{\}, \{\}\}$
   Maxima(Intents) = $\{\{\}\}$
   Add edge $\{\} \rightarrow \{b\}$

5. Current Concept = $\{\{ae\}\}$
   Border = $\{\{b\}, \{d\}, \{c\}, \{a\}\}$
   Intents = $\{\{a\}, \{\}, \{\}, \{\}\}$
   Maxima(Intents) = $\{\{a\}\}$
   Add edge $\{a\} \rightarrow \{ae\}$

6. Current Concept = $\{\{ac\}\}$
   Border = $\{\{ae\}, \{b\}, \{d\}, \{c\}\}$
   Intents = $\{\{c\}, \{a\}, \{\}, \{\}\}$
   Maxima(Intents) = $\{\{a\}, \{c\}\}$
   Add edge $\{c\} \rightarrow \{ac\}$
   Add edge $\{a\} \rightarrow \{ac\}$

7. Current Concept = $\{\{ad\}\}$
   Border = $\{\{ac\}, \{ae\}, \{b\}, \{d\}\}$
   Intents = $\{\{d\}, \{a\}, \{a\}, \{\}\}$
   Maxima(Intents) = $\{\{a\}, \{d\}\}$
   Add edge $\{d\} \rightarrow \{ad\}$
   Add edge $\{a\} \rightarrow \{ad\}$

8. Current Concept = $\{\{bc\}\}$
   Border = $\{\{ad\}, \{ac\}, \{ae\}, \{b\}\}$
   Intents = $\{\{b\}, \{c\}, \{\}, \{\}\}$
   Maxima(Intents) = $\{\{c\}, \{b\}\}$
   Add edge $\{b\} \rightarrow \{bc\}$
   Add edge $\{c\} \rightarrow \{bc\}$

9. Current Concept = $\{\{bd\}\}$
   Border = $\{\{bc\}, \{ad\}, \{ac\}, \{ae\}\}$
   Intents = $\{\{d\}, \{b\}, \{\}, \{\}\}$
   Maxima(Intents) = $\{\{b\}, \{d\}\}$
   Add edge $\{d\} \rightarrow \{bd\}$
   Add edge $\{b\} \rightarrow \{bd\}$

10. Current Concept = $\{\{bcd\}\}$
    Border = $\{\{bd\}, \{bc\}, \{ad\}, \{ac\}, \{ae\}\}$
    Intents = $\{\{bc\}, \{bd\}, \{c\}, \{d\}, \{\}\}$
    Maxima(Intents) = $\{\{bd\}, \{bc\}\}$
    Add edge $\{bc\} \rightarrow \{bcd\}$
    Add edge $\{bd\} \rightarrow \{bcd\}$

11. Current Concept = $\{\{abcde\}\}$
    Border = $\{\{ad\}, \{ac\}, \{ae\}, \{\}, \{bcd\}\}$
    Intents = $\{\{bcd\}, \{ae\}, \{ac\}, \{ad\}, \{\}\}$
    Maxima(Intents) = $\{\{bcd\}, \{ad\}, \{ac\},$
    $\{ae\}\}$
    Add edge $\{ae\} \rightarrow \{abcde\}$
    Add edge $\{ac\} \rightarrow \{abcde\}$
    Add edge $\{ad\} \rightarrow \{abcde\}$
    Add edge $\{bcd\} \rightarrow \{abcde\}$

Fig. 4. Steps performed by the border algorithm to find the covers of the concept lattice shown in Figure 5

geospatially tagged digital pictures. The digital picture context contains over 90 formal attributes.

Tests were run multiple times and the result of the final run was taken. This reduces the impact of the relational database, disk speed and other non-relevant implementation details because most information will be coming directly from RAM cache in the relational database itself. This "hot caching" is acceptable as we are mostly interested in the speed of the core algorithm which makes explicit the covering relations of the CFI. The core of the implementation was compiled with gcc 4.1.1 using the -O4 flag to turn on code optimization.

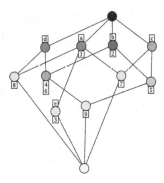

Fig. 5. Concept lattice used as example of border algorithm application

In order not to advantage either algorithm in empirical testing variants were created of each. As the intents only algorithm makes use of many temporary small sized bit sets it was implemented around `boost::dynamic_bitset<>` bit container.

As the covering edges algorithm makes extensive use of intersecting very large sets, see line 4a from Figure 1, the covering edges uses the BitMagic bitset implementation that makes use of gap encoding and Single Instruction Multiple Data (SIMD) execution. These two optimizations greatly reduce the execution time for covering edges on large data sets.

The use of SIMD carries a setup overhead and as such can actually be a disadvantage for smaller bit-sets. The choice of bit-set implementation being advantageous for each algorithm was tested by using both bit set implementations to verify empirically that the choice was optimal. For sets with 64 or less elements an implementation of intents only was created using a single 64 bit integer instead of a `boost::dynamic_bitset<>`. For covering edges a version that operates on the row reduced context was created. This implementation will be referred to as covering edges reduced.

There should be somewhat more efficient ways to perform the covering edges "set up" that is shown in the following. Regardless of potential improvements this set up time can not be avoided and the relative differences in speed shown remain.

6.1   Performance on Synthetic Data

The following use synthetic data generated with the IBM synthetic data generator [15]. Parameters include the number of transactions (ntrans), the transaction length (tlen), length of each pattern (patlen), number of patterns (npat) and number of items (nitems). The parameters were as follows ntrans=10,000, ntrans=100,000 and ntrans=1,000,000, nitems=1000, patlen=7, npats=10000. For each ntrans value, the tlen was varied between 32, 128 and 256. Some combinations of tlen and ntrans could not be generated because the implementation ran into RAM shortages.

The output of the IBM synthetic data generator is a list of *ntrans* transactions. Each transaction contains a number of items. Each item is represented by a unique integer in the range $\{1, ..., nitems\}$. Transactions were imported into a bit field column type in a PostgreSQL table for $n \in \{32, 128, 256\}$. An expression index was then created on the corresponding int array for the bit field column by using a conversion function to generate an int array from a bit field. Such an arrangement allows an RD-Tree to easily be created on the int array using the same implementation previous work [10,11]. The base data type being a bit field allows the fastest possible bulk transfers of data.

The Apriori [2] algorithm was used to generate the CFI for each dataset. Parameters for support were varied in order to obtain a desired number of CFI for each data set. The target CFI count was 1,000, 3,000 and 10,000. Results were not recorded for datasets that do not support the target number of CFI.

Shown in Table 1 is the number of CFI for each input data set along with the size of the row reduced formal context. The results of running the two algorithms against the datasets in Table 1 are shown in Table 2. Where the number of CFI is less than 3,000 the intents only method has a clear advantage. However, once the number of CFI is at 10,000 the covering edges based algorithms perform much better. The border algorithm functions very poorly on the 10,000 CFI input because the average number of concepts in the border set is very high, as shown in Table 3. As the border algorithm, shown in Fig. 2, visits each member of the border set for each concept having the average border size almost half the number of concepts makes the overall complexity of the border algorithm unacceptable.

Table 1. The number of CFI for each configuration. The reduced count is the number of transactions in an object row reduced formal context. The reduced count plays a role in the Covering Edges implementation. As can be seen the reduction process has no bearing for formal contexts with *tlen* > 32. Where the data does not support the requested number of CFI the table has blank cells.

ntrans (x1000)	tlen	|CFI| 1,000	|CFI| 3,000	|CFI| 10,000	row reduced object count (x1000)
1.0	128	931	2867	9989	1.0
10.0	32	1118	.	.	1.0
10.0	128	991	2983	9885	10.0
10.0	256	1155	3009	9767	10.0
100.0	32	906	2909	.	3.8
100.0	128	955	2880	9494	100.0

6.2 Performance on a Filesystem Data

Personal Photo Collection. The first dataset is for 2,000 photographs with metadata describing their location both semantically and quantitatively. Semantic metadata is captured by associating place names with each file. Quantitative metadata comes from both the metadata automatically recorded by a digital

Table 2. Time taken by various algorithm implementations to make the covering relations between CFI explicit

ntruns (x1000)	tlen	|CFI|	algorithm	covers time (mm:ss.d)	{m}' time (mm:ss.d)	setup X (mm:ss.d)
1	128	931	intents only	2.5		
			covering edges	4.5	0.02	1.1
		2867	intents only	13.4		
			covering edges	8.2	0.02	3.5
		9989	intents only	3:22.3		
			covering edges	30.6	0.02	12.3
10	32	1118	intents only	1.2		
			covering edges	1.4	0.01	0.8
	128	991	intents only	1.7		
			covering edges	2.6	0.2	7.0
		2983	intents only	15.9		
			covering edges	8.3	0.2	18.1
		9885	intents only	3:24.2		
			covering edges	28.9	0.2	57.7
	256	1155	intents only	2.6		
			covering edges	7.1	0.4	12.5
		3009	intents only	16.0		
			covering edges	19.1	0.4	31.0
		9767	intents only	3:20.1		
			covering edges	1:21.2	0.4	33.8
100	32	906	intents only	0.9		
			covering edges	1.4	0.9	2.8
			covering edges R	0.8	0.04	1.1
		2909	intents only	10.4		
			covering edges	5.6	0.9	6.2
			covering edges R	3.1	0.04	3.2
	128	955	intents only	1.6		
			covering edges	3.2	2.0	57.3
		2880	intents only	15.3		
			covering edges	10.5	2.0	41.5
		9494	intents only	3:12.8		
			covering edges	34.6	2.1	30.6

camera or film processing machine and also indirectly from the place name tags in the form of longitude and latitude information.

Using the border algorithm takes 0.03 seconds whereas the covering edges requires 0.49 seconds plus 0.2 seconds to set up data structures. Note that although this data was only for 2,000 images, since the border algorithm is dependent only on the CFI, this advantage would only increase proportionally with the number of images in the collection.

Two scales on 200,000 files. The term "index" can be refer to a specific implementation such as a B-Tree and also to the database as a whole. As such

Table 3. Average border size for various CFI data sets

ntrans (x1000)	tlen	|CFI|	average border size
10	256	3009	956
		9767	3399
100	32	906	331
		2909	1004
	128	955	432
		2880	1162
		9494	3991

the term `findex` is used to refer to the database as a whole which might be comprised of many database level B-Tree or spatial indices.

An `findex` was created on a Fedora Core 4 Linux machine using libferris 1.1.54 of 201,759 files in `/usr/share/`. For this findex a nominal scale was created on mimetype and a data driven scale on file modification time. A total of 141 formal attributes was created with these two scales.

With these two scales, the support was set to give two datasets with 955 and 5463 concepts. For the 955 concepts the intents only algorithm took 1.6 seconds whereas the covering edges required 7.5 seconds plus 14.8 seconds of setup time. For the larger set of 5463 concepts the intent only algorithm needed 1:07 whereas the covering edges completed with 42.5 and 25.5 seconds of setup time.

6.3 Performance on UCI Covtype Dataset

This section examines the implementations in the setting of the application of Formal Concept Analysis on a large data source. This selected application is particularly difficult due to it having 64 formal attributes as well as each formal object having a relatively large number of attributes.

The UCI covtype database consists of 581,012 tuples (formal objects) with 54 columns of data (many-valued attributes). Two ordinal columns were used: the aspect and elevation. Formal attributes were created using 32 formal attributes per ordinal scale. An example formal attribute would be created from a predicate like "elevation < 3144".

Table 4. Performance of intents only and covering edges algorithms on CFI drawn from 100,000 objects with 64 attributes

|CFI|	Average border size	Intents only (seconds)	Covering edges core (seconds)	Covering edges set-up (minutes : seconds)
1,000	522	1.8	1.7	4:08
3,000	1376	13.5	4.8	10:42

A sub-sample with only 100,000 tuples was created for testing. A support cut-off was selected to generate datasets with roughly 1,000 and 3,000 CFI. The results are shown in Table 4. It can be seen that the intents only algorithm suffers when the number of CFI increases. The intents only algorithm does not require to touch the base table in order to find X for all CFI which presents a significant set-up cost for the covering edges algorithm on such a large input dataset.

7   Conclusion

The border algorithm can suffer poor performance as the mean border set size grows. Some testing has revealed border set averages in the order of half the concept set size.

This border set issue does not significantly impact the border algorithm where the size of the CFI set is small ($< 1,000$)). The use of the border algorithm is generally advantageous when the number of concepts in the CFI is relatively small ($< 3,000$)).

The efficiency of the border algorithm is independent of the number of formal objects. Only the CFI are required for the border algorithm.

For the covering edges algorithm, as the number of formal objects increases the burden of inspecting the whole database during algorithm setup increases. The setup for the covering edges algorithm includes finding the extent of each concept and the attribute extent for every attribute in preparation for step 4a of Fig. 1.

For a large formal context where formal objects have few attributes and the set of all formal attributes A is relatively large the covering edges algorithm can be very slow. This is because step 4a of Fig. 1 will be executed on average $|C| * |A|$ times each involving a $|O|$ length bit intersection operation.

References

1. libferris visited April 2007, `http://witme.sf.net/libferris.web/`
2. Agrawal, R., Srikant, R.: Fast algorithms for mining association rules. In: Bocca, J.B., Jarke, M., Zaniolo, C. (eds.) Proc. 20th Int. Conf. Very Large Data Bases, VLDB, 12–15 1994, pp. 487–499. Morgan Kaufmann, San Francisco (1994)
3. Blake, C., Merz, C.: UCI Repository of Machine Learning Databases. In: Irvine, CA: University of California, Department of Information and Computer Science (1998), `http://www.ics.uci.edu/~mlearn/MLRepository.html`
4. Carpineto, C., Romano, G.: Concept Data Analysis. Wiley, England (2004)
5. Ganter, B., Wille, R.: Formal Concept Analysis — Mathematical Foundations. Springer–Verlag, Berlin Heidelberg (1999)
6. Goethals, B., Zaki, M.J.: Advances in frequent itemset mining implementations: Report on fimi'03. In: Goethals, B., Zaki, M.J. (eds.) Proceedings of the ICDM 2003 Workshop on Frequent Itemset Mining Implementations. CEUR Workshop Proceedings, vol. 90 (2003)

7. Grahne, O., Zhu, J.: Efficiently using prefix-trees in mining frequent itemsets. In: Goethals, B., Zaki, M.J. (eds.) Proceedings of the IEEE ICDM Workshop on Frequent Itemset Mining Implementations, Ceur (2003)

8. Han, J.: Data mining: concepts and techniques. Morgan Kaufmann Publishers, San Francisco (2001)

9. Han, J., Pei, J., Yin, Y.: Mining frequent patterns without candidate generation. In: Chen, W., Naughton, J., Bernstein, P.A. (eds.) 2000 ACM SIGMOD Intl. Conference on Management of Data, pp. 1–12. ACM Press, New York (2000)

10. Martin, B., Eklund, P.W.: Spatial Indexing for Scalability in FCA. In: Missaoui, R., Schmid, J. (eds.) Formal Concept Analysis. LNCS (LNAI), vol. 3874, pp. 205–220. Springer, Heidelberg (2006)

11. Martin, B., Eklund, P.W.: Custom asymmetric page split generalized index search trees and formal concept analysis. In: Kuznetsov, S.O., Schmidt, S. (eds.) ICFCA 2007. LNCS (LNAI), vol. 4390, Springer, Heidelberg (2007)

12. Mueller, A.: Fast sequential and parallel algorithms for association rule mining: A comparison. Technical Report CS-TR-3515, Departure of Computer Science, University of Maryland, College Park, MD (1995)

13. Pei, J.: Pattern-growth methods for frequent pattern mining, ph.d. thesis, computing science, simon fraser university (2001)

14. Pietracaprina, A., Zendolin, D.: Mining frequent itemsets using patricia tries. In: Goethals, B., Zaki, M.J. (eds.) Proceedings of the IEEE ICDM Workshop on Frequent Itemset Mining Implementations, Ceur (2003)

15. Agrawal, R., et al.: Fast discovery of association rules. In: Fayyad, U., et al. (eds.) Advances in Knowledge Discovery and Data Mining, pp. 307–328. AAAI Press, Menlo Park (1996)

16. Stumme, G., et al.: Computing iceberg concept lattices with titanic. J. on Knowledge and Data Engineering (KDE) 42, 189–222 (2002)

17. Zaki, M.J.: Scalable algorithms for association mining. Knowledge and Data Engineering 12, 372–390 (2000)

A Formal Context for Symmetric Dependencies

Jaume Baixeries

Departament de Llenguatges i Sistemes Informàtics
Universitat Politècnica de Catalunya
C/ Jordi Girona 1-3.
08034 Barcelona. Catalonia
jbaixer@lsi.upc.edu

Abstract. Armstrong and symmetric dependencies are two of the main groups of dependencies in the relational database model, both of them having their own set of axioms. The closure of a set of dependencies is the largest set of dependencies that can be calculated by the recursive application of those axioms. There are two problems related to a closure: its calculation and its characterization. Formal concept analysis has dealt with those problems in the case of Armstrong dependencies (that is, functional dependencies and alike).

In this paper, we present a formal context for symmetric dependencies that calculates the closure and the lattice characterization of a set of symmetric dependencies.

1 Introduction

Dependencies are restrictions that apply over a set of data (we assume that, generally speaking, a set of data is a set of fixed length tuples or records that have a common set of attributes). They usually indicate a relationship between sets of attributes that exist in that set of data, and may be found in different realms, as in database theory, artificial intelligence, propositional logic, knowledge discovery, algebra, etc ([1,5,7,8,22]).

Every type of dependency has a (semantical) definition that indicates the conditions that must exist in the set of data for those dependencies to hold. For instance, a functional dependency $X \xrightarrow{\text{fd}} Y$, where X and Y are sets of attributes, holds in a set of data if the values of the attributes Y can be determined by the values of the attributes X, as Example 1 shows.

Example 1. We have the set of data $\mathcal{R}$:

Name	Birth Place	Birth Year
P.D. James	Oxford	1920
de Pedrolo M.	L'Aranyó	1918
Bukowski C.	Berlin	1920
Durrell L.	Jalandhar	1912
Gombrowicz W.	Małoszyce	1904
Plenzdorf U.	Berlin	1934

R. Medina and S. Obiedkov (Eds.): ICFCA 2008, LNAI 4933, pp. 90–105, 2008.
© Springer-Verlag Berlin Heidelberg 2008

The columns are identified with the attribute names *Name, Birth Place, Birth Year* and each row represents a record or a tuple. In this case, we have that, given *Name*, we can deduce *Birth Place* and *Birth Year*. For instance, if we are given *C. Bukowski*, we know that the birth place and birth year are *Berlin* and 1920. But given *Birth Year*, we cannot deduce *Name*, since if we are given the year 1920, the author's name can be *P.D. James* or *C. Bukowski*. Neither do we have that *Birth Place* determines *Name*, since if we are given *Berlin*, the author's name may be either *C. Bukowski* or *U. Plenzdorf*. Therefore, we have that in this set of data, the functional dependency $\{Name\} \xrightarrow{fd}$ $\{Birth\ Place, Birth\ Year\}$ holds, but that $\{Birth\ Place\} \xrightarrow{fd} \{Name\}$ and $\{Birth\ Year\}$ $\xrightarrow{fd} \{Name\}$ do not hold.

For each set of dependencies Σ we take all sets of data (a possibly infinite number of them) in which all the dependencies in Σ hold: $R_1, R_2, \ldots$. Apart from all the dependencies in Σ, there may be some other dependencies not in Σ that may hold in **all** R_i as well. These dependencies, together with those in Σ are called the **closure** of Σ, and are represented as Σ^+.

The calculation of Σ^+ is closely related to the **implication problem** (also known as, membership problem), that consists in, given a set of dependencies Σ and a single dependency σ, determine if σ holds in all set of data in which all dependencies in Σ hold, or, said otherwise, if σ is a *logical consequence* of Σ ([5]). Obviously, σ is a logical consequence of Σ if and only if $\sigma \in \Sigma^+$. The calculation of Σ^+ can be performed by computing all possible sets of data R_i in which Σ hold, which is rather unfeasible since there may be an infinite number of them. Instead, we can use one set of complete and sound axioms of that type of dependency, if they exist. The axioms are a (finite) set of rules that state that a dependency holds if some conditions apply. As long as the set of dependencies Σ is finite, the axioms for those dependencies provide a finite way to calculate Σ^+.

Some important remarks concerning the comparison between the axioms for a set of dependencies compared with the definition of those same dependencies must be considered. It has been said that each type of dependency has, obviously, a semantical definition that states the conditions that must exist in a set of data for that particular dependency to hold. Unlikely the definition of a dependency, the axioms for a type of dependency are syntax-based, that is, they state what dependencies hold in a set of data if some other conditions, that **do not** depend on the set of data, exist. Therefore, the semantical definition is particular to each kind of dependency, whereas the axioms may be shared with some other dependencies. This induces a classification of dependencies according to their axioms. In database theory, there are two main of such groups: Armstrong and symmetric dependencies [8]. Functional dependencies belong to the former group, and multivalued dependencies belong to the latter, and both are well-known in the relational database model ([13]), since they are used to determine important facts on sets of data, as, for instance, if they are in some normal forms [23]. Therefore, their importance in this discipline is capital. Functional dependencies are also present in knowledge discovery [21,22].

Two important problems are important for any set of dependencies Σ: the characterization and the calculation of Σ^+. In this paper we focus on lattices, which describe one of these possible characterizations. Formal Concept Analysis has dealt with (a) the calculation and (b) the characterization of *closures* of Armstrong dependencies. Concerning (a), it is performed in [18] (Proposition 20) with a formal context such that the application of the Galois connection yields the closure of that set of dependencies, that is: $\Sigma'' = \Sigma^+$ (the set of attributes of this formal context is the set of all Armstrong dependencies that can be formed with a given set of attributes). Concerning (b), it has been studied in terms of a lattice in [18] and in [2] for Armstrong dependencies, and in [3,4] for further dependencies. This characterization was rather semantics-based and was based on a set of data. Aa lattice syntax-based characterization of Σ^+ is presented in [11].

Thus, a formal context for Armstrong dependencies provides a convenient way (a) to calculate the closure of a set of dependencies and (b) to characterize this closure. Therefore, a formal context with the same properties for other well-know dependencies may also be of interest for similar reasons. In this paper, we present a formal context, similar in nature to that for Armstrong dependencies, that calculates the closure of a set of symmetric dependencies and also its characterization in terms of a lattice.

2 Notation

We follow the notation in [18] for formal concept analysis. We define $\mathcal{U} = \{a_1, \ldots, a_n\}$ as a set of attributes. Letters $X, Y, Z, \ldots$ are used to denote sets of attributes. For a set $X \subseteq \mathcal{U}$, we represent by $(X)^C$ its complement wrt $\mathcal{U}$. Part($\mathcal{U}$) denotes the set of all possible partitions of $\mathcal{U}$. The notation of a partition $P \in \text{Part}(\mathcal{U})$ in classes is as follows: $P = [P_1 \mid \cdots \mid P_n]$, where each $P_i \subseteq \mathcal{U}$ is a class. For all $P, Q \in \text{Part}(\mathcal{U})$, we say that $P \preceq Q$ if and only if each class in Q is contained in some class of P. The relation $\preceq$ is a partial order and $\langle \text{Part}(\mathcal{U}); \preceq \rangle$ is a complete lattice ([17]). The reader can refer to [17] for more information on the lattice of the partitions of a set; for reasons that will be clear later on, the order in this paper is the dual of the order found in current literature. The bottom element is the partition $[\mathcal{U}]$, and the top element is a partition where all the classes are singletons: $[a_1 \mid \cdots \mid a_n]$. In this paper, the set notation for classes in a partition is dropped. Hence, if $\mathcal{U} = \{a, b, c, d\}$, instead of $[\{a, b\} \mid \{c\} \mid \{d\}]$ we use $[ab \mid c \mid d]$. We use the notation $\underline{X}$ in a partition, to denote that all the attributes in X are singletons in that partition. For instance, the partition $[\underline{ab} \mid cd]$ stands for $[a \mid b \mid cd]$. As it will be seen in Section 4, the definition of partitions of sets is closely related to the syntactical nature of symmetric dependencies, since partitions are the algebraic object used to describe a *dependency basis*.

Definition 1. *Let* $\mathcal{V}$ *be a lattice. For a set* $V \subseteq \mathcal{V}$, *the closure under meet of* V *is denoted as* $[V]^\wedge$.

If the set $[V]^\wedge$ has a top element, then, by Theorem 2.31 in [12], $[V]^\wedge$ is a complete lattice, since the closure under possibly infinite meets is always well defined. We say that $[V]^\wedge$ is a **closure system**. We use the notation $\mathcal{M}(\mathcal{V})$ for the set of **meet-irreducible** elements of the lattice $\mathcal{V}$.

An Armstrong and a symmetric dependency are both a binary relation between two sets of attributes, and the notation is $X \xrightarrow{\text{ad}} Y$ for the former, and $X \xrightarrow{\text{sd}} Y$ for the latter, for $X, Y \subseteq \mathcal{U}$. The closure of a set of dependencies Σ is the largest set of dependencies that hold in **all** set of data in which Σ hold, and it can be calculated by the axioms of that type of dependency. Since these axioms are syntax-based, some different kinds of dependencies may share the same axioms, although having different semantical definitions. In the relational database model, there are two main groups of dependencies: Armstrong and symmetric dependencies, each of them have their own set of axioms. Implications and functional dependencies (FD's) are **Armstrong dependencies** ([23]), whereas multivalued dependencies (MVD's), degenerated multivalued dependencies (DMVD's) and multivalued dependency-clauses (MVD-clauses), are ([8]) **symmetric dependencies** (SD's).

For a given a set of attributes $\mathcal{U}$, we define as $AD_{\mathcal{U}}$ and $SD_{\mathcal{U}}$ the set of all Armstrong dependencies and the set of all symmetric dependencies respectively that can be formed with $\mathcal{U}$. By **all possible closures of a set of symmetric dependencies**, we mean the following set:

$$\{ \Sigma^+ \mid \Sigma \subseteq SD_{\mathcal{U}} \}$$

that is, the set of all closures for all sets of dependencies that can be formed from a given set of attributes (for Armstrong dependencies, the definition is analogous, with $AD_{\mathcal{U}}$ instead of $SD_{\mathcal{U}}$).

3   Formal Context for Armstrong Dependencies

This section presents some known results mainly from [18], that characterize and calculate the closure of a set of Armstrong dependencies. These dependencies have the following axioms:

1. Reflexivity: If $Y \subseteq X$, then, $X \xrightarrow{ad} Y$ holds.
2. Augmentation: If $X \xrightarrow{ad} Y$ holds, then, $X \cup Z \xrightarrow{ad} Y \cup Z$ holds, where $Z \subseteq \mathcal{U}$.
3. Transitivity: If $X \xrightarrow{ad} Y$ and $Y \xrightarrow{ad} Z$ hold, then, $X \xrightarrow{ad} Z$ holds

As it can be seen, these axioms state the presence of an Armstrong dependency given some conditions that do not depend directly on any set of data. The calculation and characterization of Σ^+ is based on the following formal context: $\mathbb{K}_{AD}(\mathcal{U}) = (\wp(\mathcal{U}), AD_{\mathcal{U}}, \oplus)$, where the binary relation $\oplus \subseteq \wp(\mathcal{U}) \times AD_{\mathcal{U}}$ is:

Definition 2. *Let $X, Y, W \subseteq \mathcal{U}$. We say that W **respects** an Armstrong dependency $X \xrightarrow{ad} Y$ (and the notation is $W \oplus (X \xrightarrow{ad} Y)$) if and only if:*

$$W \not\supseteq X \text{ or } W \supseteq Y$$

In the context $\mathbb{K}_{AD}(\mathcal{U}) = (\wp(\mathcal{U}), AD_{\mathcal{U}}, \oplus)$, an attribute is an Armstrong dependency, and an object is a subset of $\mathcal{U}$. It may be remarked that in this case, we are dealing on the one hand with a set $\mathcal{U}$ of the attributes that are present in a hypothetical set of data, and, on the other hand, the set of attributes of the formal context which, in this case, is $AD_{\mathcal{U}}$. We also note that this formal context is only valid for a fixed set of attributes $\mathcal{U}$, since both $AD_{\mathcal{U}}$ and $\wp(\mathcal{U})$ depend on $\mathcal{U}$.

The first result concerns the calculation of the closure of all possible sets of Armstrong dependencies, and states that $\Sigma'' = \Sigma^+$ [18] (Proposition 20). Moreover, since the set of attributes in this formal context is the set $AD_{\mathcal{U}}$, the implications that hold in this formal context are between sets of Armstrong dependencies, in such a way that an implication $\Sigma \xrightarrow{\text{impl}} \Sigma''$ that holds is, in fact, characterizing the closure of Σ, that is, we have implications of the form "a set of dependencies implies its own closure (wrt Armstrong axioms)".

The second result concerns the lattice characterization of the closure of a set of Armstrong dependencies, and states that Σ' is the lattice characterization of Σ^+. This characterization is given by the following function:

$$\Upsilon_{\Sigma}(X) = \bigcap \{ Y \in \Sigma' \mid Y \supseteq X \}$$

It must be noted that Υ_{Σ} is defined wrt Σ', and characterizes all the Armstrong dependencies that are in Σ^+ in this way: $X \xrightarrow{\text{ad}} Y \in \Sigma^+$ if and only if $\Upsilon_{\Sigma}(X) = \Upsilon_{\Sigma}(X \cup Y)$.

This lattice characterization has been presented in [16,9,18,2,11,23] and some applications in database normalization can be found in [9].

4   Formal Context for Symmetric Dependencies

In this section we present a formal context for symmetric dependencies, similar in essence to that presented for Armstrong dependencies in Section 3. We first explain some peculiarities of symmetric dependencies, and then, we proceed to show our results.

4.1   Some Remarks on Symmetric Dependencies

In order to compute the closure of a set of dependencies, and as in the case of Armstrong dependencies, symmetric dependencies have their own set of axioms, which are the following:

1. Reflexivity: If $Y \subseteq X$ then, $X \xrightarrow{\text{sd}} Y$ holds.
2. Complementation: If $X \xrightarrow{\text{sd}} Y$ holds, then $X \xrightarrow{\text{sd}} (X \cup Y)^C$ holds.
3. Augmentation: If $X \xrightarrow{\text{sd}} Y$ holds and $Z \subseteq W$ then, $X \cup W \xrightarrow{\text{sd}} Y \cup Z$ holds.
4. Transitivity. If $X \xrightarrow{\text{sd}} Y$ holds and $Y \xrightarrow{\text{sd}} W$ holds, then, $X \xrightarrow{\text{sd}} (W \setminus Y)$ holds.

As a consequence, the following rules also apply([13]):

1. Right-hand side union: If $X \xrightarrow{\text{sd}} Y$ and $X \xrightarrow{\text{sd}} Z$ hold then, $X \xrightarrow{\text{sd}} Y \cup Z$ holds.
2. Right-hand side intersection: If $X \xrightarrow{\text{sd}} Y$ and $X \xrightarrow{\text{sd}} Z$ hold then, $X \xrightarrow{\text{sd}} Y \cap Z$ holds.

It is interesting to note that symmetric dependencies follow reflexivity, augmentation and transitivity just as Armstrong dependencies, with some slight differences. However, complementation does not hold for Armstrong dependencies, and, in fact, it is the axiom that from a syntactical point of view, is crucial to determine the diferences between Armstrong and symmetric dependencies.

This is a syntax-based definition, and, therefore, it can be applied to all different types of symmetric dependencies. For instance, Multivalued dependencies (MVD's), degenerated multivalued dependencies (DMVD's) and multivalued dependency clauses (MVD-cl's) are all symmetric dependencies, that is, they follow the axioms previously defined. At the same time, they all have different semantics-based definitions [4].

In order to easy our reasonings in the rest of this paper, we present the definition of the dependency basis with respect to a set of symmetric dependencies. Given a set of symmetric dependencies Σ, there is a way to condense all the information we need in order to calculate all the dependencies that are in Σ^+ that have the same left-hand side; this is the **dependency basis** ([13]). The dependency basis is a function such that given a set of attributes and a set of symmetric dependencies Σ, returns all the information about what dependencies with X in the left-hand side are in Σ^+. This information is returned as a partition of the set of the attributes. The precise definition of the dependency basis is as follows:

Definition 3. *Let $X \subseteq U$ and let Σ be a set of symmetric dependencies. The **dependency basis** of X in Σ, $DB_\Sigma(X)$ is the coarsest partition of U such that for all Y such that $X \xrightarrow{sd} Y \in \Sigma^+$, we have that Y is a union of some classes of $DB_\Sigma(X)$.*

Example 2. If the set of attributes is $U = \{a, b, c, d\}$ and $\Sigma = \{a \xrightarrow{sd} bc, a \xrightarrow{sd} d, a \xrightarrow{sd} bd, a \xrightarrow{sd} c, a \xrightarrow{sd} bcd\}$, then, all the dependencies $a \xrightarrow{sd} X$, where $X \subseteq U$, are in Σ^+. For instance, $a \xrightarrow{sd} b$ is in Σ^+ because of the intersection of the right-hand side of $a \xrightarrow{sd} bc$ and $a \xrightarrow{sd} bd$. The rest of dependencies $a \xrightarrow{sd} X$ hold because of the union of right-hand sides. Therefore:

$$DB_\Sigma(a) = [a \mid b \mid c \mid d]$$

One property of the dependency basis is:

Proposition 1. $X \xrightarrow{sd} Y \mid (X \cup Y)^C \in \Sigma^+$ *if and only if* $DB_\Sigma(X) \succeq [\underline{X} \mid Y \mid (X \cup Y)^C]$.

This proposition indicates that all the possible right-hand sides that are in $X \xrightarrow{sd} Y \mid (X \cup Y)^C$ must be characterized by $DB_\Sigma(X)$. These are Y, $(X \cup Y)^C$, as well as all the attributes of X as singletons. Therefore, $DB_\Sigma(X)$ must at least contain classes Y, $(X \cup Y)^C$ and $\underline{X}$ or coarser classes such that their union yield those sets.

The set of all the dependency bases that hold in a set of symmetric dependencies Σ is:

$$DB(\Sigma) = \{ DB_\Sigma(X) \mid X \in U \}$$

4.2   Results

We start with the definition of the formal context for symmetric dependencies $\mathbb{K}_{SD}(U) = (\mathrm{Part}(U), SD_U, \otimes)$, where the binary relation $\otimes$ is:

Definition 4. *Let* $\otimes$ *be a binary relation:* $\otimes \subseteq \mathrm{Part}(\mathcal{U}) \times SD_{\mathcal{U}}$. *Let* $P \in \mathrm{Part}(\mathcal{U})$, $X \xrightarrow{sd} Y \in SD_{\mathcal{U}}$ *and* $Z = \mathcal{U} \setminus (X \cup Y)$. *We say that the partition* P **marks** $X \xrightarrow{sd} Y$, *and we represent it as* $P \otimes (X \xrightarrow{sd} Y)$, *if and only if:*

$$\nexists P_i \in P : P_i \cap Y \neq \emptyset \text{ and } P_i \cap Z \neq \emptyset \text{ and } P_i \cap X = \emptyset$$

This definition states basically that a partition marks a symmetric dependency $X \xrightarrow{sd} Y$ if and only if there is no class in P that contains only attributes from $(Y \cup Z) \setminus X$ and at least one attribute from $Y \setminus X$ and one from Z. The *complement* of the relation $\otimes$ is $\overline{\otimes}$.

As in the preceding case for Armstrong dependencies, we prove two main results in this section that hold in the formal context $\mathbb{K}_{SD}(\mathcal{U}) = (\mathrm{Part}(\mathcal{U}), SD_{\mathcal{U}}, \otimes)$:

1. $\Sigma'' = \Sigma^+$ for any set of symmetric dependencies $\Sigma \subseteq SD_{\mathcal{U}}$.
2. Σ' is the lattice characterization of Σ^+, in terms of a closure operator (to be defined later).

We start with the first result: $\Sigma'' = \Sigma^+$ for any set $\Sigma \subseteq SD_{\mathcal{U}}$. For each axiom, we prove that all those dependencies that it can derive are in Σ''.

Proposition 2 (Reflexivity). *If* $Y \subseteq X$, *then,* $X \xrightarrow{sd} Y \in \Sigma''$.

Proof. Consider a partition $P \in \Sigma'$ and any class $P_i \in P$. Either $P_i \cap Y = \emptyset$ or $P_i \cap Y \neq \emptyset$, which means $P_i \cap X \neq \emptyset$ because $Y \subseteq X$. Whatever the case, we get $P \otimes (X \xrightarrow{sd} Y)$ for any $P \in \Sigma'$, and therefore, $X \xrightarrow{sd} Y \in \Sigma''$. ∎

Proposition 3 (Complementation). *If* $X \xrightarrow{sd} Y \in \Sigma$, *then,* $X \xrightarrow{sd} (X \cup Y)^C \in \Sigma''$.

Proof. Suppose that there is a $P \in \Sigma'$ such that $P \overline{\otimes} (X \xrightarrow{sd} (X \cup Y)^C)$. Then, there is a class in P that contains only attributes from $Y \setminus X$ and from $(X \cup Y)^C$ (and one from each, at least). But in such a case, we also have that $P \overline{\otimes} (X \xrightarrow{sd} Y)$, which contradicts the fact that $P \in \Sigma'$. ∎

Proposition 4 (Augmentation). *If* $X \xrightarrow{sd} Y \in \Sigma$ *and* $Z \subseteq W$, *then,* $X \cup W \xrightarrow{sd} Y \cup Z \in \Sigma''$.

Proof. Suppose that there is a $P \in \Sigma'$ such that $P \overline{\otimes} (X \cup W \xrightarrow{sd} Y \cup W')$. Then, P has a class that contains attributes only from $(Y \cup W') \setminus (X \cup W)$ and from $(Y \cup W')^C \setminus (X \cup W)$, and at least, one from each set. We have that

$(Y \cup W') \setminus (X \cup W)$ $=$
$(Y \cup W') \cap (X)^C \cap (W)^C$ $=$
$(Y \cap (X)^C \cap (W)^C) \cup (W' \cap (X)^C \cap (W)^C) =$ (since $W' \subseteq W$)
$(Y \cap (X)^C \cap (W)^C) \subseteq Y \setminus X$

We also have that
$(Y \cup W')^C \setminus (X \cup W)$ $=$
$(Y)^C \cap (W')^C \cap (X \cup W)^C$ $=$
$(Y)^C \cap (W')^C \cap (X)^C \cap (W)^C$ $=$ (since $(W)^C \subseteq (W')^C$)
$(Y)^C \cap (W)^C \cap (X)^C \subseteq (Y)^C \setminus X$

that is: $(Y \cup W') \setminus (X \cup W) \subseteq Y \setminus X$ and $(Y \cup W')^C \setminus (X \cup W) \subseteq (Y)^C \setminus X$. In such a case, we have that $P \overline{\otimes} (X \xrightarrow{sd} Y)$, which contradicts the fact that $P \in \Sigma'$. Since no such P can be in Σ', then, $X \cup W \xrightarrow{sd} Y \cup W' \in \Sigma''$. ∎

Proposition 5 (Transitivity). *If $X \xrightarrow{sd} Y$ and $Y \xrightarrow{sd} W \in \Sigma$, then, $X \xrightarrow{sd} W \setminus Y \in \Sigma''$.*

Proof. Suppose that there is a $P \in \Sigma'$ such that $P \otimes (X \xrightarrow{sd} W \setminus Y)$. Then, P has a class $P_i \in P$ that contains attributes only from $(W \setminus Y) \setminus X = (X)^C \cap (Y)^C \cap W$ and from $(W \setminus Y)^C \setminus X = ((X)^C \cap (W)^C) \cup ((X)^C \cap Y)$, and at least, one from each set. Since $(X)^C \cap (Y)^C \cap W \subseteq (X)^C \cap (Y)^C$, all attributes in P_i and in $(X)^C \cap (Y)^C \cap W$, are also in $(X)^C \cap (Y)^C$. Towards a contradiction, we now look at the attributes that are in P_i and in $((X)^C \cap (W)^C) \cup ((X)^C \cap Y)$. Those attributes can be all in $(X)^C \cap Y$, since $(X)^C \cap Y \subseteq ((X)^C \cap (W)^C) \cup ((X)^C \cap Y)$, but in this case, we would have $P \otimes (X \xrightarrow{sd} Y)$, which is not possible, since $P \in \Sigma'$ and $X \xrightarrow{sd} Y \in \Sigma^+$. If we have that some of the attributes (not all of them) are in $(X)^C \cap Y$, then, the remaining ones from $((X)^C \cap (W)^C) \cup ((X)^C \cap Y)$ not in $(X)^C \cap Y$ must be in $(X)^C \cap (W)^C$; besides, they cannot be in Y, because they would be then in $(X)^C \cap Y$. Therefore, they are in $(Y)^C$, that is, in $(X)^C \cap (Y)^C \cap (W)^C$, and also in $(X)^C \cap (Y)^C$, in which case, P_i contains attributes only from $(X)^C \cap (Y)^C$ and from $(X)^C \cap Y$, and then, $P \otimes (X \xrightarrow{sd} Y)$, again, a contradiction. Hence, no attribute in P_i and in $((X)^C \cap (W)^C) \cup ((X)^C \cap Y)$ can be in $(X)^C \cap Y$, that is, they all must be in $(X)^C \cap (Y)^C \cap (W)^C$. We have, then, that the attributes in P_i can be split into two classes: $P_i \cap (X)^C \cap (Y)^C \cap W$ and $P_i \cap (X)^C \cap (Y)^C \cap (W)^C$ Since $(X)^C \cap (Y)^C \cap W \subseteq (Y)^C \cap W$ and $(X)^C \cap (Y)^C \cap (W)^C \subseteq (Y)^C \cap (W)^C$, we have that P_i can be split into two non-empty classes $(Y)^C \cap W$ and $(Y)^C \cap (W)^C$, that is, classes $W \setminus Y$ and $(Y)^C \setminus W$, which means that $P \otimes (Y \xrightarrow{sd} W)$, which is also a contradiction. Hence, no such P exists in Σ' and then, $X \xrightarrow{sd} W \setminus Y \in \Sigma''$. ∎

Therefore, we have proved that Σ'' contains all the symmetric dependencies that can be derived by all the axioms of symmetric dependencies, that is, Σ^+:

Theorem 1. $\Sigma^+ \subseteq \Sigma''$ *in the formal context* $\mathbb{K}_{SD}(\mathcal{U}) = (\mathrm{Part}(\mathcal{U}), SD_{\mathcal{U}}, \otimes)$.

Proof. By Propositions 2, 3, 4 and 5. ∎

We also have that Σ'' contains a set of dependencies, and that Σ' contains a set of partitions. Before we prove that in Σ'' we have only those dependencies that can be derived by the axioms for symmetric dependencies, we study the nature of Σ' and see that, in fact, it contains all the dependency bases of Σ:

Proposition 6. $\forall X \subseteq \mathcal{U} : DB_{\Sigma}(X) \in \Sigma'$.

Proof. Note that by reflexivity, we have that $DB_{\Sigma}(X) \succeq [X \mid (X)^C]$. Suppose that $DB_{\Sigma}(X) \notin \Sigma'$. In such a case, there is a $Y \xrightarrow{sd} W \in \Sigma$ such that $DB_{\Sigma}(X) \otimes (Y \xrightarrow{sd} W)$. Then, there is a class $P_i \in DB_{\Sigma}(X)$ that contains only attributes from $W \setminus Y$ and from $(W)^C \setminus Y$, and, at least, one from each. Since $Y \xrightarrow{sd} W \in \Sigma$, then, by Proposition 4 (augmentation), $(P_i)^C \xrightarrow{sd} (W \setminus Y) \cap P_i \in \Sigma^+$. Note that since P_i must have, at least, one attribute from $(W)^C$, then, $(W \setminus Y) \cap P_i \subset P_i$. Towards a contradiction, by Theorem 1, we have that $X \xrightarrow{\text{sd}} P_i \in \Sigma''$, also that $X \xrightarrow{\text{sd}} (P_i)^C \in \Sigma''$ and finally, by Proposition 5 (transitivity) we have $X \xrightarrow{\text{sd}} (W \setminus Y) \cap P_i \in \Sigma''$. Since $X \xrightarrow{\text{sd}} (W \setminus Y) \cap P_i \in \Sigma''$, then, P_i must be formed by the union of some classes in $DB_\Sigma(X)$ which is not possible because $(W \setminus Y) \cap P_i \subset P_i$ and P_i is one class in $DB_\Sigma(X)$, which is a contradiction and then, $DB_\Sigma(X) \in \Sigma'$.   ∎

We will restrict this result further later on, but we are using it now to finally prove that Σ'' contains **only** those dependencies that can be recursively derived by the axioms of symmetric dependencies.

Theorem 2. $\Sigma'' = \Sigma^+$.

Proof. We prove that $\Sigma'' \subseteq \Sigma^+$ since in Theorem 1 it has been proved that $\Sigma^+ \subseteq \Sigma''$. By contradiction, assume that $X \xrightarrow{\text{sd}} Y \in \Sigma''$ and $X \xrightarrow{\text{sd}} Y \notin \Sigma^+$: there is a $P \in \Sigma'$ such that $P \otimes (X \xrightarrow{\text{sd}} Y)$. Since $X \xrightarrow{\text{sd}} Y \notin \Sigma^+$, we have that $DB_\Sigma(X) \not\leq [X \mid Y \mid Z]$, otherwise $X \xrightarrow{\text{sd}} Y$ could be derived by the axioms of symmetric dependencies and would belong to Σ^+. By Proposition 6, we have that $DB_\Sigma(X) \in \Sigma'$. Then, there are two attributes in a class P_i in $DB_\Sigma(X)$ that are in different classes in $[X \mid Y \mid Z]$. Since X is all singletons in $DB_\Sigma(X)$, then, $P_i \subseteq Y \cup Z$. Then, $P_i \cap Y \neq \emptyset$ and $P_i \cap Z \neq \emptyset$. In this case, we have that $DB_\Sigma(X) \otimes (X \xrightarrow{\text{sd}} Y)$, which contradicts the fact that $DB_\Sigma(X) \in \Sigma'$, and then, we have that $X \xrightarrow{\text{sd}} Y \mid Z \notin \Sigma''$.   ∎

Example 3. We define the context $\mathbb{K}_{SD}(\mathcal{U}) = (\text{Part}(\mathcal{U}), SD_{\mathcal{U}}, \otimes)$, where $\mathcal{U} = \{a, b, c, d\}$. According to the binary relation $\otimes$, the formal context is as follows:

Now, we assume that $\Sigma = \{a \xrightarrow{\text{sd}} b \mid cd, c \xrightarrow{\text{sd}} a \mid bd\}$. In order to calculate the closure of Σ, according to Theorem 2, we have that $\Sigma^+ = \Sigma''$. Therefore, we proceed with calculating Σ'' in that formal context. First, we calculate Σ' (marked in grey in the object row in the left-hand context) and finally Σ'' which is the set of dependencies that are marked in grey in the attribute column in the right-hand context:

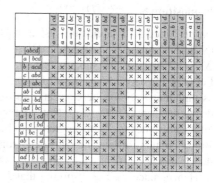

This set of symmetric dependencies happen to be Σ^+ according to the axioms for symmetric dependencies, that is, $\Sigma^+ = \{\, a \xrightarrow{\text{sd}} b \mid cd,\, c \xrightarrow{\text{sd}} a \mid b \mid d,\, ac \xrightarrow{\text{sd}} b \mid c,\, ad \xrightarrow{\text{sd}} b \mid c,\, bc \xrightarrow{\text{sd}} a \mid d,\, cd \xrightarrow{\text{sd}} a \mid b \,\}$.

So far, we have seen that the binary relation $\otimes$ has been able to discard those dependencies that are not in Σ^+. The intuition behind this binary relation will be explained later, once we see how Σ' is a partition lattice that characterizes Σ^+. We now prove some extra properties of Σ' with respect to the dependency bases of a set of dependencies. We have seen in Proposition 6 that Σ' contains all the dependency bases of Σ. Now, we further restrict this result and prove that Σ' is the set of all dependency bases of Σ closed under meet. First, we prove that Σ' is a closure system:

Proposition 7. $\Sigma' = [\Sigma']^\wedge$.

Proof. We prove that if $P, Q \in \Sigma'$, then, $P \wedge Q \in \Sigma'$ by contradiction. Assume that $P \wedge Q \notin \Sigma'$. There is a $X \xrightarrow{\text{sd}} Y \mid Z \in \Sigma$, where $Z = (X \cup Y)^C$, such that $P \wedge Q \otimes (X \xrightarrow{\text{sd}} Y \mid Z)$. That is, $P \wedge Q$ has a class P_i that contains only attributes from $Y \setminus X$ and Z. By construction of $P \wedge Q$, it can happen that $P_i \in P$ and $P_i \in Q$, but it would mean that $P \overline{\otimes} (X \xrightarrow{\text{sd}} Y \mid Z)$ and that $Q \overline{\otimes} (X \xrightarrow{\text{sd}} Y \mid Z)$, which is a contradiction. Then, P_i is not contained in partitions P, Q. We can split the attributes in P_i into two classes: $P_i \cap (Y \setminus X)$ and $P_i \cap Z$. We take all the classes in P and in Q that are (properly) contained in P_i, and assume that all those classes are contained in $P_i \cap (Y \setminus X)$ or in $P_i \cap Z$. By construction of $P \wedge Q$, then, class P_i would not be in $P \wedge Q$, since there is no class in P nor in Q that contains some attributes from $P_i \cap (Y \setminus X)$ and from $P_i \cap Z$, and hence, in $P \wedge Q$, class P_i would be split into classes $P_i \cap (Y \setminus X)$ and $P_i \cap Z$. Then, there is at least a class in P or in Q that contains some attributes from $P_i \cap (Y \setminus X)$ and from $P_i \cap Z$ which is obviously contained in $(Y \setminus X) \cup Z = (X)^C$. In this case, we have that $P \overline{\otimes} (X \xrightarrow{\text{sd}} Y \mid Z)$ or $Q \overline{\otimes} (X \xrightarrow{\text{sd}} Y \mid Z)$ which is a contradiction. ∎

As a natural consequence, we have:

Corollary 1. $[DB(\Sigma)]^\wedge \subseteq \Sigma'$.

Proof. By Proposition 6 and 7. ∎

And we now prove that the set of partitions in Σ' is exactly $[DB(\Sigma)]^\wedge$.

Proposition 8. $\Sigma' \subseteq [DB(\Sigma)]^{\wedge}$.

Proof. Since Σ' is a closure system by Proposition 7 and $\forall X \subseteq \mathcal{U} : DB(X) \in \Sigma'$ by Proposition 6, we prove that $\forall P \in \mathcal{M}(\Sigma')$, P is a dependency basis for some set $X \subseteq \mathcal{U}$. Suppose that this in not true: $P \in \mathcal{M}(\Sigma')$ and P is not a dependency basis. Let Q be the dependency basis of the set of attributes that are all singletons in P. Now, assume that $P \not\geq Q$: there are at least two attributes in a class $C \in P$ that are in two different classes $C_i, C_j \in Q$. Note that the attributes in C, C_i, C_j are not in X, because all the attributes in X are singletons in both P and Q. By Theorem 2, $X \xrightarrow{sd} C_i \in \Sigma''$. But C contains at least one attribute from C_i and one from $(X \cup C_i)^C$, which means that $P \overline{\otimes} (X \xrightarrow{sd} C_i)$, which contradicts the fact that $P \in \Sigma'$.

Hence, since we are assuming that P is not a dependency basis, we only have the possibility $P \succ Q$. Let $P = [\underline{X} \mid X' \mid C_1 \mid \cdots \mid C_n]$, where C_i are non-trivial classes. Since $Q = DB_{\Sigma}(X) \succeq [\underline{X} \mid (X)^C]$ and $P \succ Q$, and $P \neq [\underline{\mathcal{U}}]$ because $DB_{\Sigma}(\mathcal{U}) = [\underline{\mathcal{U}}]$, then, $n \geq 1$. For each C_i, we claim that $DB_{\Sigma}((C_i)^C) = [\underline{(C_i)^C} \mid C_i]$. Assume that this is not true: $DB_{\Sigma}((C_i)^C) \succ [\underline{(C_i)^C} \mid C_i]$, that is, the class C_i is split in $DB_{\Sigma}((C_i)^C)$ in, at least, two different classes, let them be $C_i^1, \ldots C_i^m$, where necessarily $m \geq 2$. It means that $(C_i)^C \xrightarrow{sd} C_i^j$, but in such a case, we have that $P \overline{\otimes} (X \xrightarrow{sd} C_i^j)$ because C_i contains all C_i^j. Since we are assuming that $P \in \Sigma'$, then, $DB_{\Sigma}((C_i)^C) = [\underline{(C_i)^C} \mid C_i]$. But in this case, we have that $P = \bigwedge\{ DB_{\Sigma}((C_i)^C) \mid C_i \in P \}$, which contradicts the fact that $P \in \mathcal{M}(\Sigma')$. Then, we have that if $P \in \mathcal{M}(\Sigma')$, then, it must be a dependency basis for some X. ∎

We now can prove the second result, that states that Σ' is the lattice characterization of Σ^+ in terms of the following a closure operator:

Definition 5. *For a given set $\Sigma \subseteq SD_{\mathcal{U}}$, where $P, Q \in \mathrm{Part}(\mathcal{U})$, we define the function $\Gamma_{\Sigma} : \mathrm{Part}(\mathcal{U}) \mapsto \mathrm{Part}(\mathcal{U})$ as:*

$$\Gamma_{\Sigma}(P) = \bigwedge\{ Q \in \Sigma' \mid Q \succeq P \}$$

We remark the structural similarity of this operator with $\Upsilon_{\Sigma}(X)$ defined in Section 3. We also note that this operator also depends on the set Σ' to compute its results. We claim that this closure operator characterizes Σ^+. We first prove that this operator computes, in fact, the dependency basis for a given set of atributes:

Proposition 9. $DB_{\Sigma}(X) = \Gamma_{\Sigma}([\underline{X} \mid (X)^C])$.

Proof. Assume that $DB_{\Sigma}(X) = [\underline{X} \mid Y_1 \mid \cdots \mid Y_n]$. We have that $DB_{\Sigma}(X) \in [DB(\Sigma)]^{\wedge}$, and, by Proposition 1, $DB_{\Sigma}(X) \succeq [\underline{X} \mid (X)^C]$, so then, $DB_{\Sigma}(X) \succeq \Gamma_{\Sigma}([\underline{X} \mid (X)^C])$. By definition of Γ_{Σ}, we have that if $\Gamma_{\Sigma}([\underline{X} \mid (X)^C]) \neq DB_{\Sigma}(X)$ it is because there is a $DB_{\Sigma}(X')$ such that $DB_{\Sigma}(X) \wedge DB_{\Sigma}(X') \succeq [\underline{X} \mid \mathcal{U}(X)^C]$. In this case, we have that all $a \in X$ are singletons in $DB_{\Sigma}(X')$. Hence, we have that $X' \xrightarrow{sd} X \in \Sigma^+$, and by transitivity, we have that $\forall Y_i \in DB_{\Sigma}(X) : X' \xrightarrow{sd} Y_i \in \Sigma^+$, which implies that every class $Y_i \in DB_{\Sigma}(X)$ must be the union of some classes of $DB_{\Sigma}(X')$, it is: every class in $DB_{\Sigma}(X')$ must be included in some class of $DB_{\Sigma}(X)$. But it would contradict that $DB_{\Sigma}(X) \wedge DB_{\Sigma}(X') \succeq [\underline{X} \mid (X)^C]$, unless $DB_{\Sigma}(X) = DB_{\Sigma}(X') = \Gamma_{\Sigma}([\underline{X} \mid (X)^C])$, as was to be shown. ∎

And, finally, we prove the characterization of the set of symmetric dependencies:

Theorem 3. $X \overrightarrow{_{sd}} Y \in \Sigma^+$ *if and only if* $\Gamma_\Sigma([\underline{X} \mid Y \cup Z]) = \Gamma_\Sigma([\underline{X} \mid Y \mid Z])$.

Proof. We have that $(X)^C = Y \cup Z$, and, obviously, Γ_Σ is a closure operator. We first note that $\Gamma_\Sigma([\underline{X} \mid Y \cup Z]) \succeq [\underline{X} \mid Y \mid Z]$ if and only if $\Gamma_\Sigma([\underline{X} \mid Y \cup Z]) = \Gamma_\Sigma([\underline{X} \mid Y \mid Z])$. The right to left implication is as follows: by extensity, we have that $\Gamma_\Sigma([\underline{X} \mid Y \mid Z]) \succeq [\underline{X} \mid Y \mid Z]$, and then, $\Gamma_\Sigma([\underline{X} \mid Y \cup Z]) \succeq [\underline{X} \mid Y \mid Z]$. The left to right implication is as follows: by monotonicity, we have that $\Gamma_\Sigma([\underline{X} \mid Y \cup Z]) \preceq \Gamma_\Sigma([\underline{X} \mid Y \mid Z])$, and by Proposition 9 we have $\Gamma_\Sigma([\underline{X} \mid Y \cup Z]) \succeq [\underline{X} \mid Y \mid Z]$. Again, by monotonicity and idempotency, we have $\Gamma_\Sigma([\underline{X} \mid Y \cup Z]) \succeq \Gamma_\Sigma([\underline{X} \mid Y \mid Z])$, and then, $\Gamma_\Sigma([\underline{X} \mid Y \cup Z]) = \Gamma_\Sigma([\underline{X} \mid Y \mid Z])$.

Finally, we have that by Proposition 1, $X \overrightarrow{_{sd}} Y \mid (X \cup Y)^C \in \Sigma^+$ if and only if $DB_\Sigma(X) \succeq [\underline{X} \mid Y \mid (X \cup Y)^C]$, that is, by Proposition 9, if and only if $\Gamma_\Sigma([\underline{X} \mid Y \cup Z]) \succeq [\underline{X} \mid Y \mid Z]$, and, as proved previously, if and only if $\Gamma_\Sigma([\underline{X} \mid Y \cup Z]) = \Gamma_\Sigma([\underline{X} \mid Y \mid Z])$. ∎

This theorem states that the lattice Σ' characterizes those symmetric dependencies that belong to the closure of Σ calculating the closure of $[\underline{X} \mid Y \cup Z]$ in the lattice Σ'. Again, we note that both closure operators Γ_Σ and Υ_Σ are structurally equivalent, and also are the algorithm that calculates Γ_Σ and Υ_Σ. The difference in the computation of this algorithm is purely the lattice over which the result is calculated, which, in turn, depends on the formal context.

Example 4. We take the same context given in Example 3. In this case, we have that if $\Sigma = \{ a \overrightarrow{_{sd}} b \mid cd, c \overrightarrow{_{sd}} a \mid bd \}$, then, $\Sigma' = \{ [a \mid b \mid c \mid d], [a \mid b \mid cd], [ac \mid b \mid d], [acd \mid b], [abc \mid d], [abcd] \}$ that forms this lattice:

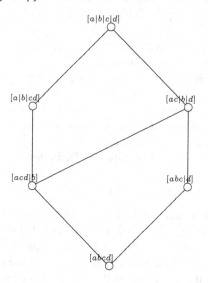

As we have stated in Theorem 3, this lattice is the lattice characterization of $\Sigma^+ = \{ a \overrightarrow{_{sd}} b \mid cd, c \overrightarrow{_{sd}} a \mid b \mid d, ac \overrightarrow{_{sd}} b \mid c, ad \overrightarrow{_{sd}} b \mid c, bc \overrightarrow{_{sd}} a \mid d, cd \overrightarrow{_{sd}} a \mid b \}$. For instance, we have that $c \xrightarrow{\text{sd}} b \mid ad$ is in Σ^+ because $\Gamma_\Sigma([c \mid abd]) = \Gamma_\Sigma([ad \mid b \mid c])$, which is $[a \mid b \mid c \mid d]$. As a negative example, we have that $ab \xrightarrow{\text{sd}} c \mid d$ is not in Σ^+, since $\Gamma_\Sigma([a \mid b \mid cd]) \neq \Gamma_\Sigma([a \mid b \mid c \mid d])$, because $\Gamma_\Sigma([a \mid b \mid cd]) = [a \mid b \mid cd]$ and $\Gamma_\Sigma([a \mid b \mid c \mid d]) = [a \mid b \mid c \mid d]$.

4.3 Discussion

We have previously deferred the discussion about the *intuition* behind the binary relation $\otimes$ until the explanation of how Σ' characterized Σ^+ was made. It has been seen in Corollary 1 and in Proposition 8 that Σ' is the closure under meet of the set of all dependency bases that hold in Σ. Therefore, the idea of the relation $\otimes$ is to reject all those partitions that (via closure under meet) would avoid a dependency from holding. Those partitions can be divided into two groups: those that avoid complementation and those that avoid transitivity (reflexivity and complementation are trivial in this discussion). This distinction is not necessary for Armstrong dependencies.

For example, let us suppose that we have the set of attributes $\mathcal{U} = \{a, b, c, d, e\}$, and a symmetric dependency $bc \xrightarrow{\text{sd}} d \mid ae \in \Sigma$. We first study what partitions avoid this dependency from holding in the lattice characterization of Σ^+. In such a case, since this dependency is in Σ, then, the dependency basis of bc in Σ would be greater or equal than $[b \mid c \mid d \mid ae]$, that is, $DB_\Sigma(bc) \succeq [b \mid c \mid d \mid ae]$ by Proposition 1. We have proved in Theorem 3 that in such a case the equality $\Gamma_\Sigma([b \mid c \mid dae]) = \Gamma_\Sigma([b \mid c \mid d \mid ae])$ must hold in the lattice characterization of Σ^+. Then, it necessarily implies that we must reject in Σ' all the partitions that avoid this equality from holding, for instance, the partition $[b \mid c \mid ad \mid e]$, because by Proposition 6 $DB_\Sigma(bc) \in \Sigma'$, and by Proposition 7 (that states that Σ' is a closure system) $[b \mid c \mid ad \mid e] \wedge DB_\Sigma(bc)$, that is, $[b \mid c \mid ade]$, would also be in Σ'. As a consequence, we would have that $\Gamma_\Sigma([b \mid c \mid dae]) = [b \mid c \mid dae]$ and that $\Gamma_\Sigma([b \mid c \mid d \mid ae]) = DB_\Sigma(bc)$, and since $DB_\Sigma(bc) \succeq [b \mid c \mid d \mid ae]$, then, the equality $\Gamma_\Sigma([b \mid c \mid dae]) = \Gamma_\Sigma([b \mid c \mid d \mid ae])$ would **not** hold, which is a violation of Theorem 3. Therefore, for any dependency $X \xrightarrow{\text{sd}} Y \mid Z$, where $Z = (X \cup Y)^C$, we have that the equality $\Gamma_\Sigma([\underline{X} \mid Y \cup Z]) = \Gamma_\Sigma([\underline{X} \mid Y \mid Z])$ must hold in Σ', and, therefore, all the partitions P such that $P \succeq [\underline{X} \mid Y \cup Z]$ and $P \not\succeq DB_\Sigma(X)$ are forbitten in Σ'. As we can see, these partitions must have all the attributes X as singletons, otherwise the condition $P \succeq [\underline{X} \mid Y \cup Z]$ would not be fulfilled, and also that there is a class that contains, at least, one attribute from Y and one attribute from Z, otherwise, the condition $P \not\succeq DB_\Sigma(X)$ would not be fulfilled. It is easy to see that those partitions are included in Definition 4:

$$\nexists P_i \in P : P_i \cap Y \neq \emptyset \text{ and } P_i \cap Z \neq \emptyset \text{ and } P_i \cap X = \emptyset$$

In fact, if those partitions were in the lattice characterization of Σ^+, they not only would avoid the dependency $X \xrightarrow{\text{sd}} Y \mid Z$ from holding, but also, all those dependencies that can be derived from $X \xrightarrow{\text{sd}} Y \mid Z$ by augmentation, which obviously must also be characterized by that same lattice. Again, we take the case of $bc \xrightarrow{\text{sd}} d \mid ae$. We have said previously that the partition $[b \mid c \mid ad \mid e]$ avoids $bc \xrightarrow{\text{sd}} d \mid ae$ from holding in the lattice characterization of Σ^+. We now take an example of a dependency that can be derived by augmentation from $bc \xrightarrow{\text{sd}} d \mid ae$ and we will see that it is not characterized by Σ' because of the presence of the partition $bc \overset{\longrightarrow}{_{\text{sd}}} d \mid ae$. We take the dependency $bce \overset{\longrightarrow}{_{\text{sd}}} d \mid a$, that be in Σ^+ because it can be derived from $bc \overset{\longrightarrow}{_{\text{sd}}} d \mid ae$ by augmentation, and should be characterized by Σ'. This is not the case because $DB_\Sigma(bce) = [a \mid b \mid c \mid d \mid e]$, which is present in the lattice and, then, we have that $\Gamma_\Sigma([b \mid c \mid ad \mid e]) = [b \mid c \mid ad \mid e]$ and that $\Gamma_\Sigma([a \mid b \mid c \mid d \mid e]) = [a \mid b \mid c \mid d \mid e]$, which again contradicts Theorem 3. This is only an example, but can easily be proved for all the dependencies that can be derived by augmentation.

The expression $(not(P \succeq [\underline{X} \mid Y \cup Z])$ and $P \not\succeq DB_\Sigma(X))$ is equivalent to that in Definition 1, which characterizes **all** the elements that are to be rejected in the lattice characterization of Σ^+ for Armstrong dependencies. However, this is not the case for symmetric dependencies. There is one more case not covered by the previous example which is that of a partition that fulfils the conditions in Definition 4 **and** such that all the attributes in X are not singletons, unlikely the previous example. This is the case of transitivity, and, again, we take a simple example in order to illustrate it. We have the dependency $bc \overset{\longrightarrow}{_{\text{sd}}} d \mid ae \in \Sigma$. It is obvious that any dependency in Σ such as $X \overset{\longrightarrow}{_{\text{sd}}} bc$ (where $X \subseteq \mathcal{U}$) forces the presence of the dependency $X \overset{\longrightarrow}{_{\text{sd}}} d \mid ae$ in Σ^+. Therefore, all those partitions that do not allow the dependency $X \overset{\longrightarrow}{_{\text{sd}}} d \mid ae$ from holding in the lattice characterization of Σ^+ cannot be present in Σ'. For instance, if we have that, apart from $bc \overset{\longrightarrow}{_{\text{sd}}} d \mid ae$, the dependency $a \overset{\longrightarrow}{_{\text{sd}}} bc \mid de$ is in Σ as well, then, we have that $a \overset{\longrightarrow}{_{\text{sd}}} d$ and $a \overset{\longrightarrow}{_{\text{sd}}} e$ are also in Σ^+ by transitivity. In this case, the *forbitten* partition is $[a \mid bc \mid de]$. It is necessary to note that this partition does not fulfil the condition $P \succeq [\underline{X} \mid Y \cup Z]$ and $P \not\succeq DB_\Sigma(X)$ as in the previous case. In general, these partitions that avoid transitivity from holding are described as partitions that have X in **one class** and also one class that contains at least one attribute from Y and one from Z.

5   Conclusions

We have presented a formal context that provides the calculation of the closure and its characterization for sets of symmetric dependencies. A similar formal context for Armstrong dependencies was presented in [18]. The interest of this formal context is twofold:

1. We can use existing FCA tools to calculate the closure of a set of symmetric dependencies and its characterization, and we can also use this context to answer the implication problem for symmetric dependencies.
2. We may use this context to answer questions for symmetric dependencies that have already been solved for Armstrong dependencies.

Concerning the first point, the formal context defined in Section 4 provides an algorithmically simple way to check whether a symmetric dependency σ belongs to the closure of a set of symmetric dependencies Σ. This computation can also be seen as to check whether the Horn clause $\Sigma \to \Sigma \cup \{\sigma\}$ holds in that context. In both cases, this task consists purely in a search throughout the context. The method for performing this computation is the same for Armstrong and symmetric dependencies. Intuitively, and comparing these methods with the algorithm in [5], we only need to perform a search, since the part that consists in calculating the dependency basis is already *contained* in the binary relation. This is why the presented formal context can be seen as a generalization for treating both Armstrong and symmetric dependencies, as well as an alternative for existing algorithms that calculate the implication problem or the closure for symmetric dependencies (efficient algorithms that handle formal contexts can be found in [18,24]).

On the other hand, concerning the second point, and parallelizing the work already started in [18,15] for Armstrong dependencies, this formal context opens the way to deduce the *basis* of a set of symmetric dependencies, that is, the minimal set of partitions that characterizes a set of dependencies. Although to check whether a dependency holds in a lattice characterization of Σ^+ is also the same algorithm for Armstrong and symmetric dependencies, we note that the definitions in [15,18] for calculating the minimal representation of a set $\Sigma \subseteq SD_U$ cannot be directly translated for symmetric dependencies, since the lattices that characterize Armstrong dependencies are different from those that have been presented in Section 3. However, the same intuitions may be used.

References

1. Baixeries, J., Balcázar, J.L.: Discrete Deterministic Data Mining as Knowledge Compilation. In: Proceedings of Workshop on Discrete Mathematics and Data Mining in SIAM International Conference on Data Mining (2003)
2. Baixeries, J.: A Formal Concept Analysis Framework To Model Functional Dependencies. In: Proceedings of Mathematical Methods for Learning (2004)
3. Baixeries, J., Balcázar, J.L.: Characterization and Armstrong Relations for Degenerate Multivalued Dependencies Using Formal Concept Analysis. Formal Concept Analysis. In: Ganter, B., Godin, R. (eds.) ICFCA 2005. LNCS (LNAI), vol. 3403, pp. 162–175. Springer, Heidelberg (2005)
4. Baixeries, J., Balcázar, J.L.: Unified Characterization of Symmetric Dependencies with Lattices. In: Contributions to ICFCA 2006. 4th International Conference on Formal Concept Analysis (2005)
5. Beeri, C.: On the Membership Problem for Functional and Multivalued Dependencies in Relational Databases. ACM Transactions on Database Systems (September 1980)
6. Beeri, C., Vardi, M.: Formal Systems for Tuple and Equality Generating Dependencies. SIAM Journal on Computing (1984)
7. Caspard, N., Monjardet, B.: The Lattices of Closure Systems, Closure Operators, and Implicational Systems on a Finite Set: a Survey. In: Proceedings of the 1998 Conference on Ordinal and Symbolic Data Analysis (OSDA-98). Discrete Applied Mathematics (2003)
8. Day, A.: A Lattice Interpretation of Database Dependencies. Semantics of Programming Languages and Model Theory (1993)
9. Demetrovics, J., et al.: Normal Form Relation Schemes: a New Characterization. Acta Cybernetica (1992)
10. Demetrovics, J., Huy, X.: Representation of Closure for Functional, Multivalued and Join Dependencies. Computers and Artificial Intelligence (1992)
11. Demetrovics, J., Libkin, L., Muchnik, I.: Functional Dependencies in Relational Databases: a Lattice Point of View. Discrete Applied Mathematics (1992)
12. Davey, B.A., Priestley, H.A.: Introduction to Lattices and Order. Cambridge University Press, Cambridge (1990)

13. Beeri, C., Fagin, R., Howard, J.H.: A Complete Axiomatization for Functional and Multivalued Dependencies in Database Relations. In: Proceedings of the 1977 ACM SIGMOD International Conference on Management of Data, August 3-5, 1977, Toronto, Canada (1977)
14. Fagin, R.: Armstrong Databases. In: Proc. 7th IBM Symposium on Mathematical Foundations of Computer Science, Kanagawa, Japan (1982)
15. Duquenne, V., Guigues, J.L.: Familles Minimales d'Implications Informatives Resultant d'un Tableau de Donées Binaires. Mathematics and Social Sciences (1986)
16. Gottlob, E., Libkin, L.: Investigations on Armstrong Relations, Dependency Inference, and Excluded Functional Dependencies. Acta Cybernetica (1990)
17. Grätzer, G.: General Lattice Theory. Academic Press, London (1978)
18. Ganter, B., Wille, R.: Formal Concept Analysis: Mathematical Foundations. Springer, Heidelberg (1999)
19. Kuznetsov, S.O., Obiedkov, S.A.: Comparing performance of algorithms for generating concept lattices. J. Exp. Theor. Artificial Intelligence 14(2–3) (2004)
20. Kuznetsov, S.O., Obiedkov, S.A.: Algorithms for the Construction of Concept Lattices and Their Diagram Graphs. In: Siebes, A., De Raedt, L. (eds.) PKDD 2001. LNCS (LNAI), vol. 2168, Springer, Heidelberg (2001)
21. Lopes, S., Petit, J.-M., Lakhal, L.: Functional and Approximate Dependency Mining: Database and FCA Points of View. Journal of Experimental and Theoretical Artificial Intelligence 14(2-3) (2002)
22. Lopes, S., Petit, J.-M., Lakhal, L.: Efficient Discovery of Functional Dependencies and Armstrong Relations. In: Zaniolo, C., et al. (eds.) EDBT 2000. LNCS, vol. 1777, Springer, Heidelberg (2000)
23. Ullman, J.D.: Principles of Database Systems. Computer Science Press (1982)
24. Valtchev, P., Missaoui, R., Lebrun, P.: A Fast Algorithm for Building the Hasse Diagram of a Galois Lattice. Actes du Colloque LaCIM (2000)

The Number of Plane Diagrams of a Lattice

Christian Zschalig

Institut für Algebra, TU Dresden, Germany
Christian.Zschalig@tu-dresden.de

Abstract. In this work we want to clarify, how many non-similar plane diagrams a planar lattice can have. In the first part demonstrate how to find all these diagrams by specifying all realizers, i.e. all pairs of linear orders whose intersection equals to the lattice order. The tools we use to achieve that goal are *Ferrers-graphs* [DDF84, Reu89] and *left-relations on contexts* [Zsc07]. Finally we determine the set of numbers which can occur as the number of plane diagrams of a planar lattice.

1   Introduction

A lattice $\underline{\mathfrak{V}}$ is called *planar* if it possesses a *diagram*[1] $\mathrm{pos}(\underline{\mathfrak{V}})$ without *edge crossings*. There exist several characterizations of planar lattices and many authors worked on this topic, see [BFR71, Cog82, DDF84, DM41, KR75, Pla76, Spi82] for instance. Some of the results allow to specify a plane diagram in polynomial time (w.r.t. the size of the lattice) by considering the underlying graph [Pla76] or the incomparability graph of the lattice [Spi82].

In this work we are interested to find all plane diagrams of a lattice. Since their number is either zero or uncountable, we need to classify diagrams. This is done by the notion of *similarity* [KR75].

The base for our consideration is the *Ferrers-graph* [DDF84] of the *standard context* [GW99] of a lattice. Its deep interrelation to planarity [DDF84, Zsc07] and the usage of *left-relations* [Zsc05] will help us to find an algorithm that determines the number of all plane diagrams in polynomial time. Thereby we also determine the number of *realizers* [DM41] of a lattice of dimension two.

Finally we want to answer the question, which finite numbers can occur as the number of plane diagrams of a planar lattice.

2   Preliminaries

In this work, all considered structures will be finite. A lattice $\underline{\mathfrak{V}}$ possesses a set of $\bigvee$-irreducibles $J(\underline{\mathfrak{V}})$ and a set of $\bigwedge$-irreducibles $M(\underline{\mathfrak{V}})$ (which are the sets of irreducible objects and attributes respectively in terms of FCA). By the word context we will understand the standard context $(J(\underline{\mathfrak{V}}), M(\underline{\mathfrak{V}}), \leq)$ of $\underline{\mathfrak{V}}$. Sloppily we will sometimes write J and M instead of $J(\underline{\mathfrak{V}})$ and $M(\underline{\mathfrak{V}})$ respectively.

---

[1] We sloppily use this term for the notion of *line diagram*. That is the image of a function pos mapping lattice elements to points in the Euclidian plane and elements of the neighborhood relation of $\underline{\mathfrak{V}}$ to straight line segments connecting their end points. For a formal definition of that notion, see for instance [KR75, Zsc05].

R. Medina and S. Obiedkov (Eds.): ICFCA 2008, LNAI 4933, pp. 106–123, 2008.

2.1   Dimension and Conjugate Orders

Dushnik and Miller [DM41] were the first to introduce these two concepts to order theory. The *dimension* of a poset $\underline{P} = (P, \leq)$ is the smallest number of linear orders intersecting in $\leq$. A *conjugate order* on a poset is a strict order[2] on P whose elements are pairs of incomparable elements of P.

Definition 1. *[DM41] Let $\underline{P} = (P, \leq)$ be an ordered set. The incomparability relation in $\underline{P}$ is denoted by $\|$.*

1. *We call L_c conjugate relation if $L_c \cup L_c^{-1} = \|$.*
2. *We call L_c conjugate order if additionally L_c is a strict order.*

Definition 2. *[DM41] Let $\underline{P} = (P, \leq)$ be an ordered set. A family $\{R_i\}_{i \in I}$ of linear orders on P is called* realizer *of $\underline{P}$ if $\leq = \bigcap_{i \in I} R_i$. The dimension $dim(\underline{P})$ is the smallest cardinal number m, s.t. there exists an m-elemental realizer of $\underline{P}$.*

They also discovered the dependency between both terms if $\underline{P}$ is a lattice and $dim(\underline{P}) = 2$:

Lemma 1. *[DM41] Let $\underline{\mathfrak{V}} = (\mathfrak{V}, \leq)$ be a lattice possessing a conjugate order L. Then the set $\{L \cup <, L^{-1} \cup <\}$ is a realizer.*

   Let $\underline{\mathfrak{V}}$ be a lattice of dimension 2 with a realizer $\{L, R\}$. Then both $L_c := L \setminus \leq$ and $R_c := R \setminus \leq$ are conjugate orders. In particular, $L_c^{-1} = R_c$.

The importance of both concepts to the characterization of planar lattices is clarified in the following theorem.

Theorem 1. *[BFR71, Bir67, DM41] Let $\underline{\mathfrak{V}}$ be a lattice. Then the following are equivalent:*

1. *$\underline{\mathfrak{V}}$ is planar.*
2. *There exists a conjugate order on $\underline{\mathfrak{V}}$.*
3. *$dim(\underline{\mathfrak{V}}) \leq 2$.*

2.2   Left-Relations on Lattices

A *sorting relation* on a lattice is just the union of strict linear orders on incomparable $\wedge$-irreducibles sharing a common upper neighbor:

Definition 3. *[Zsc05] Let $\underline{\mathfrak{V}}$ be a finite lattice and M be the set of its $\wedge$-irreducible elements. A strict order $L_a \subseteq M \times M$ is called* sorting relation *if the following condition[3] holds for all elements $m, n \in M$:*

$$m^* = n^* \iff m \, L_a \, n \text{ or } n \, L_a \, m.$$

---

[2] i.e. an irreflexive, asymmetric and transitive binary relation.

[3] With m^* we denote the unique upper neighbor of a $\wedge$-irreducible m.

Definition 4. *[Zsc05] Let $\mathfrak{V}$ be a finite lattice with a given sorting relation L_a. For arbitrary lattice elements v and w, we define*

$$M(v,w) = \{(v',w') \subset M \times M \mid v \leq v', w \leq w', v \parallel w', w \parallel v'\}.$$

We define the relation $L \subseteq \mathfrak{V} \times \mathfrak{V}$ according to:

$$v\,L\,w : \Longleftrightarrow \begin{cases} v\,L_a\,w, & v,w \in M, v^* = w^* \\ \exists (m,n) \in M(v,w) : m\,L\,n, & else \end{cases}$$

L is called left-relation *and $R := L^{-1}$ is called* right-relation *on the lattice $\mathfrak{V}$. If L is additionally a strict order, we will call it* left-order.

A sorting relation uniquely defines a left-relation. Instead of considering only some $\bigwedge$-irreducible elements, the left-relation states for all pairs of incomparable lattice elements (v,w), whether v is left or right of w. The comparable elements are understood to be above or below each other. This extension is done iteratively: v is left of w if there exist two $\bigwedge$-irreducibles m and n above v and w respectively, s.t. m is left of n:

The interrelation to conjugate orders is revealed in the following proposition. Namely every conjugate order is a left-order, i.e. can be constructed from a sorting relation.

Proposition 1. *[Zsc05] Let L be a relation on a finite lattice $\mathfrak{V}$. Then*

$$L \text{ is a conjugate order } \iff L \text{ is a left-order.}$$

Since a left-relation ise determined by a sorting relation defined only on some pairs of $\bigwedge$-irreducibles it is easier to handle than a conjugate order.

2.3   Ferrers-Graphs

Definition 5. *[Cog82, GW99] A* Ferrers-relation *F is a relation $F \subseteq J \times M$ meeting*

$$j_1 F m_1 \wedge j_2 F m_2 \implies j_1 F m_2 \vee j_2 F m_1.$$

for all $j_1, j_2 \in J$ and $m_1, m_2 \in M$. The Ferrers-dimension *$fdim(R)$ of a relation $R \subseteq J \times M$ is the smallest number of Ferrers-Relations $F_t \subseteq J \times M$, $t \in T$, whose intersection is equal to R, i.e. $R = \bigcap_{t \in T} F_t$.*

In a cross table representing a relation F we notice that F is a Ferrers-relation if and only if the configuration depicted on the right does not occur.

The inverse $\overline{F}$ of a Ferrers-relation is again a Ferrers-relation. Hence the Ferrers-dimension of a relation R is the smallest cardinality of a set of Ferrers-relations $\{F_t\}_{t \in T}$ covering the empty cells of its cross table [GW99], i.e. $\overline{R} := (J \times M) \setminus R = \bigcup_{t \in T} F_t$.

The next result connects the Ferrers-dimension and the order dimension. Since a lattice is planar if and only if its order dimension is at most two (see Theorem 1), this gives us the key to our further considerations.

Theorem 1. [4] *[GW99] Let $\mathfrak{V}$ be a lattice. Then $fdim(\leq) = dim(\mathfrak{V})$.*

Although the calculation of the Ferrers-dimension in general is $\mathcal{NP}$-complete [GW99], it is treatable in the case (≤ 2) that we are interested in. For that purpose we introduce the notion of a *Ferrers-graph*. Its nodes are the empty cells of a context and its edges indicate which vertices cannot belong to the same Ferrers-relation F_t. See Figure 1 for an example.

Definition 6. *[DDF84, Reu89] Let $R \subseteq J \times M$ be a relation. We define the Ferrers-graph $\tilde{\Gamma}(R)$ as an undirected simple graph with vertex set V and edge set E as follows:*

$$V := \overline{R} \qquad E := \{\{(j_1, m_2), (j_2, m_1)\} \mid (j_1, m_1), (j_2, m_2) \in R\}.$$

The bare *Ferrers-graph $\Gamma(R)$ is obtained from $\tilde{\Gamma}(R)$ by deleting all isolated vertices. By the* Ferrers-graph $\tilde{\Gamma}$ *of a lattice $\mathfrak{V}$ we denote the Ferrers-graph of its standard context $\mathbb{K} = (J(\mathfrak{V}), M(\mathfrak{V}), \leq)$.*

$\mathbb{K}$	m_1	m_2	m_3	m_4
j_1	×	×	•	•
j_2	•	×	×	•
j_3	•	•	×	×

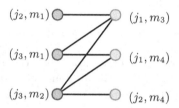

Fig. 1. A context $\mathbb{K}$ given by a cross table (left) and its appropriate Ferrers-graph Γ (right)

The following theorem connects the Ferrers-dimension to a property of the Ferrers-graph, namely the bipartiteness[5]. A constructive proof (finding a plane diagram of the respective lattice) is provided in [Zsc07].

Theorem 2. *[DDF84] A relation R has the Ferrers-dimension of at most two if and only if its Ferrers-graph is bipartite.*

2.4   Left-Relations on Contexts

A *left-relation on a context* can be understood as a restriction of a lattice's left-relation on the incomparable pairs (j, m) of a $\bigvee$-irreducible j and a $\bigwedge$-irreducible m. The relation can be filled into the empty cells of the standard context of $\mathfrak{V}$:

---

[4] Originally this assertion was given more generally for posets by Cogis in [Cog82].

[5] A graph (V, E) is *bipartite* if their exists a partition of V into two classes V_1 and V_2, s.t. there is no edge in the subgraphs induced by V_1 and V_2.

Definition 7. *Let $\mathfrak{V}$ be a lattice and $\mathbb{K} = (J, M, \leq)$ its standard context.*

1. *The relation* $\|_\mathbb{K} = (J \times M) \cap \|$ *is called* incomparability relation on $\mathbb{K}$.
2. *A relation* $\mathfrak{L} \subseteq \|_\mathbb{K}$ *is called* left-relation on $\mathbb{K}$. *We denote* $\mathfrak{R} := \|_\mathbb{K} \setminus \mathfrak{L}$.

Let $\mathfrak{V}$ be a lattice with Ferrers-graph $\tilde{\Gamma}$. If $\tilde{\Gamma}$ is bipartite, it is evident to consider the bipartition classes $\tilde{\mathfrak{L}}$ and $\tilde{\mathfrak{R}}$ of $\tilde{\Gamma}$ as left- and right-relation on the standard context. More precisely:

$$\mathfrak{L} := \tilde{\mathfrak{L}} \setminus \{(j, m) \mid j > m\} \text{ and } \mathfrak{R} := \tilde{\mathfrak{R}} \setminus \{(j, m) \mid j > m\}$$

are the respective left- and right- relations. They *induce* a relation $\tilde{L} \subseteq M \times M$ by

$$m \, \tilde{L} \, n : \Longleftrightarrow m \parallel n \text{ and } (\exists j \in J : m \geq j \, \mathfrak{L} \, n \text{ or } n \geq j \, \mathfrak{R} \, m).$$

Furthermore, if $\tilde{\Gamma}$ consists of components $\tilde{\Gamma}_k, k \in K$ we partition the vertex classes $\mathfrak{L}$ and $\mathfrak{R}$ by $\mathfrak{L}_k := \mathfrak{L} \cap V(\tilde{\Gamma}_k)$ and $\mathfrak{R}_k := \mathfrak{R} \cap V(\tilde{\Gamma}_k)$. We introduce induced relations $\tilde{L}_k, k \in K$ by

$$m \, \tilde{L}_k \, n \Longleftrightarrow m \parallel n \text{ and } (\exists j \in J : m \geq j \, \mathfrak{L}_k \, n \text{ or } n \geq j \, \mathfrak{R}_k \, m).$$

We observe that for a bipartite graph Γ and $\bigwedge$-irreducibles $m \parallel n$ the following equivalence holds:

$$\exists j \in J : m \geq j \, \mathfrak{L}_k \, n \Longleftrightarrow \exists h \in J : n \geq h \, \mathfrak{R}_k \, m.$$

The following Lemma shows that the introduction of the induced relation $\tilde{L}$ "almost" guarantees the existence of a conjugate order on $\mathfrak{V}$ if the Ferrers-graph Γ is bipartite. Unfortunately the transitivity of $\tilde{L}$ is much harder to treat than the asymmetry.

Lemma 2. *[Zsc07] Let $\mathfrak{V}$ a lattice and Γ its bipartite Ferrers-graph with vertex classes $\mathfrak{L}$ and $\mathfrak{R}$. Let $\tilde{L}$ be the relation induced by $\mathfrak{L}$. Then $\tilde{L}$ is asymmetric and connex on pairs of incomparable attributes. Furthermore the following equivalence holds:*

$$\tilde{L} \text{ is transitive} \Longleftrightarrow \tilde{L} \text{ can be extended}[6] \text{ to a left-order.}$$

2.5   The Geometry of Lattice Diagrams

Kelly and Rival gave a possibility to categorize diagrams of ordered structures by means of *similarity*. That concept is comparable to the *topological equivalence* used in graph theory [Die96].

Definition 8. *[KR75] Let $\mathfrak{V} = (\mathfrak{V}, \leq)$ be a lattice and pos($\mathfrak{V}$) a plane line diagram of it. Let $\lambda^* \subseteq \mathfrak{V} \times \mathfrak{V}$ be a relation defined as follows: $v \, \lambda^* \, w$ holds if and only if v and w have a common upper neighbor v^* and the straight line*

---

[6] i.e. there exists a sorting relation $L_a \subseteq \tilde{L}$, s.t. the left-relation L induced by L_a contains $\tilde{L}$.

segment connecting v and v^* is left[7] of the straight line segment connecting w and v^*. Two diagrams are called similar *if their respective λ^* relations are the same. The* left-relation $\lambda \subseteq \mathfrak{V} \times \mathfrak{V}$ *induced by* $\mathrm{pos}(\mathfrak{V})$ *is defined by*

$$v \,\lambda\, w : \Longleftrightarrow \quad v \parallel w \text{ and}$$
$$\exists v' \geq v, w' \geq w : v', w' \prec (v \vee w) \text{ with } v' \,\lambda^*\, w'.$$

The following theorem states that we can obtain all non-similar plane diagrams of a lattice $\mathfrak{V}$ by calculating all left-orders on $\mathfrak{V}$. By additionally reminding Lemma 1 and Proposition 1, we conclude that every realizer $\{L, R\}$ corresponds to exactly two plane diagrams possessing the left-relations L and R respectively.

Theorem 3. *[KR75, Zsc05] Let $\mathfrak{V}$ be a finite lattice. The following statements are equivalent.*

1. *There exists a plane diagram* $\mathrm{pos}(\mathfrak{V})$ *with the left-relation L.*
2. *L is a left-order on $\mathfrak{V}$.*

It is well known that for triconnected[8] planar graphs every two diagrams are equivalent [Die96], this also holds for diagrams of triconnected digraphs [BDMT98].

   If we are interested in all plane diagrams, we have to be able to handle lattices $\mathfrak{V} = (\mathfrak{V}, \leq)$ having several triconnected components in their corresponding graph $(\mathfrak{V}, \prec)$. We therefore use the following notion:

Definition 9. *[KR75] Let $\mathfrak{V}$ be a finite lattice and let $a < b$ for two elements of $\mathfrak{V}$. An $< a, b >$-component is a connected component of the graph $((a, b), \prec)$. An $< a, b >$-component C is called* proper *if $y \leq x \implies y \leq a$ and $y \geq x \implies y \geq b$ holds for all elements $x \in C$ and $y \in \mathfrak{V} \setminus C$.*

Later we will see later that "reflecting" and "permuting" proper components will supply all the plane diagrams of a lattice. An appropriate characterization was already given in [KR75].

3 How to Find All Plane Diagrams of a Lattice

3.1 Foundation

We know that a lattice is planar if its Ferrers-graph is bipartite [DDF84]. A bipartition $V(\Gamma) = \mathfrak{L} \,\dot{\cup}\, \mathfrak{R}$ of Γ determines a conjugate order $\hat{L}$ on $\mathfrak{V}$ [Zsc07]. Moreover, a left-order L defines uniquely a bipartition on the bare graph Γ by restricting L to incomparable pairs of $J \times M$. This follows from Lemma 2.

---

[7] We think that the meaning of "left" is intuitively clear. We omit the precise definition (given in [KR75]) due to better readability of the paper.

[8] A graph is *triconnected* if removing two arbitrary nodes keeps the graph connected and non-empty.

Hence the possible candidates for plane diagrams can be obtained by considering the set of bipartitions of Γ. If Γ consists of κ components, we find at most 2^κ plane diagrams and $2^{\kappa-1}$ realizers of the lattice order $\leq$ (see Theorem 3 and Lemma 1).

3.2 About the Structure of the Bare Ferrers-Graph Γ

Our first result clarifies the connection between the bare Ferrers-graph of a lattice and the incomparability relation on its standard context $\mathbb{K}$. The Ferrers-graph $\tilde{\Gamma}$ is defined on all pairs (j, m) satisfying $j > m$ or $j \parallel m$. For the bare Ferrers-graph Γ however $V(\Gamma) = \parallel_{\mathbb{K}}$ holds. That means that a bipartition $\mathfrak{L} \dot{\cup} \mathfrak{R}$ of Γ exactly defines two left-relations $\mathfrak{L}$ and $\mathfrak{R}$ on $\mathbb{K}$.

Proposition 2. *[Zsc07] Let $\mathfrak{V}$ be a lattice and $\tilde{\Gamma}$ its Ferrers-graph. A vertex (j, m) of Γ is isolated if and only if $j > m$.*

Now we want to observe which vertices of Γ are in the same component for a given context $\mathbb{K}$. It turns out that sequences of $\bigvee$-irreducibles and $\bigwedge$-irreducibles of the form

$$m_0 \geq j_1 \leq m_1 \geq j_2 \leq m_2 \geq \ldots \geq j_r \leq m_r$$

play an important role for connectivity.

Definition 10. *[Zsc07] Let $\mathbb{K} = (J, M, \leq)$ be a context and $[\underline{v}, \overline{v}]$ be an interval in $\mathfrak{V}$. A sequence $p = m_0 \geq j_1 \leq m_1 \geq j_2 \ldots \geq j_r \leq m_r$ of $\bigvee$-irreducibles j_i and $\bigwedge$-irreducibles m_i is called* connection *of m_0 and m_r in $[\underline{v}, \overline{v}]$ if*

$$\forall i \in \{0, \ldots, r\} : m_i \not\geq \overline{v} \quad and \quad \forall i \in \{1, \ldots, r\} : j_i \not\leq \underline{v}$$

If the condition $j_i \leq m_k \implies k \in \{i, i-1\}$ holds for all $i \in \{1, \ldots, r\}$ as well then p is called shortest connection.

Every connection p between m_0 and m_r contains a shortest connection q [Zsc07].

Definition 11. *[Zsc07] Let $\mathbb{K}$ be the standard context of a lattice $\mathfrak{V}$ and $[\underline{v}, \overline{v}]$ be an interval in $\mathfrak{V}$.*

1. *A $\bigwedge$-irreducible $m \in [\underline{v}, \overline{v}]$ is called* bound *if there exists a $\bigwedge$-irreducible $n \notin [\underline{v}, \overline{v}]$ and a connection $p = m \ldots n$ in $[\underline{v}, \overline{v}]$.*
2. *Two $\bigwedge$-irreducibles $m, n \in [\underline{v}, \overline{v}]$ are called* connected *if there exists a connection $p = m \ldots n$ in $[\underline{v}, \overline{v}]$.*
3. *Three pairwise incomparable $\bigwedge$-irreducibles m_1, m_2, m_3 are called* free triple *if none is bound and no two are connected in $[m_1 \wedge m_2 \wedge m_3, m_1 \vee m_2 \vee m_3]$.*

Definition 12. *[Zsc07] Let $[\underline{v}, \overline{v}]$ be an interval in the lattice $\mathfrak{V}$. Let $U[\underline{v}, \overline{v}]$ denote the set of non-bound $\bigwedge$-irreducibles in $[\underline{v}, \overline{v}]$. Let $m \in U[\underline{v}, \overline{v}]$. The set*

$$U_m[\underline{v}, \overline{v}] := \{n \in M \mid m \text{ and } n \text{ are connected}\}$$

is called the m-component *of $[\underline{v}, \overline{v}]$.*

Obviously "connected" in $[\underline{v}, \overline{v}]$ is an equivalence relation. Therefore the set of m-components in $[\underline{v}, \overline{v}]$ is a partition of $U[\underline{v}, \overline{v}]$. An equivalence class $U_m[\underline{v}, \overline{v}]$ contains exactly the $\bigwedge$-irreducibles of one proper component of $[\underline{v}, \overline{v}]$.

Definition 13. *Let $\mathbb{K} = (J, M, \leq)$ be the standard context of a lattice $\mathfrak{V}$, Γ its Ferrers-graph and $m \parallel n \in M$. By $\Gamma_{m,n}$ we denote that component of Γ containing all edges between m and n, i.e. all edges[9] of the form $\{(j, m), (h, n)\} \in E(\Gamma)$.*

The next two results of this section point out that the free triples are the only problematic case for finding transitive induced left-relations $\tilde{L}$ out of a bipartition $\mathfrak{L} \mathbin{\dot{\cup}} \mathfrak{R}$ of the bare Ferrers-graph Γ.

Corollary 1. *[Zsc07] Let $\mathfrak{V}$ be a lattice with bipartite Ferrers-graph Γ. Let $\tilde{L}$ be the induced relation by an appropriate bipartition and $m_1, m_2, m_3 \in M(\mathfrak{V})$. If $m_1 \mathbin{\tilde{L}} m_2 \mathbin{\tilde{L}} m_3 \mathbin{\tilde{L}} m_1$ then (m_1, m_2, m_3) is a free triple.*

Lemma 3. *[Zsc07] Let $\mathfrak{V}$ be a lattice with Ferrers-graph Γ. Let m_1, m_2, m_3 be $\bigwedge$-irreducible elements. Then (m_1, m_2, m_3) is a free triple if and only if the components Γ_{m_i, m_k} ($i \neq k \in \{1, 2, 3\}$) are pairwise disjoint.*

Finally we remind an obvious observation about bipartite graphs. If some of the components of a graph possessing a bipartition are "turned around", one again obtains a bipartition.

Lemma 4. *[Zsc07] Let $\Gamma = (V, E)$ be a bipartite graph with vertex classes X and Y and let Γ_k, $k \in K$ its components. Let $X_k = X \cap V(\Gamma_k)$ and $Y_k = Y \cap V(\Gamma_k)$ be the vertex classes of the appropriate components Γ_k. Let $R_k \in \{X_k, Y_k\}$ for all $k \in K$. Then the sets $R = \bigcup_{k \in K} R_k$ and $V(\Gamma) \setminus R$ are a bipartition of Γ.*

3.3 Components of Γ and m-Components of $\mathfrak{V}$

The strategy to find all plane diagrams of a lattice $\mathfrak{V}$ is as follows: We classify the components of the bare Ferrers-graph Γ into two types. A component Γ_k of type 1 acts on one proper component (i.e. for all vertices (j, m) of that component we find a proper $[\underline{v}, \overline{v}]$-component with $j, m \in [\underline{v}, \overline{v}]$). In contrary, a component Γ_k of type 2 acts on two proper $[\underline{v}, \overline{v}]$-components C_1, C_2, i.e. for each vertice (j, m) either $j \in C_1, m \in C_2$ or $j \in C_2, m \in C_1$ holds. Then we show (by considering the appropriate components of Γ) that all plane diagrams of a planar lattice can be obtained by reflecting one proper component or by permuting proper components over one interval $[\underline{v}, \overline{v}]$. However, instead of proper components C we use to consider the respective m-components U_m, which are just a restriction to the contained $\bigwedge$-irreducible elements. Before we need some preliminary observations.

Definition 14. *Let $\mathfrak{V}$ be a lattice with Ferrers-graph Γ. Let Γ_k be a component of Γ. By*

$$M(\Gamma_k) := \{m \in M \mid \exists j \in J : (j, m) \in V(\Gamma_k)\}$$

---

[9] It is easy to see that indeed all these edges are in one component of Γ (see [Zsc07]).

we denote the set of all $\bigwedge$-irreducibles which belong to an edge of Γ_k. Dually we define the set $J(\Gamma_k)$. The interval $I(\Gamma_k) = [\underline{v}, \overline{v}]$ of a component Γ_k is defined by

$$\underline{v} := \bigwedge M(\Gamma_k) \qquad and \qquad \overline{v} := \bigvee M(\Gamma_k).$$

Lemma 5. *Let $\mathfrak{V}$ be a lattice with Ferrers-graph Γ. Let Γ_k be a component possessing the interval $[\underline{v}, \overline{v}] = I(\Gamma_k)$ with $|V(\Gamma_k)| > 1$. Then*

$$M(\Gamma_k) \subseteq (\underline{v}, \overline{v}) \qquad and \qquad J(\Gamma_k) \subseteq (\underline{v}, \overline{v})$$

Proof. 1. Let $m \in M(\Gamma_k)$. Since Γ_k consists of at least two vertices, we find elements $n \in M(\Gamma_k)$ and $j, h \in J(\Gamma_k)$, s.t. $((j, m), (h, n))$ is an edge in Γ_k. By Definition 6 we note $m \parallel n$ and therefore $\underline{v} \leq m \wedge n < m < m \vee n \leq \overline{v}$.

2. Let $j \in J(\Gamma_k)$. Since Γ_k consists of at least two vertices, we find elements $m, n \in M(\Gamma_k)$ and $h \in J(\Gamma_k)$, s.t. $((j, m), (h, n))$ is an edge in Γ_k. By Definition 6 we note $j \leq n$ and $j \parallel m$ and find with the first statement of Lemma 5 $j \leq n < \overline{v}$ and $\underline{v} < m$, i.e. $j \not\geq \underline{v}$.

We finally have to prove $j \not\parallel \underline{v}$. It is a basic result of lattice theory that $j = \bigwedge \{ \tilde{n} \in M \mid m \geq j \}$. Hence we have to show $\forall \tilde{n} \in M : \tilde{n} \geq j \implies \tilde{n} \geq \underline{v}$ only. Let $\tilde{n}$ be an arbitrary $\bigwedge$-irreducible satisfying $j \leq \tilde{n}$. Since $\tilde{n} \not\leq m$ (that would imply $j \leq m$) we have either $\tilde{n} > m > \underline{v}$ or $\tilde{n} \parallel m$. The latter case implies the existence of a $\bigvee$-irreducible $\tilde{h}$ satisfying $\tilde{n} \parallel \tilde{h} \leq m$. Hence $\{(\tilde{h}, \tilde{n}), (j, m)\}$ is an edge of Γ_k. Therefore $\tilde{n} \in M(\Gamma_k)$ and hence $\tilde{n} > \underline{v}$.

Lemma 6. *Let $\mathfrak{V}$ be a lattice with Ferrers-graph Γ. Let*

$$(j_0, n_0) E(j_1, n_1) E \ldots E(j_{r-1}, n_{r-1}) E(j_r, n_r)$$

be an edge sequence in a component Γ_k of the Ferrers-graph. Then n_0 is connected to either n_{r-1} or n_r in the interval $[\underline{v}, \overline{v}]$ of Γ_k.

Proof. By Definition 10 we note that $p := n_0 \geq j_1 \leq n_2 \geq j_3 \leq n_4 \ldots \leq n_s$ is a connection in $\mathfrak{V}$ with either $s = r - 1$ (if r is odd) or $s = r$ (if r is even). Let $[\underline{v}, \overline{v}]$ be the interval of Γ_k. With Lemma 5 we know that $j_i > \underline{v}$ and $n_i < \overline{v}$ holds for all $i \in \{0, \ldots, r\}$. Hence p is a connection of n_0 and n_s in $[\underline{v}, \overline{v}]$.

See the picture on the right for a visualization, the edge sequence E is represented by the graph in the diagonal and the connection p by the thick crosses.

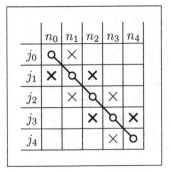

Lemma 7. *Let $\mathfrak{V}$ be a lattice with Ferrers-graph Γ. Let $\Gamma_{m_1, m_2} = \Gamma_{m_3, m_4}$ be a component of Γ possessing the interval $[\underline{v}, \overline{v}]$. Then m_1, m_2 are connected in $[\underline{v}, \overline{v}]$ if and only if m_3, m_4 are connected in $[\underline{v}, \overline{v}]$.*

Proof. Since $\Gamma_{m_1,m_2} = \Gamma_{m_3,m_4}$, there exists an edge sequence

$$(h_1, m_1)E(h_2, m_2)E\ldots E(h_3, m_3)E(h_4, m_4)$$

By Lemma 6 this refers w.l.o.g. to a connection between m_1 and m_3 and m_2 and m_4 in $[\underline{v}, \overline{v}]$ respectively. If m_1 and m_2 are connected, we find a connection

$$p := m_3 \ldots m_1 \ldots m_2 \ldots m_4$$

in $[\underline{v}, \overline{v}]$. Since $m_3, m_4 \in M(\Gamma_{m_3,m_4})$, both $\bigwedge$-irreducibles are connected. Dually, if m_3, m_4 are connected, then m_1, m_2 too.

Definition 15. *Let Γ be the Ferrers-graph of a lattice $\mathfrak{V}$. A component $\Gamma_{m,n}$ is of type 1 if m and n are connected in $I(\Gamma_{m,n})$. Otherwise $\Gamma_{m,n}$ is of type 2.*

Definition 15 is well-defined, in particular a component is either of type 1 or of type 2. This is a direct consequence of Lemma 7.

In the following we prepare, as emphasized in the beginning of that section, for a result stating that every component of a Ferrers-graph acts on one or on two m-components.

Lemma 8. *Let $\mathfrak{V}$ be a lattice with Ferrers-graph Γ. Let $m, m_1, m_2 \in M(\mathfrak{V})$, s.t. there is a connection between m and m_1 in $[\underline{v}, \overline{v}] := [m_1 \wedge m_2, m_1 \vee m_2]$. Then $\Gamma_{m_1,m_2} = \Gamma_{m,m_1}$ or $\Gamma_{m_1,m_2} = \Gamma_{m,m_2}$.*

Proof. Let $q = n_0 \geq h_1 \ldots \geq h_r \leq n_r$ be a shortest connection in $[\underline{v}, \overline{v}]$ for adequate chosen $\bigvee$-irreducibles h_i and $\bigwedge$-irreducibles n_i between $n_0 := m$ and $n_r := m_1$. Let $\{(j_1, m_2), (j_2, m_1)\}$ be an edge of Γ_{m_1,m_2}. We distinguish the following cases:

1. Let $j_2 \not\leq n_i$ for all $i \in \{0, \ldots, r\}$ and $h_i \not\leq m_2$ for all $i \in \{1, \ldots, r\}$. Then

   $$(j_2, n_0)E(h_1, m_2)E(j_2, n_1)E(h_2, m_2)E\ldots E(h_r, m_2)E(j_2, n_r),$$

   is an edge sequence and hence $\Gamma_{m_1,m_2} = \Gamma_{m,m_2}$. Note that this case always occurs if Γ_{m_1,m_2} is of type 2.
2. Let $j_2 \leq n_k$ for one $k \in \{0, \ldots, r\}$ and $h_r \not\leq n_k$. Then $h_r \leq n_r$ implies $k < r$. Then

   $$(j_1, m_2)E(j_2, n_r)E(j_1, n_k)E(h_k, n_r)E(j_1, n_{k-1})E(h_{k-1}, n_r)E$$
   $$\ldots E(h_1, n_r)E(j_1, n_0)$$

   is an edge sequence and hence $\Gamma_{m_1,m_2} = \Gamma_{m,m_1}$.
3. Let $h_k \leq m_2$ for one $k \in \{1, \ldots, r\}$. From $h_k \not\leq m_1 \wedge m_2$ we conclude $k < r$. Then

   $$(j_2, m_1)E(j_1, m_2)E(h_k, n_r)E(j_1, n_{k-1})E(h_{k-1}, n_r)E\ldots E(h_1, n_r)E(j_1, n_0)$$

   is an edge sequence and hence $\Gamma_{m_1,m_2} = \Gamma_{m,m_1}$.

4. Let k be an index, s.t. $j_2, h_r \leq n_k$. Note $k < r$, since $(j_2, m_1) \in V(\Gamma)$ and hence $j_2 \parallel m_1$. Since q is a shortest connection, we observe $k = r - 1$. Since $n_{r-1} \not\geq m_1 \vee m_2$ we conclude w.l.o.g. $n_{r-1} \not\geq m_2$. Hence we find a $\vee$-irreducible j satisfying $j \leq m_2$ and $j \not\leq n_{r-1}$. Then

$$(j_1, m_2)E(j_2, m_1)E(h_r, m_2)E(j, n_{r-1})E(h_{r-1}, m_2)E \ldots E(h_1, m_2)E(j_2, n_0)$$

is an edge sequence and hence $\Gamma_{m_1, m_2} = \Gamma_{m, m_2}$.

Lemma 9. *Let $m \in M(\Gamma_{m_1, m_2})$, s.t. m is not comparable to m_1 or m_2. Then $\Gamma_{m_1, m_2} = \Gamma_{m, m_1}$ or $\Gamma_{m_1, m_2} = \Gamma_{m, m_2}$.*

Proof. If there is a connection between m and m_1 or m_2 in $I := [m_1 \wedge m_2, m_1 \vee m_2]$ then we can apply Lemma 8 to prove the claim. Since $m \in M(\Gamma_{m_1, m_2})$, we find (for adequate $\vee$-irreducibles h_i and $\wedge$-irreducibles n_i) an edge sequence $K := (h_1, n_1) E(h_2, n_2) E \ldots E(h_r, n_r)$ with $n_1 = m_1$, $n_2 = m_2$ und $n_r = m$. W.l.o.g. (in case of $2 \nmid r$)

$$p = n_1 \geq h_2 \leq n_3 \geq \ldots \geq h_{r-1} \leq n_r \text{ and } q = h_1 \leq n_2 \geq h_3 \leq \ldots \leq n_{r-1} \leq h_r$$

are connections in $I(\Gamma_{m_1, m_2})$. Consider the sequence

$$S = (h_r, n_r), (h_{r-1}, n_{r-1}), (h_{r-2}, n_r), (h_{r-1}, n_{r-3}), (h_{r-4}, n_r),$$
$$\ldots, (h_{r-1}, n_2), (h_1, n_r).$$

1. In case of $h_i \parallel n_r$ for all $i \in \{1, \ldots, r\} \setminus \{r - 1\}$ and $h_{r-1} \parallel n_j$ for all $j \in \{1, \ldots, r\} \setminus \{r, r - 2\}$, S is an edge sequence and $\Gamma_{n_1, n_2} = \Gamma_{n_{r-1}, n_r} = \Gamma_{n_2, n_r}$.

2. Otherwise let t be the smallest index satisfying $h_t \leq n_r$ or $h_{r-1} \leq n_t$ (remind that $h_t > n_r \implies h_{r-1} \leq n_{t-1}$ and $h_{r-1} > n_t \implies h_{t-1} \leq n_r$).
   (a) Let $h_{r-1} \leq n_t$:
   (a1) $t \in \{1, 2\}$. First we observe $h_{r-1} \leq n_2 \iff h_{r-1} \leq n_1$, otherwise we had a connection $n_t \geq h_{r-1} \leq n_r$ in I. By precondition $n_r \not\leq n_1 \wedge n_2$. Therefore we find a $\vee$-irreducible $j \leq n_r$ satisfying w.l.o.g. $j \not\leq n_1$. If $j \not\leq n_{r-1}$ then we extend K by two nodes and gain the sequence $\tilde{K} = (h_1, n_1)E \ldots E(h_r, n_r)E(j, n_{r-1}E(h_r, n_r)$. For the appropriate sequence

$$\tilde{S} = (h_r, n_r), (j, n_{r-1}), (h_{r-2}, n_r), (j, n_{r-3}), (h_{r-4}, n_r),$$
$$\ldots, (j, n_2), (h_1, n_r)$$

   we can apply one of the other cases.
   If otherwise $j \leq n_{r-1}$ then we find in case of $h_2 \leq n_{r-1}$ an edge sequence $(j, n_1)E(h_2, n_r)E(h_{r-1}, n_{r-1})$ (or a connection $m_1 \geq h_2 \leq n_r$ in I) and hence $\Gamma_{n_1, n_r} = \Gamma_{n_r, n_{r-1}} = \Gamma_{n_1, n_2}$. Finally assume $h_2 \not\leq n_{r-1}$, then $(h_2, n_r)E(j, n_1)E(h_2, n_{r-1})$ is an edge sequence, i.e. $\Gamma_{n_1, n_r} = \Gamma_{n_1, n_{r-1}}$. Moreover, by the construction of K, also n_{r-1} is an element of $M(\Gamma_{n_1, n_2})$. We observe that $n_1 \geq h_{r-1} \leq n_r$ is a connection in $[n_r \wedge n_{r-1}, n_r \vee n_{r-1}]$. With Lemma 8 we conclude $\Gamma_{n_r, n_{r-1}} = \Gamma_{n_1, n_r}$ or $\Gamma_{n_r, n_{r-1}} = \Gamma_{n_1, n_{r-1}}$ which both implies $\Gamma_{n_1, n_2} = \Gamma_{n_1, n_r}$.

(a2) $t > 2$: Then

$$(h_t, n_t)E(h_{r-1}, n_{t-1})E(h_{t-2}, n_r)E(h_{r-1}, n_{t-3})E \ldots E(h_{r-1}, n_y)E(h_x, n_r)$$

with $y - 1 = t \pmod 2$ and $x = 3 - y$ is an edge sequence in $\Gamma_{m,m_2} = \Gamma_{m_1,m_2}$ since $(h_t, n_t) \in V(\Gamma_{m_1,m_2})$.

(b) Let $h_t \leq n_r$: In case of $t \leq 2$ we find the connection $m_{3-t} \geq h_t \leq m$ in I. Otherwise $t > 2$ and

$$(h_t, n_t)E(h_{t-1}, n_r)E(h_{r-1}, n_{t-2})E(h_{t-3}, n_r)E \ldots E(h_y, n_r)E(h_{r-1}, n_x)$$

is an edge sequence with $y - 1 = t \pmod 2$ and $x = 3 - y$ in Γ_{m_1,m_2}.

Lemma 10. *Let $\mathfrak{V}$ be a lattice with Ferrers-graph Γ. Let m, m_1, m_2 be $\bigwedge$-irreducibles satisfying $m \in U_{m_1}(I(\Gamma_{m_1,m_2}))$. Then $m \in M(\Gamma_{m_1,m_2})$.*

Proof. If $m \in U_{m_1}[m_1 \wedge m_2, m_1 \vee m_2]$ then the claim follows from Lemma 8.

Let otherwise $q = m \geq h_1 \leq n_1 \geq \ldots \geq h_r \leq m_1$ be a shortest connection containing a $\bigvee$-irreducible $h_k \leq m_1 \wedge m_2$ or a $\bigwedge$-irreducible $n_k \geq m_1 \vee m_2$. Then h_k or n_k respectively are elements of $I(\Gamma_{m_1,m_2})$ and according to Definition 14 we find a $\bigwedge$-irreducible $\tilde{m} \in M(\Gamma_{m_1,m_2})$ incomparable to h_k or n_k. According to Lemma 9 let w.l.o.g. $\Gamma_{m_1,m_2} = \Gamma_{\tilde{m},m_1}$. We observe that h_k or n_k respectively are elements of $I(\Gamma_{\tilde{m},m_1})$ and hence q is a connection in $I(\Gamma_{\tilde{m},m_1})$. Applying Lemma 8 we find $m \in M(\Gamma_{\tilde{m},m_1}) = M(\Gamma_{m_1,m_2})$.

Corollary 2. *Let $\mathfrak{V}$ be a lattice with Ferrers-graph Γ. Let Γ_{m_1,m_2} be a component of Γ possessing the interval $I := I(\Gamma_{m_1,m_2})$.*

1. *If Γ_{m_1,m_2} is of type 1 then $M(\Gamma_{m_1,m_2}) = U_{m_1}(I) = U_{m_2}(I)$.*
2. *If Γ_{m_1,m_2} is of type 2 then $M(\Gamma_{m_1,m_2}) = U_{m_1}(I) \mathbin{\dot\cup} U_{m_2}(I)$.*

Proof. Lemma 10 states $M(\Gamma_{m_1,m_2}) \supseteq U_{m_1}(I) \cup U_{m_2}(I)$.

Let $m \in M(\Gamma_{m_1,m_2})$. Then there exists an edge sequence

$$(j, m)E \ldots E(j_1, m_1)E(j_2, m_2)$$

for some adequate $j, j_1, j_2 \in J$. With Lemma 6 we conclude that m is connected to m_1 or m_2, i.e. $m \in U_{m_1}(I) \cup U_{m_2}(I)$. Hence $M(\Gamma_{m_1,m_2}) \subseteq U_{m_1}(I) \cup U_{m_2}(I)$.

Finally, by Definition 15 we immediately see $U_{m_1}(I) = U_{m_2}(I)$ for type 1 and $U_{m_1}(I) \cap U_{m_2}(I) = \emptyset$ for type 2.

Corollary 2 gives an explanation of the type numbers of components of Γ. The ones of type one act on exactly one proper component whereas the ones of type 2 act on exactly two.

Lemma 11. *Let $\mathfrak{V}$ be a lattice with bare Ferrers-graph Γ. Let $V(\Gamma) = \mathfrak{L} \mathbin{\dot\cup} \mathfrak{R}$ be a bipartition of Γ, s.t. the induced relation $\tilde{L}_1$ can be extended to a left-order L_1. Furthermore let Γ_k be a component of Γ of type 1. Then the bipartition*

$$V(\Gamma) = ((\mathfrak{L} \setminus \mathfrak{L}_k) \cup \mathfrak{R}_k) \mathbin{\dot\cup} ((\mathfrak{R} \setminus \mathfrak{R}_k) \cup \mathfrak{L}_k)$$

is again a bipartition inducing a relation $\tilde{L}_2$ that can be extended to a left-order L_2 distinct from L_1.

Proof. We have to evidence the following claims

1. $((\mathfrak{L} \setminus \mathfrak{L}_k) \cup \mathfrak{R}_k) \;\dot{\cup}\; ((\mathfrak{R} \setminus \mathfrak{R}_k) \cup \mathfrak{L}_k)$ is again a bipartition of the vertex set of Γ. This is intuitively clear since we just "turned around" one of its components. See Lemma 4 for a proof.

2. The induced relation $\tilde{L}_2$ can be extended to a left-order: We know (see Lemma 2) that $\tilde{L}_2$ is asymmetric and connex since it is induced by a bipartition. Now assume $\tilde{L}_2$ not to be transitive. Hence we have three $\bigwedge$-irreducibles satisfying $m_1 \, \tilde{L}_2 \, m \, \tilde{L}_2 \, m_2 \, \tilde{L}_2 \, m_1$. Since $\tilde{L}_1$ is transitive by precondition, let w.l.o.g. $m_1 \, \tilde{L}_1 \, m_2$. In particular this means $\Gamma_k = \Gamma_{m_1,m_2}$ and $m \notin M(\Gamma_k)$. Therefore we conclude $I(\Gamma_k) \subset [m \wedge m_1 \wedge m_2, m \vee m_1 \vee m_2]$.

   On the other hand we observe applying Corollary 1 that m, m_1, m_2 is a free triple. Hence m_1, m_2 are not connected in $[m \wedge m_1 \wedge m_2, m \vee m_1 \vee m_2] \supset I(\Gamma_k)$. This contradicts our precondition that Γ_k is of type 1.

3. $L_1 \neq L_2$: The component Γ_k contains at least one edge, i.e. $\Gamma_k = \Gamma_{m_1,m_2}$ for some $m_1, m_2 \in M$. W.l.o.g. let $m_1 \, \tilde{L}_1 \, m_2$ and $m_2 \, \tilde{L}_2 \, m_1$. This extends to the respective left-relations, i.e. $m_1 \, L_1 \, m_2$ and $m_2 \, L_2 \, m_1$. Since both left-orders are asymmetric, we conclude $L_1 \neq L_2$.

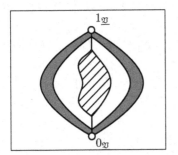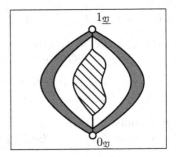

Fig. 2. In a plane diagram, turning a type 1 component results again in a plane diagram. The corresponding proper component (depicted diagonally striped) is thereby reflected.

Lemma 12. *Let $\mathfrak{V}$ be a lattice with Ferrers-graph Γ. Let $V(\Gamma) = \mathfrak{L} \,\dot{\cup}\, \mathfrak{R}$ be a bipartition of Γ, s.t. the induced relation $\tilde{L}_1$ can be extended to a left-order. Furthermore let $\Gamma_{[\underline{v},\overline{v}]} := \{\Gamma_1, \dots, \Gamma_r\}$ be the set of components of type 2 possessing the interval $[\underline{v}, \overline{v}]$. Then*

1. $r = \dbinom{s}{2}$ *for some natural number $s \geq 2$. There exist exactly s m-components $U_{m_1}[\underline{v},\overline{v}], \dots, U_{m_s}[\underline{v},\overline{v}]$ in the interval $[\underline{v},\overline{v}]$.*

2. *Let $K \subseteq \{1, \dots, r\}$. The relation $\tilde{L}$ induced by the bipartition*

$$V(\Gamma) = ((\mathfrak{L} \setminus \bigcup_{k \in K} \mathfrak{L}_k) \cup \bigcup_{k \in K} \mathfrak{R}_k) \;\dot{\cup}\; ((\mathfrak{R} \setminus \bigcup_{k \in K} \mathfrak{R}_k) \cup \bigcup_{k \in K} \mathfrak{L}_k) \tag{1}$$

can be extended to a left-order if and only if there exists a linear order $\sqsubseteq$ on the $\bigwedge$-irreducibles $m_1, \ldots, m_s$, s.t. $m_i \sqsubseteq m_j \iff m_i \hat{L} m_j$ holds for all $i, j \in \{1, \ldots, s\}$.

3. *There exist exactly $s!$ bipartitions of the form of Equation (1) whose induced relations can be extended to a left-order.*

Proof. 1. Let $U_m[\underline{v}, \overline{v}] \neq U_n[\underline{v}, \overline{v}]$ be arbitrary m-components. Then $M(\Gamma_{m,n})$ is a subset of $[\underline{v}, \overline{v}]$ and therefore $[\underline{v}, \overline{v}]$ is the interval of $\Gamma_{m,n}$. Furthermore m, n are not connected in that interval, i.e. $\Gamma_{m,n}$ is of type 2. That means that for each pair of proper components in $[\underline{v}, \overline{v}]$, the component $\Gamma_{m,n}$ is an element of $\Gamma_{[\underline{v}, \overline{v}]}$. We conclude $\Gamma_{m,n} \in \Gamma_{[\underline{v}, \overline{v}]} \iff U_m[\underline{v}, \overline{v}] \neq U_n[\underline{v}, \overline{v}]$ with Corollary 2. Therefore r is the number of pairs of proper components over $[\underline{v}, \overline{v}]$. Hence we have s such components, with $r = \binom{s}{2}$.

2. If such a linear order $\sqsubseteq$ does not exist, we find a triple of $\bigwedge$-irreducibles fulfilling $m_1 \hat{L} m_2 \hat{L} m_3 \hat{L} m_1$. Therefore $\hat{L}$ can not be extended to a left-order.

   Let $\sqsubseteq$ be a linear order fulfilling $m_i \sqsubseteq m_j \iff m_i \hat{L} m_j$ for all pairs of $\bigwedge$-irreducibles $m_i, m_j \in [\underline{v}, \overline{v}]$. We assume to find a triple $m_1, m_2, m_3 \in M(\mathfrak{V})$ satisfying $m_1 \hat{L} m_2 \hat{L} m_3 \hat{L} m_1$. By the precondition we know that at least one of the three $\bigwedge$-irreducibles is not an element of $[\underline{v}, \overline{v}]$, since they are ordered linearly. Additionally we know by applying Corollary 1 that m_1, m_2, m_3 is a free triple. Therefore $m_1 \notin [\underline{v}, \overline{v}]$ and $m_2, m_3 \in [\underline{v}, \overline{v}]$ is not the case since then we find $\underline{v} = m_i \wedge m_j > m \wedge m_i \wedge m_j$. Finally the case $m_1, m_2 \notin [\underline{v}, \overline{v}]$ does not occur since then all involved components $\Gamma_{m_1, m_2}, \Gamma_{m_1, m_3}$ and Γ_{m_2, m_3} are not in the set $\Gamma_{[\underline{v}, \overline{v}]}$ and are therefore not changed. This contradicts our assumption and hence the relation $\hat{L}$ transitive. With Lemma 2 we conclude that $\hat{L}$ can be extended to a left-order.

3. This is obvious: there exist $s!$ linear orders on an s-elemental set and therefore also $s!$ orientations of the involved components of Γ inducing a relation $\hat{L}$ that can be extended to a left-order.

The previous two lemmas allow to characterize all non-similar plane diagrams of a lattice:

Theorem 4. *Let $\mathfrak{V}$ be a lattice with the bare Ferrers-graph Γ.*

1. *If Γ is not bipartite then $\mathfrak{V}$ is not planar.*
2. *If Γ is bipartite with κ components of type 1 and $\mu = \mu_1 + \ldots + \mu_t$ components of type 2, s.t. μ_i is the number of components of the set $\Gamma_{[\underline{v}, \overline{v}]}$ for some interval $[\underline{v}, \overline{v}]$ (containing proper components). Then $\mathfrak{V}$ possesses $2^\kappa \cdot \prod_{i=1}^{t} \mu_i!$ non-similar plane diagrams.*

Proof. 1. This is due to Theorem 2.
2. Follows from Lemma 11 and Lemma 12.

Corollary 3. *Let $\mathfrak{V}$ be a lattice with the bipartite bare Ferrers-graph Γ as previously described in Theorem 4. Then $\mathfrak{V}$ possesses $2^{\kappa-1} \cdot \prod_{i=1}^{t} \mu_i!$ realizers.*

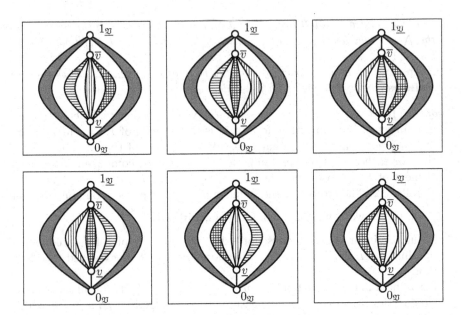

Fig. 3. In a plane diagram, turning type 2 components results again in a plane diagram if their respective ordering corresponds to a linear ordering of the involved proper components. In this picture we have 3 proper components U_{m_1} (horizontally striped), U_{m_2} (vertically striped) and U_{m_3} (crosshatched). Therefore there exist 3 components of type 2, namely Γ_{m_1,m_2}, Γ_{m_1,m_3} and Γ_{m_2,m_3}. Of the eight possible orientations of that components, six supply a plane diagram.

Proof. We concluded from Lemma 1, Proposition 1 and Theorem 3 that every two plane diagrams of $\mathfrak{V}$ correspond to exactly one realizer of $\mathfrak{V}$. The claim follows then directly from Theorem 4 and Lemma 1.

In [DDF84] a characterization of posets of dimension 2 possessing a unique realizer is given. Applying our consideration we can derive this result, restricted to lattices, too:

Corollary 4. *[DDF84] Let $\mathfrak{V}$ be a lattice with the bipartite bare Ferrers-graph Γ. Then $\mathfrak{V}$ possesses a unique realizer of size two if and only if Γ is connected.*

Proof. Follows immediately from Corollary 3.

Finally we can specify the subset of natural numbers that can be described as the number of non-similar plane diagrams of a lattice:

Corollary 5. *Let α be a natural number. There exists a lattice possessing α non-similar plane diagrams if and only if $\alpha = 0$ or $\alpha = \prod_{i=1}^{t} \alpha_i!$ for natural numbers $\alpha_1, \ldots, \alpha_t$.*

Proof. The case $\alpha = 0$ refers to non-planar lattices. The number of non-similar plane diagrams of a planar lattice is a product of factorials; this is implied by Theorem 4 if one reminds $2 = 2!$.

Otherwise for any number $\alpha = \prod_{i=1}^{t} \alpha_i!$, the lattice consisting of a parallel composition of lattices M_{α_i} has exactly α plane diagrams: The bare Ferrers-graph of the lattice M_n[10] possesses $\binom{n}{2}$ components of size 2, which all share the interval $[0_{M_n}, 1_{M_n}]$ and has therefore $n!$ plane diagrams. Additionally, the bare Ferrers-graph Γ of the parallel composition of two lattices $\mathfrak{A}_1$ and $\mathfrak{A}_2$ is exactly the (disjoint) union of the bare Ferrers-graphs of its parts. This is due to the fact that all pairs of the form (j, m) with $j \in J(\mathfrak{A}_1)$ and $m \in M(\mathfrak{A}_2)$ or $j \in J(\mathfrak{A}_2)$ and $m \in M(\mathfrak{A}_1)$ fulfill either $j \le m$ or $j \ge m$. Both cases imply $(j, m) \notin V(\Gamma)$.

3.4   An Algorithm for Finding All Plane Diagrams of a Lattice

With the help of the previous statements we can design an algorithm that specifies all left-orders, i.e. all non-similar plane diagrams, of a lattice $\mathfrak{A}$. it consists of the following steps:

1. Calculate the bare Ferrers-graph Γ of $\mathfrak{A}$.
2. Decide, whether Γ is bipartite. If yes, assign a bipartition $\mathfrak{L} \cup \mathfrak{R}$ to the vertices of Γ.
3. Determine all components of Γ.
4. Calculate the interval to each component Γ_k and determine, whether Γ_k is of type 1 or type 2.
5. Subsume the components of type 2 in equivalence classes based on their interval.
6. Calculate the m-components of each interval $[\underline{v}, \overline{v}]$ defining an equivalence class in terms of step 5.
7. Find all bipartitions inducing a relation $\tilde{L}$ that can be extended to a left-order.

The algorithm finds exactly all bipartitions that define a left-order. This is assured by Theorem 4. To gain a plane diagram from that, one can use for instance the method described in [Zsc05] or more general in [Zsc08].

Unfortunately that algorithm is not polynomial (both in terms of $\bigwedge$-irreducibles and of lattice elements) in general due to the last step. We may find up to $|M(\mathfrak{A})|!$ non-similar plane diagrams. The other steps are polynomial however, as we want to clarify in the following.

1. The first step needs a complexity of $\mathcal{O}((|J| \cdot |M|)^2)$ by a naive calculation.
2. Finding a bipartition of a graph (V, E) can be done in $\mathcal{O}(|E|) = \mathcal{O}((|J| \cdot |M|)^2)$ by a standard algorithm [Jun94].
3. Finding all components of the bare graph can be done in $\mathcal{O}(|E|) = \mathcal{O}((|J| \cdot |M|)^2)$ [Jun94].

---

[10] The lattice M_n consists of an n-elemental antichain completed by top and bottom element.

4. In a planar lattice every element v can be represented as the infimum of at most two $\wedge$-irreducibles and as the supremum of at most two $\vee$-irreducibles [Zsc08]. Therefore the calculation of the interval can be done in $\mathcal{O}(|M(\Gamma_k)|)$ for each component Γ_k, that is $\mathcal{O}(|M|^3)$ altogether.

The destinction between type 1 and type 2 components can be done easily in this process, as the following lemma states:

Lemma 13. *Let $\Gamma_{m,n}$ be a component of the Ferrers-graph of a lattice and $I(\Gamma_{m,n}) = [\underline{v}, \overline{v}]$.*

1. *If $\Gamma_{m,n}$ is of type 2 then $m \wedge n = \underline{v}$.*
2. *If $\Gamma_{m,n}$ is of type 1 then there exists an edge $\{(j_1, m_1), (j_2, m_2)\}$ in $\Gamma_{m,n}$ with $m_1 \wedge m_2 > \underline{v}$.*

Proof. 1. Let $m \wedge n > \underline{v}$. Then we find a $\vee$-irreducible j satisfying $j \leq m \wedge n$ and $j \not\leq \underline{v}$. Hence m and n are connected in $[\underline{v}, \overline{v}]$, i.e. $\Gamma_{m,n}$ is not of type 2.

2. The second fact is shown by induction over the length of a shortest connection p between m and n in $[\underline{v}, \overline{v}]$.

If p is of length one, i.e. $p = m \geq j \leq n$ then $m \wedge n \geq j \not\leq \underline{v}$, hence $m \wedge n > \underline{v}$.
If p is of length $r > 1$, i.e. $p = n_0 \geq h_1 \leq n_1 \geq \ldots \geq h_r \leq n_r$ (with $m = n_0$ and $n = n_r$) then n_1 is connected to m in $[m \wedge n, m \vee n]$. We conclude with Lemma 8 that $\Gamma_{m,n} = \Gamma_{m,n_1}$ or $\Gamma_{m,n} = \Gamma_{n,n_1}$. In both cases the connections between m and n_1 and n and n_1 respectively are shorter than p. Hence we find a pair of $\wedge$-irreducibles meeting the requirements of the claim.

5. This step is obviously done in $\mathcal{O}(\mu)$ where $\mu \leq |M|^2$ is the number of components of type 2.
6. Exactly the pairs of $\wedge$-irreducibles m, n connected by an edge of the form $\{(j, m), (h, n)\}$ are in different m-components. Hence this step takes a complexity of $\mathcal{O}(|M(\Gamma_k)|)$ for each component Γ_k and $\mathcal{O}(|M|^3)$ altogether.

By this consideration we observe that the first six steps of the algorithm need a time complexity of

$$\max\{\mathcal{O}((|J| \cdot |M|)^2), \mathcal{O}(|M^3|)\} \leq \max\{\mathcal{O}((|J|^4), \mathcal{O}(|M^4|)\} = \mathcal{O}(|\mathfrak{V}|^2).$$

See [Zsc05, Zsc08] for the equality between the second and third term.

4   Conclusion

In this work we gave a possibility to specify all plane diagrams and all realizers of an arbitrary finite lattice. As far as we know, such a characterization was not done so far. The algorithm described in Section 3.4 allows to calculate the number of all plane diagrams and the number of all realizers in polynomial time. More precisely the complexity of the method is $\mathcal{O}(|\mathfrak{V}|^2)$ for a lattice $\mathfrak{V} = (\mathfrak{V}, \leq)$. Actually writing down the respective left-orders can not be done in polynomial time since their number can reach $|M(\mathfrak{V})|!$, where $|M(\mathfrak{V})|$ is the number of $\wedge$-irreducibles in $\mathfrak{V}$.

Our approach may be a useful instrument to draw nice diagrams of concept lattices. Given a context $\mathbb{K} = (G, M, I)$, one can search for large subcontexts $\tilde{\mathbb{K}} = (G, \tilde{M}, \tilde{I})$ (with $\tilde{M} \subseteq M$ and $\tilde{I} = I \cap G \times \tilde{M}$), s.t. $\mathfrak{B}(\tilde{\mathbb{K}})$ is planar. One can provide all non-similar plane diagrams of $\mathfrak{B}(\tilde{\mathbb{K}})$ and add the images of the principal ideals $(\mu m]$ of the remaining attributes $m \in M \setminus \tilde{M}$. There exists an algorithm [Sch02] that adds a principal ideal and thereby maximizes the *conflict distance*[11] between inserted and already existing nodes and lines.

References

[BDMT98] Bertolazzi, P., et al.: Optimal upward planarity testing of single-source digraphs. SIAM J. COmput. 27, 132–169 (1998)
[BFR71] Baker, K.A., Fishburn, P.C., Roberts, F.S.: Partial orders of dimension 2. Networks 2, 11–28 (1971)
[Bir67] Birkhoff, G.: Lattice theory, third ed., Amer. Math. Soc (1967)
[Cog82] Cogis, O.: On the ferrers-dimension of a digraph. Discrete Math. 38, 47–52 (1982)
[DDF84] Doignon, J.P., Ducamp, A., Falmagne, J.C.: On realizable biorders and the biorder dimension of a relation. J. of. Math. Psychology 28, 73–109 (1984)
[Die96] Diestel, R.: Graphentheorie. Springer, Heidelberg (1996)
[DM41] Dushnik, B., Miller, E.W.: Partially ordered sets. Amer. J. Math 63, 600–610 (1941)
[Gan04] Ganter, B.: Conflict avoidance in order diagrams. J. of Univ. Comp. Sc. 10(8), 955–966 (2004)
[GW99] Ganter, B., Wille, R.: Formal concept analysis - mathematical foundations. Springer, Heidelberg (1999)
[Jun94] Jungnickel, D.: Graphen, Netzwerke und Algorithmen, third ed., BI-Wiss.-Verl. (1994)
[KR75] Kelly, D., Rival, I.: Planar lattices. Can. J. Math. 27(3), 636–665 (1975)
[Pla76] Platt, C.R.: Planar lattices and planar graphs. J. Combinatorial Theory Ser. B, 30–39 (1976)
[Reu89] Reuter, K.: Removing critical pairs, Tech. Report 1241, TU Darmstadt (1989)
[Sch02] Schmidt, B.: Ein Algorithmus zum Zeichnen von Liniendiagrammen, Master's thesis, TU Dresden (2002)
[Spi82] Spinrad, J.: Two dimensional partial orders, Ph.D. thesis, Princeton Univ (1982)
[Zsc05] Zschalig, C.: Planarity of lattices - an approach based on attribute additivity. In: Ganter, B., Godin, R. (eds.) ICFCA 2005. LNCS (LNAI), vol. 3403, pp. 391–402. Springer, Heidelberg (2005)
[Zsc07] Bipartite ferrers-graphs and planar concept lattices. In: Kuznetsov, S.O., Schmidt, S. (eds.) ICFCA 2007. LNCS (LNAI), vol. 4390, pp. 313–327. Springer, Heidelberg (2007)
[Zsc08] Left-relations, A.: unified approach to characterize planar lattices, Ph.D. thesis, TU Dresden (to appear, 2008)

---

[11] That is the minimal distance between a node and a non-incident line, see [Gan04] for a formal definition.

Spectral Lattices of $\overline{\mathbb{R}}_{\max,+}$-Formal Contexts

Francisco J. Valverde-Albacete and Carmen Peláez-Moreno*

Dpto. de Teoría de la Señal y de las Comunicaciones.
Universidad Carlos III de Madrid
Avda. de la Universidad, 30. Leganés 28911. Spain
fva,carmen@tsc.uc3m.es

Abstract. In [13] a generalisation of Formal Concept Analysis was introduced with data mining applications in mind, $\mathcal{K}$-Formal Concept Analysis, where incidences take values in certain kinds of semirings, instead of the standard Boolean carrier set. Subsequently, the structural lattice of such generalised contexts was introduced in [15], to provide a limited equivalent to the main theorem of $\mathcal{K}$-Formal Concept Analysis, resting on a crucial parameter, the degree of existence of the object-attribute pairs φ. In this paper we introduce the spectral lattice of a concrete instance of $\mathcal{K}$-Formal Concept Analysis, as a further means to clarify the structural and the $\mathcal{K}$-Concept Lattices and the choice of φ. Specifically, we develop techniques to obtain the join- and meet-irreducibles of a $\overline{\mathbb{R}}_{\max,+}$-Concept Lattice independently of φ and try to clarify its relation to the corresponding structural lattice.

1 Motivation: The Analysis of Confusion Matrices with $\mathcal{K}$-Formal Concept Analysis

Consider sets of *entities* G and *patterns* M with $|G| = g \in \mathbb{N}$, $|M| = m \in \mathbb{N}$ and a device called a *classifier* accepting a *characterisation* of an entity i, $0 \leq i \leq g$, normally a vector of features, and returning the index of a pattern j, $0 \leq j \leq m$.

A *confusion matrix* or *contingency table* $C \in \mathbb{N}^{g \times n}$ tries to capture at a glance the performance of such classifier: for each classification act we increase C_{ij} by one, tallying classification hits and errors, which makes C a *semiring-valued matrix*. With the aim of better understanding the performance of the classifier we would like to find a way to analyse the geometry of the spaces associated to matrices with properties similar to those of C .

For that purpose, in [13] a generalisation of Formal Concept Analysis was introduced that allows incidences to take values in *dioids*, or idempotent semirings: for $g, m \in \mathbb{N}$, given two sets of objects $G = \{g_i\}_{i=1}^{g}$, and attributes $M = \{m_j\}_{j=1}^{m}$, let $\mathcal{K}$, be a complete, idempotent semifield [2,13], and a $\mathcal{K}$-valued matrix, $R \in K^{g \times m}$, the triple $(G, M, R)_{\mathcal{K}}$ is called a $\mathcal{K}$-*formal context*. We interpret $R_{ij} = \lambda$ as "object g_i has attribute m_j in degree λ" or, dually, "attribute m_j is manifested in object g_i to degree λ".

---

* This work has been partially supported by a Spanish Government-Comisión Interministerial de Ciencia y Tecnología project TEC2005-04264/TCM.

R. Medina and S. Obiedkov (Eds.): ICFCA 2008, LNAI 4933, pp. 124–139, 2008.
© Springer-Verlag Berlin Heidelberg 2008

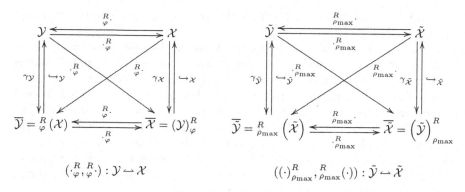

Fig. 1. Diagrams depicting the structures in the Galois connection of Eq. 1 (left), and Corollary 10.1, (right)

Now, for each (multiplicatively) invertible $\varphi \in K$, call $(\mathcal{K}, \varphi)$ a *reflexive idempotent semiring* if the following maps define a Galois connection, with $Y = K^{1 \times g}$, $X = K^{m \times 1}$ and the bracket $\langle \cdot \mid \cdot \rangle : Y \times X \to K, (y, x) \mapsto \langle y \mid x \rangle = yRx$, ([3,13] and §2.1 below):

$$\overset{R}{\underset{\varphi}{\cdot}} : Y \to X \qquad y_{\varphi}^{R} = \bigvee\{x \in X \mid \langle y \mid x \rangle \leq \varphi\} \qquad (1)$$

$$\overset{R}{\underset{\varphi}{\cdot}} : X \to Y \qquad \overset{R}{\underset{\varphi}{}}x = \bigvee\{y \in Y \mid \langle y \mid x \rangle \leq \varphi\} .$$

in which case we call them the *φ-polars* of the $\mathcal{K}$-formal context $(G, M, R)_{\mathcal{K}}$. Under such conditions:

1. The images $\overline{\mathcal{Y}} = \overset{R}{\underset{\varphi}{}}(\mathcal{X})$ and $\overline{\mathcal{X}} = (\mathcal{Y})_{\varphi}^{R}$ are dually inverse complete subsemimodules of $\mathcal{Y}$ and $\mathcal{X}$, respectively. They are obtained from the original semimodules by the closure operators: $\gamma_{\mathcal{Y}} : Y \to Y, y \mapsto \gamma_{\mathcal{Y}}(y) = \overset{R}{\underset{\varphi}{}}((y)_{\varphi}^{R})$ and $\gamma_{\mathcal{X}} : X \to X, x \mapsto \gamma_{\mathcal{X}}(x) = (\overset{R}{\underset{\varphi}{}}(x))_{\varphi}^{R}$.
2. A (formal) *φ-concept* of the formal context $(G, M, R)_{\mathcal{K}}$ is a pair $(a, b) \in Y \times X$ such that $a_{\varphi}^{R} = b$ and $\overset{R}{\underset{\varphi}{}}b = a$. We call a the *φ-extent* and b the *φ-intent* of the concept (a, b), and φ its *(maximum) degree of existence*.
3. If $(a_1, b_1), (a_2, b_2)$ are φ-concepts of a context, they are ordered by the relation $(a_1, b_1) \leq (a_2, b_2) \iff a_1 \leq a_2 \overset{op}{\iff} b_1 \leq b_2$, called the *hierarchical order*. The set of all concepts ordered in this way is called the *φ-concept lattice*, $\underline{\mathfrak{B}}^{\varphi}(G, M, R)_{\mathcal{K}}$, of the $\mathcal{K}$-valued context $(G, M, R)_{\mathcal{K}}$.

In [14] a preliminary application of $\mathcal{K}$-Concept Lattices to data mining was described to characterise the behaviour of n-class classifiers. Two sources of complexity associated to trying to understand such lattices were detected therein: first, the potentially vast size of $\mathcal{K}$-Concept Lattices, and second, the need to sweep over parameter $\varphi \in K$ to find all possible lattices which prove only slightly different for similar φ's, pairwise considered.

To overcome the first difficulty the *structural lattice* of a $\mathcal{K}$-formal context was introduced in [15] as a sort of skeleton for it. This had the supplementary benefit of providing a (limited) second half for the fundamental theorem of $\mathcal{K}$-Concept Lattices.

Discouragingly, nothing conclusive was found in the same paper with regard to the role of φ in the series of lattices generated by sweeping over the parameter apart from a non-monotone relationship on the number of concepts.

Therefore, in this paper we introduce another lattice related to a $\mathcal{K}$-Formal Context which we obtain independently of any φ, the *spectral lattice of a $\mathcal{K}$-Formal Context*. Furthemore, for gaining a more concrete understanding of the problem, we apply the results stated so far to the well-known maxplus $\mathbb{R}_{\max,+}$ and minplus $\mathbb{R}_{\min,+}$ semirings [1]. This not only provides concrete examples for all abstract notions we have been manipulating so far but also enables to leverage powerful techniques specially developed for such semirings.

For that purpose, we introduce these concrete algebras and a notation to be able to handle expressions mixing them in Section 2.1. With this toolkit we easily introduce in Section 2.2 the notion of the spectrum of any square, $\mathbb{R}_{\max,+}$-valued matrix. As the main contribution of this paper, in Section 3 we find dually order isomorphic lattices related to the spectra of the projectors onto the image subsemimodules of the Galois connection, $\overline{\mathcal{Y}}$ and $\overline{\mathcal{X}}$, implied in the main theorem. Finally, in Section 4 we show a more involved application to the analysis of confusion matrices.

2 $\overline{\mathbb{R}}_{\mathbf{max},+}$ Spectral Theory

2.1 $\overline{\mathbb{R}}_{\mathbf{max},+}$ and $\overline{\mathbb{R}}_{\mathbf{min},+}$ Algebra

Idempotent semirings. A *semiring*[1] $\mathcal{S} = \langle S, \oplus, \otimes, \epsilon, e \rangle$ is an algebra whose additive structure, $\langle S, \oplus, \epsilon \rangle$, is a commutative monoid and whose multiplicative structure, $\langle S\backslash\{\epsilon\}, \otimes, e \rangle$, is a monoid whose multiplication distributes over addition from right and left and whose neutral element w.r.t. $\oplus$ is absorbing for $\otimes$, i.e. $\forall a \in S,\ \epsilon \otimes a = \epsilon$. On any semiring $\mathcal{S}$ left and right multiplications can be defined: $L_a : S \rightarrow S, b \mapsto L_a(b) = ab$, and $R_a : S \rightarrow S, b \mapsto R_a(b) = ba$. A *commutative semiring* is one whose multiplicative structure is commutative.

A *semifield* $\mathcal{K}$ is a semiring whose multiplicative structure $\langle K\backslash\{\epsilon\}, \otimes \rangle$ is a group, that is, there is an operation, $\cdot^{-1} : K\backslash\{\epsilon\} \rightarrow S\backslash\{\epsilon\}$ such that $\forall a \in K, a \otimes a^{-1} = a^{-1} \otimes a = e$. For commutative semifields whose multiplicative structure is a commutative group we have $(a \otimes b)^{-1} = a^{-1} \otimes b^{-1}$.

An *idempotent semiring or dioid* (for double monoid) $\mathcal{D}$ is a semiring whose addition is idempotent, $\forall a \in D, a \oplus a = a$, that is, whose additive structure $\langle D, \oplus, \epsilon \rangle$ is an *idempotent semigroup* . Compared to a ring, an idempotent semiring crucially lacks additive inverses. All idempotent commutative monoids $\langle D, \oplus, \epsilon \rangle$ are endowed with a *natural order*, $\forall a, b \in D, a \preceq b \iff a \oplus b = b$, which turns them into $\vee$-semilattices with least upper bound defined as $a \vee b = a \oplus b$. Moreover, the neutral element for the additive structure of semiring $\mathcal{D}$ is the infimum for this natural order, $\epsilon = \bot$. Hence all dioids are sup-semilattices

---

[1] Henceforth $\mathcal{S}$ will be a generic semiring, $\mathcal{K}$ a semifield, and $\mathcal{D}$ an idempotent semiring.

$\langle D, \preceq \rangle$ with a bottom element. A dioid whose multiplicative structure is a group is an *idempotent semifield*. The formula for the infimum of two elements in such case was already given by Dedekind [4]: the meet law is: $a \wedge b = a \otimes (a \oplus b)^{-1} \otimes b$, hence idempotent semifields are already lattices.

A semiring $\mathcal{S}$ is *complete*, if for any index set I including the empty set, and any $\{a_i\}_{i \in I} \subseteq \mathcal{S}$ the (possibly infinite) summations $\bigoplus_{i \in I} a_i$ are defined and the distributivity conditions: $\left(\bigoplus_{i \in I} a_i \right) \otimes c = \bigoplus_{i \in I} (a_i \otimes c)$ and $c \otimes \left(\bigoplus_{i \in I} a_i \right) = \bigoplus_{i \in I} (c \otimes a_i)$, are satisfied. Note that for $c = e$ the above demand that infinite sums have a result. In complete semirings one can define the *Kleene star* of an element, $a \in \mathcal{S}$, $a^* = \sum_{i=0}^{\infty} a^i$, and also $a^+ = \sum_{i=1}^{\infty} a^i$, with $a^+ = a \otimes a^*$ and $a^* = e \oplus a^+$.

A dioid $\mathcal{D}$ is *complete*, if it is complete as a naturally ordered set $\langle D, \preceq \rangle$ and left (L_a) and right (R_a) multiplications are lower semicontinuous, that is join-preserving.

Example 1 (The Maxplus and Minplus semifields)

1. *The **Maxplus semifield**, $\mathbb{R}_{\max,+} = \langle \mathbb{R} \cup \{-\infty\}, \max, +, -\infty, 0 \rangle$ with inverse $\cdot^{-1} := -\cdot$ is an idempotent commutative semifield. It is incomplete because its bottom has no inverse: $\forall a \in \mathbb{R} \cup \{-\infty\}$, $a + (-\infty) = -\infty \neq 0$.*
2. *The **Minplus semifield**, $\mathbb{R}_{\min,+} = \langle \mathbb{R} \cup \{\infty\}, \min, +, \infty, 0 \rangle$ is an idempotent commutative semifield, with the same inverse as the previous example. It is incomplete for a similar reason: $\forall a \in \mathbb{R} \cup \{\infty\}$, $a + \infty = \infty \neq 0$.*

Top Completion of idempotent semifields ([8,10,11,12]). A non-trivial idempotent semifield $\mathcal{D} \neq \{\epsilon, e\}$ (that is, non-isomorphic to $\mathcal{B}$) cannot contain a top element, $\top$, hence it cannot be a complete dioid. In [14] a procedure is described whereby one can obtain from any (incomplete) idempotent semiring $\mathcal{D}$ a completion as follows.

For any lattice-ordered group $\mathcal{G} = \langle G, \preceq, \otimes \rangle$: adjoin two elements $\bot$ and $\top$ to G to obtain $\overline{G} = G \cup \{\bot, \top\}$ and extend the order to $\overline{G}$ as $\bot \preceq a \preceq \top, \forall a \in \overline{G}$. Then extend the product to two different operations, *upper*, $\overset{\cdot}{\otimes}$, and *lower*, $\otimes$, *multiplications*:

$$ a \overset{\cdot}{\otimes} b = \begin{cases} \bot & \text{if } a, b \in G \cup \{\bot, \top\}, \text{with } a = \bot, \text{ or } b = \bot . \\ \top & \text{if } a, b \in G \cup \{\top\}, \text{with } a = \top, \text{ or } b = \top . \\ a \otimes b & \text{if } a, b \in G . \end{cases} \tag{2} $$

$$ a \overset{\cdot}{\otimes} b = \begin{cases} \top & \text{if } a, b \in G \cup \{\bot, \top\}, \text{with } a = \top, \text{ or } b = \top . \\ \bot & \text{if } a, b \in G \cup \{\bot\}, \text{with } a = \bot, \text{ or } b = \bot . \\ a \otimes b & \text{if } a, b \in G . \end{cases} \tag{3} $$

to obtain the structure $\overline{\mathcal{G}} = \langle \overline{G}, \preceq, \overset{\cdot}{\otimes}, \otimes \rangle$, known as the *canonical enlargement* of $\mathcal{G} = \langle G, \preceq, \otimes \rangle$. In this structure, $\otimes$ and $\overset{\cdot}{\otimes}$ are associative and commutative over $\overline{G}$, as the original $\otimes$ was over G, and the isotony of the product with respect to the natural order extends to $\overline{G}$. Furthermore, if e is the unit element of $\langle G, \otimes \rangle$, it is similarly the unit of $\langle \overline{G}, \dot{\otimes} \rangle$ and $\langle \overline{G}, \otimes \rangle$. The *top completion* of a dioid $\mathcal{D}$ is another dioid $\overline{\mathcal{D}} = \langle \overline{D}, \oplus, \otimes, \epsilon, e \rangle$ where: $\overline{D} = D \cup \{\top\}$ and in which $\otimes$ coincides with its definition above when $\mathcal{D}$ is considered as bearing a lattice-ordered (multiplicative semi-)group, and we extend $\oplus$ with the extra top-element:

$$a \dot{\oplus} b = \begin{cases} \top & \text{if } a = \top \text{ or } b = \top . \\ a \oplus b, & \text{if } a, b \in D . \end{cases} \tag{4}$$

Given an (incomplete) idempotent semifield $\mathcal{D}$, on its top enlargement as above, $\overline{\mathcal{D}}$, we extend the *notation* for the inverse with the following conventions: $\epsilon^{-1} = \top$, $\top^{-1} = \epsilon$. In that way we have two related *completed idempotent semifield* structures:

- a complete lattice for the natural order $\langle \overline{D}, \preceq \rangle$, the one we have been focusing on, $\overline{\mathcal{D}} = \langle \overline{D}, \oplus = \vee, \otimes, \bot, e \rangle$, and
- a complete lattice for the *dual of the natural order*, $\langle \overline{D}, \preceq^d \rangle = \langle \overline{D}, \succeq \rangle$, $\overline{\mathcal{D}}^d = \langle \overline{D}, \dot{\oplus} = \wedge, \dot{\otimes}, \top, e \rangle$ where the meet is defined (on $\mathcal{D}$) by Dedekind's formula and the definition of $\dot{\otimes}$ follows equation (3).

Example 2. *Using the procedure above, we have that:*

- *The top completion of* $\mathbb{R}_{\max,+}$ *is* $\overline{\mathbb{R}}_{\max,+} = \langle \mathbb{R} \cup \{-\infty \, \infty\}, \max, +, -\infty, 0 \rangle$, *the* completed Maxplus semifield.
- *The top completion of* $\mathbb{R}_{\min,+}$ *is* $\overline{\mathbb{R}}_{\min,+} = \langle \mathbb{R} \cup \{-\infty, \infty\}, \min, \dot{+}, \infty, 0 \rangle$ *the* completed Minplus semifield .

Note that in this notation we have $-\infty + \infty = -\infty$ *and* $-\infty \dot{+} \infty = \infty$, *which solves several issues in dealing with the separately completed dioids, as promised.*

In the completed structure, which we prefer to denote by $\overline{\mathcal{K}}$, we have the following De Morgan-like relations between the multiplications, their residuals and inversion:

Proposition 1 ([12], lemma 2.2). *In the top enlargement* $\overline{\mathcal{K}}$ *of any commutative semifield* $\mathcal{K}$ *we have:*

$$(a \oplus b)^{-1} = a^{-1} \dot{\oplus} b^{-1} \qquad\qquad (a \dot{\oplus} b)^{-1} = a^{-1} \oplus b^{-1} \tag{5}$$

$$(a \otimes b)^{-1} = a^{-1} \dot{\otimes} b^{-1} \qquad\qquad (a \dot{\otimes} b)^{-1} = a^{-1} \otimes b^{-1}$$

If $\overline{\mathcal{K}}$ is the completion of an idempotent semifield, its upper and lower residuals:

$$a \otimes b \preceq c \Leftrightarrow b \preceq a \backslash c \Leftrightarrow a \preceq c / b \qquad a \dot{\otimes} b \preceq^d c \Leftrightarrow b \preceq^d a \dot{\backslash} c \Leftrightarrow a \preceq^d c \dot{/} b \quad (6)$$

can be expressed in terms of the multiplications and inversion as:

$$a \backslash c = a^{-1} \dot{\otimes} c = (a \otimes c^{-1})^{-1} \qquad c / a = c \dot{\otimes} a^{-1} = (c^{-1} \otimes a)^{-1} \qquad (7)$$

$$a \dot{\backslash} c = a^{-1} \otimes c = (a \dot{\otimes} c^{-1})^{-1} \qquad c \dot{/} a = c \otimes a^{-1} = (c^{-1} \dot{\otimes} a)^{-1}$$

Although associativity of $\oplus$ with respect to $\dot{\oplus}$ would be desirable, the farther we can get is:

Proposition 2 ([8], proposition 3.c). *For all $x, y, z \in \overline{\mathcal{K}}$:*

$$(x \dot{\oplus} y) \oplus z \le x \dot{\oplus} (y \oplus z) \qquad (8)$$

Example 3 (Residuation in $\overline{\mathbb{R}}_{\max,+}$, $\overline{\mathbb{R}}_{\min,+}$). *The residuals of $+$ and $\dot{+}$ are:*

$$a \backslash c := (-a) \dot{+} c = -(a + (-c)) \qquad c / a := c \dot{+} (-a) = -((-c) + a)$$

$$a \dot{\backslash} c := (-a) + c = -(a \dot{+} (-c)) \qquad c \dot{/} a := c + (-a) = -((-c) \dot{+} a)$$

Idempotent semimodules of matrices. A semimodule over a semiring is defined in a similar way to a module over a ring [3,7,6]: a *left $\mathcal{S}$-semimodule*, $\mathcal{Y}$ over a semiring $\mathcal{S}$ is an additive commutative monoid $\langle Y, \oplus, \epsilon_Y \rangle$ endowed with a map $(\lambda, y) \mapsto \lambda \odot y$ such that $\forall \lambda, \mu \in S, \; y, z \in Y$. Following the convention of dropping the symbols for the scalar action and semiring multiplication we have:

$$(\lambda\mu)y = \lambda(\mu y) \qquad\qquad \epsilon_S \odot y = \epsilon_Y \qquad (9)$$
$$\lambda(y \oplus z) = \lambda y \oplus \lambda z \qquad\qquad e_S \odot y = y$$

The definition of a *right $\mathcal{S}$-semimodule* $\mathcal{X}$ follows the same pattern with the help of a *right action*, $(\lambda, x) \mapsto x \odot \lambda$ and similar axioms to those of (9). A $(\mathcal{K}, \mathcal{S})$-*semimodule* is a set M endowed with left $\mathcal{K}$-semimodule and a right $\mathcal{S}$-semimodule structures, and a $(\mathcal{K}, \mathcal{S})$-*bisemimodule* a $(\mathcal{K}, \mathcal{S})$-semimodule such that the left and right multiplications commute. For a left $\mathcal{S}$-semimodule, $\mathcal{Y}$, the left and right multiplications are defined as: $L_\lambda^{\mathcal{S}} : Y \to Y, y \mapsto L_\lambda^{\mathcal{S}}(y) = \lambda y$, and $R_y^{\mathcal{Y}} : S \to Y, \lambda \mapsto R_y^{\mathcal{Y}}(\lambda) = \lambda y$. And similarly, for a right $\mathcal{S}$-semimodule.

Example 4. *Each semiring, $\mathcal{K}$, is a left (right) semimodule over itself, with the semiring product as left (right) action. Therefore, it is a $(\mathcal{K}, \mathcal{K})$-bisemimodule over itself, because both actions commute by associativity. Such is the case for the Boolean $(\mathcal{B}, \mathcal{B})$-bisemimodule, the Maxplus and the Minplus bisemimodules. These are all complete and idempotent.*

Example 5 (Finite matrix semirings and semimodules). *Let S be a semiring. $\mathcal{M}_n(S) = \langle S^{n \times n}, \oplus, \otimes, \mathcal{E}, E \rangle$ is semiring of (square) matrices over S with $S^{n \times n}$ denoting the set of square matrices over the semiring, matrix operations $(A \oplus B)_{ij} = A_{ij} \oplus B_{ij}, 0 \leq i, j \leq n$ and $(A \otimes B)_{ij} = \bigoplus_{k=1}^n A_{ik} \otimes B_{kj}, 0 \leq i, j \leq n$, null element the matrix $\mathcal{E}$, $\mathcal{E}_{ij} = \epsilon, 0 \leq i, j \leq n$ and unit E, $E_{ii} = e, 0 \leq i \leq n$, $E_{ij} = \epsilon, 0 \leq i, j \leq n, i \neq j$. Such semirings are not commutative in general even if S is, except for $\mathcal{M}_1(S) = S$. They are complete and idempotent if S is, in which case, the Kleene star of a square matrix, $A \in \mathcal{M}_n(S)$, can be calculated efficiently: $A^* = \mathcal{E} \oplus A \oplus A^2 \dots A^n$.*

For $g, m \in \mathbb{N}$, the semimodule of finite matrices $\mathcal{M}_{g \times m}(S) = \langle S^{g \times m}, \oplus, \mathcal{E} \rangle$ is a $(\mathcal{M}_g(S), \mathcal{M}_m(S))$-bisemimodule, with matrix multiplication-like left and right actions and componentwise addition. Special cases of it are:

- *the bisemimodules of column vectors $\mathcal{M}_{m \times 1}(S)$ and row vectors $\mathcal{M}_{1 \times g}(S)$.*
- *the semiring of square matrices $\mathcal{M}_g(S)$ with $g = m$, also a bisemimodule.*

If $S \equiv D$ is idempotent (resp. complete), then all are idempotent (resp. complete) with the component-wise partial order as their natural order. If $\overline{D}$ is a completed semifield, then matrix multiplications read for appropriate A, B and summations:

$$(A \otimes B)_{ij} = \bigoplus_{k=1}^n A_{ik} \otimes B_{kj} \qquad (A \dot{\otimes} B)_{ij} = \bigoplus_{k=1}^n A_{ik} \dot{\otimes} B_{kj}$$

For the completed semifields $\overline{\mathbb{R}}_{\max,+}$ and $\overline{\mathbb{R}}_{\min,+}$, we have:

$$(A \otimes B)_{ij} := \max_{k=1}^n (A_{ik} + B_{kj}) \qquad (C \dot{\otimes} D)_{ij} := \min_{k=1}^n (C_{ik} \dot{+} D_{kj})$$

Residuation in matrix semimodules. A left D-semimodule $\mathcal{Y}$ over an idempotent semiring D inherits the idempotent law: $\forall v \in Y, v \oplus v = v$, which induces a *natural order* on the semimodule: $\forall v, w \in Y, v \leq w \iff v \oplus w = w$, whereby it becomes a $\vee$-semilattice, with $\epsilon_{\mathcal{Y}}$ its minimum. In the following we systematically equate left (respectively right) idempotent D-semimodules and row (respectively column) semimodules over an idempotent semiring D . When D is a complete idempotent semiring, a left D-semimodule $\mathcal{Y}$ is *complete (in its natural order)* if it is complete as a naturally ordered set and its left and right multiplications are lower semicontinuous. Trivially, it is also a complete lattice, with join and meet operations given by: $v \leq w \iff v \vee w = w \iff v \wedge w = v$. This extends naturally to right- and bisemimodules.

As in the semiring case, because of the natural order structure, the actions of idempotent semimodules admit residuation: given a complete, idempotent left D-semimodule, $\mathcal{Y}$, we define for all $y, z \in Y, \lambda \in D$ the residuals are: $\left(L_\lambda^{D}\right)^{\#} : Y \to Y, \left(L_\lambda^{D}\right)^{\#}(z) = \lambda \backslash z$ and $\left(R_y^{\mathcal{Y}}\right)^{\#} : Y \to D, \left(R_y^{\mathcal{Y}}\right)^{\#}(z) = z/y$ and likewise for a right semimodule.

If D is idempotent (resp. complete), then finite matrix semimodules are idempotent (resp. complete) with the componentwise partial order as their natural order. Therefore we can define residuated operations as ([2], p. 196): let $\mathcal{D}$ be a complete dioid in which $\wedge$ exists, and $A \in D^{m \times n}, B \in D^{m \times p}, C \in D^{n \times p}$, then their left, $A \backslash B$, and right B / C residuals are:

$$(A \backslash B)_{ij} = \bigwedge_{k=1}^{m} (A_{ki} \backslash B_{kj}) \qquad (B / C)_{ij} = \bigwedge_{k=1}^{p} (B_{ik} / C_{jk}) \qquad (10)$$

For $\overline{\mathcal{K}}$ a completed idempotent semifield as in subsection 2.1, the left and right residuals of $\otimes$ and $\dot{\otimes}$ are (with the appropriate summations):

$$(A \backslash B)_{ij} = \bigoplus_{k=1}^{m} \left(A_{ki}^{-1} \dot{\otimes} B_{kj} \right) \qquad (A \dot{\backslash} B)_{ij} = \bigoplus_{k=1}^{m} \left(A_{ki}^{-1} \otimes B_{kj} \right) \qquad (11)$$

$$(B / C)_{ij} = \bigoplus_{k=1}^{p} \left(B_{ik} \dot{\otimes} C_{jk}^{-1} \right) \qquad (B \dot{/} C)_{ij} = \bigoplus_{k=1}^{p} \left(B_{ik} \otimes C_{jk}^{-1} \right)$$

To pave the way for some results in Section 3 we have:

Proposition 3 (Adapted from [5], §5.3.3 and 5.4). *For u, v, w in the appropriate S-semimodules, $(u \backslash v) \otimes w \leq u \backslash (v \otimes w)$ and equality holds when $w \in S$ is invertible or $w \in \mathcal{M}_{g \times m}(S)$ has at least one finite component in every row and column.*

Definition 6 (Conjugations). *For $\mathcal{Y} \cong \mathcal{K}^{1 \times n}, \mathcal{X} \cong \mathcal{K}^{n \times 1}$ left and right semimodules, respectively, over an idempotent reflexive semifield $(\mathcal{K}, \varphi)$ and bracket $\langle \cdot \mid \cdot \rangle : Y \times X \to K$, $\langle y \mid x \rangle = y \dot{\otimes} x$ [13] we define a conjugation to be the Galois connection obtained from the maps in eq. (1): $y^{\circledast} = y \backslash e_{\mathcal{D}}$, $^{\circledast}x = e_{\mathcal{D}} / x$, and we write simply: $(\cdot^{\circledast}, {}^{\circledast} \cdot) : \mathcal{Y} \hookrightarrow \mathcal{X}$. For any other invertible element $\varphi \in K$ we have the φ-conjugations: $y_{\varphi}^{\circledast} = y \backslash \varphi = y \backslash (e_{\mathcal{D}} \dot{\otimes} \varphi) = y^{\circledast} \dot{\otimes} \varphi$ and $_{\varphi}^{\circledast}x = \varphi \dot{\otimes} {}^{\circledast}x$.*

For instance, the conjugations in $\overline{\mathbb{R}}_{\max,+}$ are: $y^{\circledast} := -y^t,{}^{\circledast} x := -x^t$, where $\cdot^t :$ $Y \to Y$ stands for transposition. We also define without further ado: $y^{-1} = (y^t)^{\circledast} = (y^{\circledast})^t$ and similarly for right semimodules.

For adequate invertible unitary matrices, $E_{\mathcal{M}_n(\mathcal{D})}, (\mathcal{M}_n(\mathcal{D}), E_{\mathcal{M}_n(\mathcal{D})})$ is reflexive hence the conjugations of Def. (6) exist for $R \in D^{g \times m}$:

$$R^{\circledast} = R \backslash E_{\mathcal{M}_g(\mathcal{D})} \qquad\qquad {}^{\circledast}R = E_{\mathcal{M}_m(\mathcal{D})} / R \qquad (12)$$

and we can write analogues of Prop. 1 compactly:

Proposition 4. *In the top completion, $\overline{\mathcal{D}}$, of an idempotent semifield the following De Morgan-like laws hold:*

$$(A \oplus B)^{\circledast} = A^{\circledast} \dot{\oplus} B^{\circledast} \qquad\qquad (A \dot{\oplus} B)^{\circledast} = A^{\circledast} \oplus B^{\circledast} \qquad (13)$$

$$(A \otimes B)^{\circledast} = B^{\circledast} \dot{\otimes} A^{\circledast} \qquad\qquad (A \dot{\otimes} B)^{\circledast} = B^{\circledast} \otimes A^{\circledast}$$

the following residuation laws hold:

$$A \backslash_{\bullet} C = A^{\circledast} \dot{\otimes} C = (C^{\circledast} \otimes A)^{\circledast} \qquad A \dot{\backslash} C = A^{\circledast} \otimes C = (C^{\circledast} \dot{\otimes} A)^{\circledast} \qquad (14)$$

$$C /_{\bullet} A = C \dot{\otimes} A^{\circledast} = (A \otimes C^{\circledast})^{\circledast} \qquad C \dot{/} A = C \otimes A^{\circledast} = (A \dot{\otimes} C^{\circledast})^{\circledast}$$

and similarly for left conjugates.

2.2   Spectra of Reducible and Irreducible Matrices

Graphs related to a matrix. Consider a digraph $\Gamma = (V, E)$, with V a set of vertices and $E \subseteq V^2$ a set of edges. If there is a walk from a vertex i to a vertex j in Γ we say that i *has access to* j, $i \rightsquigarrow j$. This relation is transitive and reflexive. The *access equivalent classes* of Γ are the equivalence classes of the transitive, symmetric and reflexive closure of the access relation, $i \leftrightsquigarrow j \Leftrightarrow i \rightsquigarrow j \wedge j \rightsquigarrow i$. Γ is *strongly connected* if it only has one class. When $C, C' \in V/\leftrightsquigarrow$, we say that a class C *has access to* a class C', if some vertex of C has access to some vertex of C', and we say that it is *final* if it has only access to itself.

Now consider a matrix with values in a semiring, $A \in D^{n \times n}$. The *digraph* $\Gamma(A)$ *associated to this matrix* consists of the set of vertices $V = \{1, \ldots n\}$ and a set of edges, $E = \{(i, j) \mid A_{ij} \neq \epsilon_D\}$. The *classes of a matrix* A are the (access equivalent) classes of $\Gamma(A)$, hence we say that the matrix A is *irreducible* if $\Gamma(A)$ is strongly connected, and *reducible* otherwise.

A *walk* in $\Gamma(A)$ is a sequence of edges pairwise sharing an element $w = (v_1, v_2), (v_2, v_3), \ldots, (v_{k-1}, v_k)$. The weight of a walk is $|w|_A = A_{v_1 v_2} \otimes A_{v_2 v_3} \otimes \ldots \otimes A_{v_{k-1} v_k}$, and its *length* is $|w| = k - 1$. Call a *cycle* a walk with $v_1 = v_k$ and its *cycle mean* the ratio of weight-to-length. Therefore the *maximal cycle mean*, $\rho_{\max}(A)$, is the maximum of the cycle means over all cycles of $\Gamma(A)$:

$$\rho_{\max}(A) = \max_{c \text{ cycle of } \Gamma(A)} \frac{|c|_A}{|c|} \qquad (15)$$

A cycle that attains such a maximum is called a *critical cycle*. Call the union of the critical cycles the *critical digraph*, $\Gamma_c(A)$, and its vertices, the *critical vertices*, V_c . Also, call the (access equivalent) classes of the critical digraph $\Gamma_c(A)$ the *critical classes of* A.

Eigenvalues and eigenvectors in idempotent semimodules. Let $\overline{D}$ be a completed dioid. An *eigenvector* of $A \in D^{n \times n}$ is a vector $x \in D^n \backslash \{\epsilon\}$ such that $A \otimes x = \lambda \otimes x$ for some $\lambda \in D$ which is called the *(geometric) eigenvalue corresponding to* x. If λ is an eigenvalue of A then the *eigenspace* of A for the eigenvalue λ is the set of vectors, $eig(A, \lambda) = \{x \in \overline{D}^n \mid A \otimes x = \lambda \otimes x\}$.

To put a concrete example, the $\overline{\mathbb{R}}_{\max,+}$ spectral theory shows notorious differences with normal spectral theory. For $\overline{D} := \overline{\mathbb{R}}_{\max,+}$ the eigenvalue equation becomes:

$$\max_{1 \leq j \leq n} \left\{ A_{ij} + x_j \right\} = \lambda + x_i, \; \forall 1 \leq i \leq n \tag{16}$$

Now, if we define the *normalised matrix* as $\tilde{A} = \rho_{\max}(A)^{-1}A$ when D is a semifield, the following facts all refer to irreducible $A \in \mathbb{R}^{n \times n}_{\max,+}$ [1]:

Property 7 (Spectra of irreducible $\overline{\mathbb{R}}_{\max,+}$-matrices)

1. *For any matrix A, $\rho_{\max}(A)$ is an eigenvalue of A, and any eigenvalue of A is less than or equal to $\rho_{\max}(A)$.*
2. *An eigenvalue of A associated with an eigenvector in $\overline{\mathbb{R}}^n_{\max,+}$ must be equal to $\rho_{\max}(A)$.*
3. *If A is irreducible, then $\rho_{\max}(A) > \epsilon$ and it is the only eigenvalue of A .*
4. *For all critical vertices $i \in V_c(A)$, the column $\tilde{A}^*_{\cdot i}$ is an eigenvector of A for the eigenvalue $\rho_{\max}(A)$.*
5. *If i and j belong to the same critical class, then $\tilde{A}^*_{\cdot i} = \tilde{A}^*_{\cdot j} \otimes \tilde{A}^*_{ji}$.*
6. *(Eigenspace for the eigenvalue $\rho_{\max}(A)$). Let $\{C_t\}_{t=1}^s$ be the set of critical classes of A. Arbitrarily select one vertex i_t from each class. The columns $\tilde{A}^*_{\cdot i_s}, t = 1 \ldots s$ span the eigenspace of A for the maximal cycle mean $\rho_{\max}(A)$,*
$$eig(A, \rho_{\max}(A)) = span\left(\{\tilde{A}^*_{\cdot i_s}\}_{t=1}^s \right) .$$

The most notable difference here is the existence of a single eigenvalue $\rho_{\max}$ per irreducible matrix. In fact in such situations we drop the specification of the eigenvalue from the eigenspace notation $eig(A) = eig(A, \rho_{\max})$ thereby implying that A is irreducible.

Now, denote by $A[C, C]$ the submatrix of A selected by the vertices in class C and call a class C of A *basic* if $\rho_{\max}(A[C, C]) = \rho_{\max}(A)$. The following facts relate to reducible matrices[2]:

Property 8 (Spectra of reducible $\overline{\mathbb{R}}_{\max,+}$-matrices)

1. *A scalar $\lambda \neq \epsilon$ is an eigenvalue of A if and only if there is at least one class of A such that $\rho_{\max}(A[C, C]) = \lambda$ and $\rho_{\max}(A[C, C]) \geq \rho_{\max}(A[C', C'])$ for all classes C' that have access to D . The spectrum of A, $spec(A)$, is the set of such eigenvalues, which is essentially the union of the spectra of some of its irreducible blocks.*
2. *$A \in \mathbb{R}^{n \times n}$ has an eigenvector in $\mathbb{R}^n$ iff all its final classes are basic.*
3. *(Eigenspace for eigenvalue λ.) Let $\{C^k\}_{k=1}^m$ denote all the classes of A such that if $\rho_{\max}(A[C^k, C^k]) = \lambda_k$ then $\rho_{\max}(A[C', C']) \leq \lambda_k$ for all classes C' that have access to C^k . For every $1 \leq k \leq m$, let $\{C_t^k\}_{t=1}^{s_k}$ denote the critical classes of the matrix $A[C^k, C^k]$. For each $1 \leq k \leq m, 1 \leq t \leq s_k$, choose an arbitrary $j_{k,t} \in C_t^k$. Then the columns of the λ-normalized columns $eig(A, \lambda) = span(\{(\lambda \setminus A)^*_{\cdot j_{k,t}} \mid 1 \leq k \leq m, 1 \leq t \leq s_k, j_{k,t} \in C_t^k\})$ span the eigenspace of A for λ and any spanning family of this eigenspace contains a scalar multiple of every one of these.*

---

[2] We mention in passing that there are algorithms for transforming a reducible matrix into an upper or lower block-triangular form.

Again the extra requisites on the spectral eigenvalues related to the order of the reachable classes is a deviation from "standard" spectral theory.

Calculating such spectra is specially easy in a certain kind of matrices:

Definition 9. *Let $A \in \mathcal{M}_n(\overline{\mathbb{R}}_{\max,+})$. After [9], we call A definite if its maximal cycle mean is $\rho_{\max}(A) = e$ and its diagonal entries equal $A_{ii} = e$.*

We have the following:

Proposition 5 ([9], prop. 7). *If A is a definite matrix, then:*

1. *It has a unique eigenvalue $\lambda = e = \rho_{\max}(A)$.*
2. *$eig(A) = span(A^*)$.*

The important thing about definite matrices is that the very complex eigenvalue-eigenvector calculation is reduced to the calculation of a star operation. We prove in passing the next result to be used later implying that the left and right residuals of any rectangular matrix are halfway to being a definite matrix:

Proposition 6. *Let $\overline{K}$ be the top completion of an idempotent semifield. For $R \in \mathcal{M}_{g \times m}(\overline{K})$, the diagonal entries of $R \backslash R \in \mathcal{M}_m(\overline{K})$ and $R / R \in \mathcal{M}_g(\overline{K})$ equal e iff at least some row, or column of $\dot{R}$ is finite.*

Proof. Call $P = R \backslash R \in \mathcal{M}_m(\overline{\mathcal{D}})$. Recall that $R \backslash R = R^{\circledast} \dot{\otimes} R$, so for each

$$1 \leq i \leq m, P_{ii} = \dot{\bigoplus}_{1 \leq i \leq m} R_{ij}^{\circledast} \dot{\otimes} R_{ji} \text{ . Now, } R_{ij}^{\circledast} = R_{ji}^{-1} \text{ . Hence, for } R_{ji} \in \mathcal{D}$$

this means $R_{ij}^{\circledast} \dot{\otimes} R_{ji} = R_{ji}^{-1} \dot{\otimes} R_{ji} = e$, and for $R_{ji} \in \{\bot, \top\}$, $R_{ji}^{-1} \dot{\otimes} R_{ji} = \top$. If at least one of the elements is finite, then the total sum, being an inf, becomes e . The proof for R / R is the same.   □

3   The Spectral Lattice of an $\overline{\mathbb{R}}_{\max,+}$-Context

Consider the *right* semimodules $\mathcal{Y} \cong \overline{K}^{g \times 1}$, $\mathcal{X} \cong \overline{K}^{m \times 1}$ and the bracket $\langle y \mid x \rangle = y^t \dot{\otimes} R \dot{\otimes} x$ where we have switched to consider columns as vectors as customary in data mining and signal processing applications[3]. We can give algebraic expressions for the φ-polars in the completed semifield:

Proposition 7. *The φ-polars have the algebraic form: $y_\varphi^R = R^{\circledast} \dot{\otimes} y^{-1} \dot{\otimes} \varphi$, ${}_\varphi^R x = \varphi \dot{\otimes} x^{\circledast} \dot{\otimes} R^{\circledast}$.*

Proof. This is straightforward using the maxplus/minplus algebra developed in section 2.1:

$$y_\varphi^R = (y^t \dot{\otimes} R) \backslash \varphi \qquad\qquad {}_\varphi^R x = \varphi \dot{/} (R \dot{\otimes} x) \qquad\qquad (17)$$

$$= R^{\circledast} \dot{\otimes} y^{-1} \dot{\otimes} \varphi \qquad\qquad = \varphi \dot{\otimes} x^{\circledast} \dot{\otimes} R^{\circledast}$$

---

[3] This will only entail minimal tinkering with the notation.

This suggests that we call $\tilde{x} = x \, \dot{/} \, \varphi = x \otimes \varphi^{\circledast}$ and $\tilde{y}^{t} = \varphi \, \dot{\backslash} \, y^{t} = \varphi^{\circledast} \, \dot{\otimes} \, y^{t}$,

$(\tilde{y} = y \, \dot{/} \, \varphi^{t} = y^{t} \otimes \varphi^{-1})$, so that the *normalised semimodules (wrt. φ)* are:

$$\tilde{\mathcal{Y}}^{t} = \{ \tilde{y}^{t} \mid y^{t} \in \mathcal{Y}^{t} \} \qquad \tilde{\mathcal{Y}} = \{ \tilde{y} \mid y \in \mathcal{Y} \} \qquad \tilde{\mathcal{X}} = \{ \tilde{x} \mid x^{t} \in \mathcal{X} \} \qquad (18)$$

Then we have the following:

Proposition 8 (Decoupled eigenequations)

1. With $P_{\mathcal{Y}^{t}} = R \, \dot{\otimes} \, R^{\circledast} \in \mathcal{M}_{g}(\overline{\mathbb{R}}_{\max,+})$, we have $\tilde{y}^{t} \otimes P_{\mathcal{Y}^{t}} = \tilde{y}^{t}$.

2. With $P_{\mathcal{X}} = R^{\circledast} \, \dot{\otimes} \, R \in \mathcal{M}_{m}(\overline{\mathbb{R}}_{\max,+})$ we have $P_{\mathcal{X}} \otimes \tilde{x} = \tilde{x}$.

3. With $P_{\mathcal{Y}} = (P_{\mathcal{Y}^{t}})^{t} = (R \, \dot{\otimes} \, R^{\circledast})^{t} = R^{-1} \, \dot{\otimes} \, R^{t}$ we have $P_{\mathcal{Y}} \otimes \tilde{y} = \tilde{y}$.

Proof The equation for the concepts can be written as:

$$\underset{\varphi}{\overset{R}{}} x = y^{t} \qquad\qquad y \underset{\varphi}{\overset{R}{}} = x \qquad (19)$$

Therefore, equating Eqs. (17) and (19):

$$\varphi \, \dot{\otimes} \, x^{\circledast} \, \dot{\otimes} \, R^{\circledast} = y^{t} \qquad\qquad R^{\circledast} \, \dot{\otimes} \, y^{-1} \, \dot{\otimes} \, \varphi = x$$

hence from $x^{\circledast} = \varphi^{\circledast} \otimes y^{t} \otimes R$ and $y^{-1} = R \otimes x \otimes \varphi^{\circledast}$ we get:

$$\varphi \, \dot{\otimes} (\varphi^{\circledast} \otimes y^{t} \otimes R) \, \dot{\otimes} \, R^{\circledast} = y^{t} \qquad\qquad R^{\circledast} \, \dot{\otimes} (R \otimes x \otimes \varphi^{\circledast}) \, \dot{\otimes} \, \varphi = x$$

whence, for invertible φ:

$$(\varphi^{\circledast} \otimes y^{t} \otimes R) \, \dot{\otimes} \, R^{\circledast} = \varphi \, \dot{\backslash} \, y^{t} \qquad\qquad R^{\circledast} \, \dot{\otimes} (R \otimes x \otimes \varphi^{\circledast}) = x \, \dot{/} \, \varphi$$

$$(\tilde{y}^{t} \otimes R) \, \dot{\otimes} \, R^{\circledast} = \tilde{y}^{t} \qquad\qquad R^{\circledast} \, \dot{\otimes} (R \otimes \tilde{x}) = \tilde{x}$$

Finally, by Prop. 3 we have:

$$\tilde{y}^{t} \otimes (R \, \dot{\otimes} \, R^{\circledast}) = \tilde{y}^{t} \qquad\qquad (R^{\circledast} \, \dot{\otimes} \, R) \otimes \tilde{x} = \tilde{x} \qquad (20)$$

For the third proposition we write $P_{\mathcal{Y}} = (P_{\mathcal{Y}^{t}})^{t} = (R \, \dot{\otimes} \, R^{\circledast})^{t} = R^{-1} \, \dot{\otimes} \, R^{t}$ and then transpose the whole equation for $\tilde{y}^{t}$ in Eq. (20). $\qquad \square$

Note that although it is apparently a major unbalance, the eigenvalue equation for $P_{\mathcal{Y}}$ allows us to write the very balanced:

$$\begin{bmatrix} R^{-1} \, \dot{\otimes} \, R^{t} & 0_{g \times m} \\ 0_{m \times g} & R^{\circledast} \, \dot{\otimes} \, R \end{bmatrix} \otimes \begin{bmatrix} \tilde{y} \\ \tilde{x} \end{bmatrix} = \begin{bmatrix} \tilde{y} \\ \tilde{x} \end{bmatrix} \qquad (21)$$

For $\varphi^{\circledast} = \varphi^{-1}$ and $\tilde{z} = [\tilde{y}^t \ \tilde{x}^t]^t = z \otimes \varphi^{\circledast}$ we can write $C \otimes \tilde{z} = \tilde{z}$, and call it the *extended eigenvalue equation*, which shows that the normalised formal concepts are also the fixpoint of some sort of matrix operator.

Another practical advantage of using P_y is to be able to refer all results to column semimodules. This is what we will do hence.

Consider now $\overline{\mathcal{K}} = \mathbb{R}_{max,+}$. With regard to the eigenspaces of these projections we have the following proposition:

Proposition 9. P_y and P_x are definite matrices.

Proof. The proof that they are matrices with their diagonals set to e_D is in Prop. 6. Now consider any of the equations in Proposition 8. These are clearly equations for the eigenvalue $\lambda = e_D$. From Property 7.2 this eigenvalue has to be ρ_{max}. □

Proposition 10. *1. P_y and P_x are closure operators in matrix form over their respective normalised semimodules.*
2. *$P_y = P_y^*$ and $P_x = P_x^*$.*

Proof. From Props. 2 and 2, $eig(\overline{\mathcal{Y}}^t, \rho_{max}) = span\left(\left(\overline{\mathcal{Y}}^t\right)^*\right) = span(\overline{\mathcal{X}})$. □

Then we have the following easy corollaries:

Corollary 11 (The spectral Galois connection)

1. *P_{y^t} and P_x are the closure operators in matrix form of the Galois connection $((\cdot)_{\rho_{max}}^R, \frac{R}{\rho_{max}}(\cdot)) : \tilde{\mathcal{Y}} \leftharpoonup \tilde{\mathcal{X}}$*
2. *The subsemimodule $\overline{\mathcal{Y}}$ is the eigenspace $eig(P_y) = span(P_y)$ and the subsemimodule $\overline{\mathcal{X}}$ is the eigenspace $eig(P_x) = span(P_x)$.*

Proof. For the first subproposition, consider the polars of the generic Galois connection and rewrite: $y_\varphi^R = R^{\circledast} \dot{\otimes} (\varphi^{\circledast} \otimes y)^{\circledast} = (\tilde{y}^t \otimes R) \dot{\setminus} e = \tilde{y}_e^R = \tilde{y}_{\rho_{max}}^R$, and similarly n $_\varphi^R x = \frac{R}{\rho_{max}} \tilde{x}$. By ([3], Th. 42) this is a Galois connection, whose closure operators by the proof of Proposition 8 are exactly the matrices pointed to above. For the second proposition, combine Propositions 5.2 and 10.2 to get the closure lattices, $\overline{\mathcal{Y}}$ and $\overline{\mathcal{X}}$. □

This suggests that $\varphi = \rho_{max} = e$ for both matrices is a special choice, so we give it its right status:

Definition 10. *For a set G of objects and a set M of attributes, of widths $g, m \in \mathbb{R}$ respectively, with $R \in \mathcal{M}_{g \times m}(\overline{\mathbb{R}}_{max,+})$ building the $\overline{\mathbb{R}}_{max,+}$-formal context $\mathbb{K} = (G, M, R)_{\overline{\mathbb{R}}_{max,+}}$ the spectral lattice, $\mathfrak{B}^{\rho_{max}}(G, M, R)_{\overline{\mathbb{R}}_{max,+}}$ is the lattice of ρ_{max}-formal concepts of the connection $((\cdot)_{\rho_{max}}^R, \frac{R}{\rho_{max}}(\cdot)) : \tilde{\mathcal{Y}} \leftharpoonup \tilde{\mathcal{X}}$*

Indeed, this is the Galois connection depicted to the right of Figure 1. The next proposition paves the way for a more familiar representation, the structural lattice:

Proposition 12. *1. The join irreducibles $\mathcal{J}\left(\mathfrak{B}^{\rho_{\max}}(G, M, R)_{\overline{\mathbb{R}}_{\max,+}}\right)$ are the pairs (a_i, b_i) such that i ranges over the columns of $P_{\mathcal{Y}}$ and $b_i = (a_i)^R_{\rho_{\max}}$.*

2. The meet irreducibles $\mathcal{M}\left(\mathfrak{B}^{\rho_{\max}}(G, M, R)_{\overline{\mathbb{R}}_{\max,+}}\right)$ are the pairs (a_j, b_j) such that j ranges over the columns of $P_{\mathcal{X}}$ and $a_j = {}^R_{\rho_{\max}}(b_j)$.

Proof. The Galois connection between the closure lattices $\overline{\overline{\mathcal{Y}}}$ and $\overline{\overline{\mathcal{X}}}$ ensures that the columns of each of the projectors are the basis of the eigenspaces, that is the join-irreducibles of each lattice. The join-irreducibles of $\overline{\overline{\mathcal{Y}}}$ generate the join-irreducibles of $\mathfrak{B}^{\rho_{\max}}(G, M, R)_{\overline{\mathbb{R}}_{\max,+}}$ by applying the polars of the Galois connection. However, because of the inversion for the second domain, the join-irreducibles of $\overline{\overline{\mathcal{X}}}$ generate the *meet-irreducibles* of $\mathfrak{B}^{\rho_{\max}}(G, M, R)_{\overline{\mathbb{R}}_{\max,+}}$.   □

Once we have both meet- and join-irreducibles it is easy to obtain the structural (concept) lattice of the spectral lattice by the procedure described in [13].

4   Application: The Analysis of Confusion Matrices

To illustrate the calculations behind the spectral lattice we retake now the problem of analysing confusion matrices. Figure 2 illustrates one such matrices with the usual hypothesis in pattern recognition, $g = m$.

For simplification's sake, consider every row and column in C to have at least one non-null entry and call D_G and D_M those diagonal matrices such that their diagonal elements are the sums of rows and columns respectively, $(D_G)_{ii} = \sum_{j=1}^{m} C_{ij}$, $(D_M)_{jj} = \sum_{i=1}^{g} C_{ij}$, $(D_G)_{ij} = (D_M)_{ij} = 0, i \neq j$. Therefore, D_G, D_M are invertible so the matrix $R = \log[(D_G)^{-1}C(D_M)^{-1}]$ is defined and has entries in $\mathbb{R} \cup \{-\infty\}$. In this case, R happens to be irreducible.

$$M = \begin{bmatrix} 5 & 3 & 0 \\ 2 & 3 & 1 \\ 0 & 2 & 11 \end{bmatrix} \qquad R = \begin{bmatrix} 3.821457e & 2.852357 & \varepsilon \\ -0.7378621 & 2.272438 & -5.856696 \\ \varepsilon & -1.249387 & 2.796319 \end{bmatrix} \cdot 10^{-01}$$

Fig. 2. The confusion matrix, M, its version as a $\mathbb{R}_{\max,+}$ matrix, R

Now consider $P_{\mathcal{Y}}$ and $P_{\mathcal{X}}$ as per the definitions in Eq. (8). These are both definite and irreducible hence their eigenvectors are all of their columns. The structural spectral Formal Context and its Concept Lattice are shown in Fig. 3. Interestingly, this trivial lattice already justifies the asymmetric treatment of real and recognised classifier tags. It questions Pattern Recognition approaches to confusion matrix analysis that impose a symmetrical structure on these.

As a further example we introduce the (abridged) analysis of the performance of an automatic speech recognizer for Spanish in figure (4). Its confusion matrix (and $\overline{\mathbb{R}}_{\max,+}$-Formal Context), illustrates the baseline performance for a certain type of recognition technology: most of the vowels can be adequately decoded, the less so nasals. However, approximants (soft /b/,/d/,/g/ in vocalic context) cannot be told apart in the spectral lattice at all, a weakness of this recognizer.

Fig. 3. Structural spectral context and Concept lattice for matrix R in Figure 2

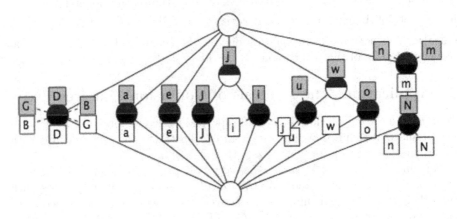

Fig. 4. Confusion matrix for an automatic speech recognizer for Spanish. Objects: real phonemes; attributes: recognised phonemes (SAMPA). /J/ stands for the phoneme of letter "ñ".

5   Conclusion

In this paper, we have tried to justify the importance of a particular value of the exploration parameter φ to obtain structural lattices [15] for the concrete case of $\overline{\mathbb{R}}_{\max,+}$-Formal Concept Analysis, viz, the case where we consider it to be an eigenvalue of the projectors onto the closure lattices in the Galois connection. The latter can be readily obtained as the left and right residuals of the $\overline{\mathbb{R}}_{\max,+}$-valued incidence in the completed semiring, which makes the spectral estimation a very light process computationally speaking.

References

1. Akian, M., Bapat, R., Gaubert, S.: Handbook of Linear Algebra. In: Max-Plus Algebra, CRC Press, Boca Raton (2006)
2. Baccelli, F., et al.: Synchronization and Linearity. Wiley, Chichester (1992)
3. Cohen, G., Gaubert, S., Quadrat, J.–P.: Duality and separation theorems in idempotent semimodules. Linear Algebra and Its Applications 379, 395–422 (2004)

4. Erné, M.: Adjunctions and Galois connections: Origins, History and Development. In: Mathematics and Its Applications, vol. 565, pp. 1–138. Kluwer Academic, Dordrecht (2004)

5. Gaubert, S.: Théorie des systèmes linéaires dans les dioïdes. Thèse, École des Mines de Paris (July 1992)

6. Golan, J.S.: Power Algebras over Semirings. With Applications in Mathematics and Computer Science. In: Mathematics and its applications, vol. 488, Kluwer Academic Publishers, Dordrecht (1999)

7. Golan, J.S.: Semirings and Their Applications. Kluwer Academic Publishers, Dordrecht (1999)

8. Moreau, J.J.: Inf-convolution, sous-additivité, convexité des fonctions numériques. J. Math. pures et appl. 49, 109–154 (1970)

9. Sergeev, S.: Max-plus definite matrix closures and their eigenspaces. Linear Algebra and its Applications 421(2-3), 182–201 (2007)

10. Singer, I.: Abstract Convex Analysis. Monographs and Advanced Texts. Wiley-Interscience (1997)

11. Singer, I. (*,s)-dualities. Journal of Mathematical Sciences 115(4), 2506–2541 (2003)

12. Singer, I.: Some relations between linear mappings and conjugations in idempotent analysis. Journal of Mathematical Sciences 115(5), 2610–2630 (2003)

13. Valverde-Albacete, F.J., Peláez-Moreno, C.: Towards a generalisation of Formal Concept Analysis for data mining purposes. In: Missaoui, R., Schmidt, J. (eds.) Formal Concept Analysis. LNCS (LNAI), vol. 3874, pp. 161–176. Springer, Heidelberg (2006)

14. Valverde-Albacete, F.J., Peláez-Moreno, C.: Further Galois connections between semimodules over idempotent semirings. In: Diatta, J., Eklund, P. (eds.) Proceedings of the 4th Conference on Concept Lattices and Applications (CLA 2007), October 2007, Montpellier, pp. 199–212 (2007)

15. Valverde-Albacete, F.J., Peláez-Moreno, C.: Galois connections between semimodules and applications in data mining. In: Kuznetsov, S.O., Schmidt, S. (eds.) ICFCA 2007. LNCS (LNAI), vol. 4390, pp. 181–196. Springer, Heidelberg (2007)

About Keys of Formal Context and Conformal Hypergraph

Pierre Colomb and Lhouari Nourine

LIMOS - CNRS UMR 6158
Université Blaise Pascal, Clermont-Ferrand
{colomb,nourine}@isima.fr

Abstract. In this paper we study the problem of generating all keys of a formal context as well as a hypergraph. We show that computing the maximum size of a key is NP-complete. Consequently, there is no polynomial time algorithm that decides if a hypergraph is k-conformal, unless P=NP. We also present an algorithmic framework based on decomposition to enumerates all keys of a hypergraph. As example we propose a decomposition of a hypergraph into conformal hypergraphs. Computing a minimal decomposition of an arbitrary hypergraph into conformal hypergraphs remains open in this paper.

1 Introduction

In Formal Concept Analysis (FCA), keys and implications are viewed as information implicitly present in a context. Extract a such information represents a crucial part in KDD process and it is strongly connected to minimal transversals of a hypergraph [1].

Generating all keys of a context or all minimal transversals of a hypergraph is a very popular problem. More recently, keys and minimal transversals have been intensively studied in several areas, such as database theory, artificial intelligence, machine learning, graph theory [1, 2, 3]. This problem has two independent sources of complexity; (1) The number of keys can be exponential in the size of the input, and (2) Time complexity needed for the generating algorithm is unknown. The best known algorithm is due to Fredman and Khachiyan [4], which has a quasi-polynomial time complexity in the size of input and output.

Several authors aimed at identifying special cases for which minimal transversals (or keys) can be generated in polynomial total time. For example bounded edge-intersections [5], bounded dimension [6], acyclicity [1]. (see also [3, 7, 8]). In [9] authors design an efficient algorithm to compute minimal transversal of $k-$conformal hypergraphs when k is constant.

In this paper we study the problem of generating all keys of a context as well as a hypergraph. We show that computing the maximum size of a key is NP-complete; as a consequence there is no polynomial time algorithm that decides if a hypergraph is k-conformal, unless P=NP. We introduce the notion

R. Medina and S. Obiedkov (Eds.): ICFCA 2008, LNAI 4933, pp. 140–149, 2008.

of conformal decomposition of a hypergraph into conformal hypergraphs. Then we present an algorithm that decomposes a given hypergraph into conformal hypergraphs and enumerates all its keys. Computing a minimal decomposition of an arbitrary hypergraph into conformal hypergraphs remains open in this paper.

The content of this paper is structured as follows. Section 2 presents the basic notions from Formal Concept Analysis and hypergraphs. Section 3 presents the link between keys and $k-$conformal hypergraphs. NP-Completeness for the largest key and k-conformal hypergraph recognition are shown. Section 4 describes an approach to compute keys based of a conformal decomposition.

2 Preliminaries

This paper is written using Formal Concept Analysis [10], and hypergraphs [11] terminology. We recall some notions and results concerning hypergraphs and keys of a formal context.

Let $\mathbb{K} = (G, M, I)$ be a reduced formal context where G is the set of objects, M the set of attributes and $I \subseteq G \times M$. A subset $Q \subseteq M$ is said a *super-key* of $\mathbb{K}$ if there is no object $g \in G$ such that $Q \subseteq g'$ (where g' denote the set of attributes owned by the object g). A super-key is called a *key* if it is (inclusionwise) minimal. $\Sigma(\mathbb{K})$ denotes the set of all keys of $\mathbb{K}$. The notion of keys is strongly related to minimal transversals of a hypergraph (see for example [12]).

A *hypergraph* $\mathcal{H}$ consists of a finite collection of subsets over a finite set V. The elements of $\mathcal{H}$ are called *hyperedges*, or simply *edges*. A hypergraph $\mathcal{H}$ is said *simple* (or Sperner family) if it has no pair of hyperedges E and E' such that E is properly contained in E'. A *transversal* (or *hitting set*) of $\mathcal{H}$ is a set $T \subseteq V$ that intersects every edge of $\mathcal{H}$, i.e. for all $E \in \mathcal{H}$, $T \cap E \neq \emptyset$. A transversal is *minimal* if it does not contain any other transversal as a subset. The set $Tr(\mathcal{H})$ of all minimal transversals of $\mathcal{H}$ is also a hypergraph on V, which is called the *transversal hypergraph* of $\mathcal{H}$. It's well known that a simple hypergraph satisfies $Tr(Tr(\mathcal{H})) = \mathcal{H}$ [13].

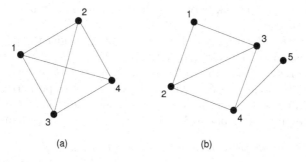

(a)   (b)

Fig. 1. The 2-section graph of hypergraphs $\mathcal{H}_a = \{123, 124, 34\}$ and $\mathcal{H}_b = \{123, 234, 45\}$

The complementary hypergraph of a hypergraph $\mathcal{H}$ is denoted by $\overline{\mathcal{H}} = \{V \setminus E \mid E \in \mathcal{H}\}$. When hyperedges have size at most 2, $\mathcal{H}$ is known as an *undirected graph* (or *graph*).

In the following, we consider only simple hypergraphs without isolated vertex, i.e. $V = \bigcup_{E \in \mathcal{H}} E$. The *2-section* graph $G(\mathcal{H})$ of a hypergraph $\mathcal{H}$ is a graph where vertices are the set $X = \bigcup_{E \in \mathcal{H}} E$ and edges are all pairs $u, v \in V$ that are contained in a hyperedge of $\mathcal{H}$.

The *clique hypergraph* of a graph G, denoted by $\mathcal{C}(G)$, is a hypergraph with the same vertices as G and whose hyperedges are the vertex sets of maximal cliques of G. A hypergraph $\mathcal{H}$ is said *conformal* if it is the clique hypergraph of its 2-section graph $G(\mathcal{H})$ [14]. We can check that the hypergraph $\mathcal{H}_a$ in Figure 1 is not conformal, but the hypergraph $\mathcal{H}_b$ is conformal.

Berge [11] introduce the notion $k-$conformal hypergraphs which can be seen as a generalization of conformal hypergraphs. A hypergraph $\mathcal{H}$ is said $k-$conformal if and only if for every vertex set $X \subseteq V$ the following conditions are equivalents:

(i) There exists $E \in \mathcal{H}$ such that $X \subseteq E$
(ii) For all $S \subseteq X, \mid S \mid \leq k$, there exists $E \in \mathcal{H}$ such that $S \subseteq E$

Note that (i) implies always (ii)
For example the hypergraph $\mathcal{H}_a$ considered in Figure 1, is 3-conformal. Notice that a conformal hypergraph is 2-conformal.

3 k-Conformal Hypergraph Recognition

In this section, we consider the problem of recognizing if a given hypergraph is k-conformal, and show that is coNP-complete. More formally, this problem is specified as follows:

Problem: CONFORMALITY DEGREE (CD)
Instance: $\mathcal{H} \subseteq 2^V$ a hypergraph on a finite set V, a positive integer $k \leq \mid V \mid$.
Question: Is $\mathcal{H}$ k-conformal?

First, we highlight the link between k-conformality in a hypergraph and keys of a formal context.

Let $\mathbb{K} = (G, M, I)$ be a reduced formal context. Consider the hypergraph $\mathcal{H}_{\mathbb{K}}$ defined on the set M and whose hyperedges are $\{g' \mid g \in G, g'\text{is (inclusion wise)}$ maximal$\}$. We remark that a key of $\mathbb{K}$ is a subset $Q \subseteq M$ such that there is no hyperedge $E \in \mathcal{H}_{\mathbb{K}}$ with $Q \subseteq E$ and it is minimal with this property. Proposition 1 shows that minimal transversals of $\overline{\mathcal{H}_{\mathbb{K}}}$ are exactly keys of $\mathbb{K}$ (see Example 1).

In the remaining of this paper, We will use $\mathbb{K}$ and $\mathcal{H}_{\mathbb{K}}$ interchangeably.

Proposition 1. *[12] Let $\mathbb{K}$ be a formal context. Then $\Sigma(\mathbb{K}) = Tr(\overline{\mathcal{H}_{\mathbb{K}}})$.*

Example 1. Consider the following formal context:

$\mathbb{K}$	1	2	3	4	5
a	1	0	0	0	1
b	0	0	0	1	1
c	0	1	1	0	1
d	1	0	1	1	0
e	0	0	1	0	1

Its associated hypergraph is $\mathcal{H}_{\mathbb{K}} = \{15, 45, 134, 235\}$. From $\mathcal{H}_{\mathbb{K}}$ we can construct the set of all keys of the formal context $\mathbb{K}$: $\Sigma(\mathbb{K}) = Tr(\overline{\mathcal{H}_{\mathbb{K}}}) = \{12, 24, 345, 145\}$

Berge [11] gives a characterization of a $k-$conformal hypergraph using minimal transversal as follows:

Theorem 1. *[11] Let $\mathcal{H}$ be a hypergraph. Then $\overline{\mathcal{H}}$ is $k-$conformal iff for all $T \in Tr(\mathcal{H})$, $| T | \leq k$.*

This useful corollary follows immediately from Proposition 1 and the notion of keys.

Corollary 1. *A hypergraph $\mathcal{H}$ is $k-$conformal if and only if the size of every key of $\mathcal{H}$ is at most k.*

Corollary 1 tells us that the problem CD is strongly connected with the problem of computing the size of the largest key of a hypergraph. This last problem can be stated as follows:

Problem: MAXIMUM KEY (MK)
Instance: $\mathcal{H} \subseteq 2^V$ a hypergraph on a finite set V, a positive integer $k \leq | V |$.
Question: Is there a key Q of $\mathcal{H}$ such that $| Q | \geq k$?

Notice that the problem for computing the minimum size of a key in a hypergraph is NP-complete ([15]: SP8). Surprising, it turns out that the problem of computing the largest key of a given hypergraph is also NP-Complete.

Theorem 2. *Maximum Key is NP-Complete.*

Proof. Let Q be a subset of V. We can check in polynomial time if Q is a key of $\mathcal{H}$ and it's size k. Thus MK is in NP.

To show NP-Hardness, we consider the problem of dominating set which is NP-complete ([15]: GT2). A dominating set in a graph $G = (V, E)$ is a set $D \subset V$ such that for all $x \in V \setminus D$ there exists $y \in D$ for which $xy \in E$. From [15], we learn that the following problem remains NP-Complete if D is required to be both a dominating set and an independent set.

Problem: INDEPENDENT DOMINATING SET (IDS)
Instance: A graph $G = (V, E)$, a positive integer $k \leq | V |$.
Question: Is there an independent dominating set D such that $| D | \leq k$ for G?

Consider a graph $G = (V, E)$ and a positive integer $k \leq |V|$. We construct a hypergraph $\mathcal{H}$ such that D is an independent dominating set of G of size k iff $V \setminus D$ is a key of $\mathcal{H}$ of size $|V| - k$. The vertices of H are elements of V and the set of hyperedges is $\mathcal{H} = \{V \setminus \{u, v\} \mid uv \in E\}$.

A subset $D \subseteq V$ is an independent dominating set of G iff D is a maximal independent set of G iff $V \setminus D$ is a minimal transversal of G. According to the bijection between minimal transversal and keys, we have $V \setminus D$ is a key of $\mathcal{H}$ iff D is an independent dominating set of G.

We conclude that MK is NP-complete, since the construction of $\mathcal{H}$ is polynomial. □

As a consequence, there is no polynomial time algorithm, unless $P = NP$, to compute the largest key of a given hypergraph. Therefore, the problem CD is intractable.

Corollary 2. *Conformal Degree is CoNP-Complete*

4 Decomposition of a Hypergraph into Conformal Hypergraphs

We now turn to the issue of generating all keys of a given hypergraph. We first define an algorithmic framework to generate all keys of a given hypergraph. This framework is based on decomposition of hypergraph into sub-hypergraphs. As example, we propose a decomposition of a given hypergraph into conformal hypergraph.

Let us first recall the following classical operations on two hypergraphs $\mathcal{H} = \{E_1, ..., E_m\}$ and $\mathcal{G} = \{F_1, ..., F_l\}$

- *Union* : $\mathcal{H} \cup \mathcal{G} = \{E_1, ..., E_m, F_1, ..., F_l\}$.
- *Product* : $\mathcal{H} \vee \mathcal{G} = \{E_i \cup F_j \mid i = 1, ...m \text{ and } j = 1, ...l\}$.

The following proposition can be found in [11] in term of minimal transversals.

Proposition 2. *Let $\mathcal{H}$ and $\mathcal{G}$ be simple hypergraphs. Then $\Sigma(\mathcal{H} \cup \mathcal{G})$ is the (inclusionwise) minimal sets of $\Sigma(\mathcal{H}) \vee \Sigma(\mathcal{G})$.*

Proof. Let $Q_1 \in \Sigma(\mathcal{H})$ and $Q_2 \in \Sigma(\mathcal{G})$. Then $Q_1 \cup Q_2$ is a superkey of $\mathcal{H} \cup \mathcal{G}$.

Conversely, consider $Q \in \Sigma(\mathcal{H} \cup \mathcal{G})$. There exist $Q_1 \in \Sigma(\mathcal{H})$ and $Q_2 \in \Sigma(\mathcal{G})$ such that $Q_1 \cup Q_2 \subseteq Q$. Moreover $Q_1 \cup Q_2 = Q$ since $Q_1 \cup Q_2$ is a superkey. □

From Proposition 2 we obtain a general decomposition-based algorithm to compute keys of a given hypergraph.

Algorithm 1. Generating keys of a hypergraph

Data : An hypergraph $\mathcal{H} = \{E_1, ..., E_m\}$ on a set V.

Result: $\Sigma(\mathcal{H})$ the set of all keys of $\mathcal{H}$.

(1) Let $\mathbb{H} = \{\mathcal{H}_1, ..., \mathcal{H}_k\}$ be a decomposition of $\mathcal{H}$.

(2) Compute $\Sigma(\mathcal{H}_i)$ for each $i \in [1, k]$.

(3) Output minimal sets of $\Sigma(\mathcal{H}_1) \vee ... \vee \Sigma(\mathcal{H}_k)$.

The fundamental point of this framework is to use a decomposition in which keys of each component of the decomposition can be computed efficiently. The famous algorithm of Berge [11] which computes minimal transversals can be viewed as an instance of Algorithm 1. It uses a trivial decomposition where each component contains only one hyperedge. In this case we have $\Sigma(\mathcal{H}_i = E_i) = \{v \mid v \notin E_i\}$ as shown in Example 2.

Example 2. Let us illustrate Berge algorihm on $\mathcal{H} = \{123, 124, 34, 25, 45, 135\}$:

$$\Sigma(\mathcal{H}) = \Sigma(123) \vee \Sigma(124) \vee \Sigma(34) \vee \Sigma(25) \vee \Sigma(45) \vee \Sigma(135)$$

$$\Sigma(\mathcal{H}) = \{4, 5\} \vee \{3, 5\} \vee \{1, 2, 5\} \vee \{1, 3, 4\} \vee \{1, 2, 3\} \vee \{2, 4\}$$

$$\Sigma(\mathcal{H}) = \{125, 145, 134, 235, 234, 245, 345\}$$

Now we propose a decomposition technique based on the notion of conformal hypergraph. Given a non conformal hypergraph $\mathcal{H}$, it's always possible to decompose it into a collection $\{\mathcal{H}_1..., \mathcal{H}_k\}$ of conformal hypergraphs. Each component of the collection can be represented by its 2-section graph. And $\mathcal{H}$ can be seen as the union of clique hypergraphs of these graphs.

For example, consider the hypergraph $\mathcal{H} = \{123, 124, 34\}$. Collections $\{\mathcal{H}_1 = \{123, 124\}, \mathcal{H}_2 = \{34\}\}$, $\{\mathcal{H}_1 = \{123, 34\}, \mathcal{H}_2 = \{124\}\}$ and $\{\mathcal{H}_1 = \{124, 34\}, \mathcal{H}_2 = \{123\}\}$ are possible decompositions into conformal hypergraphs.

Remark 1. Although we consider hypergraphs without isolated vertex. It's possible that such vertex appear in sub hypergraphs during decomposition process. Consequently, 2-section graph of a conformal sub-hypergraph can't own all vertices of the given hypergraph.

The problem that find the minimum k such that a such decomposition exists can be stated as follows:

Problem: MINIMUM CONFORMAL DECOMPOSITION (MCD)
Instance: $\mathcal{H} \subseteq 2^V$ a hypergraph on a finite set V, a positive integer $k \leq |\mathcal{H}|$.
Output: Is there a decomposition of $\mathcal{H}$ into k conformal hypergraphs?

The problem MCD is motivated by the following proposition which makes link between decomposition of a hypergraph into conformal hypergraph and k−conformality.

Proposition 3. *Let $\mathcal{H}$ be a hypergraph. If $\mathcal{H}$ can be decomposed in k conformal hypergraphs then $\mathcal{H}$ is at most $2k-$conformal.*

Proof. If a hypergraph $\mathcal{H}$ can be decomposed into k conformal hypergraphs, then each key $Q \in \Sigma(\mathcal{H})$ can be written as a union of k sets of size 2. Thus the size of the largest key is at most $2k$. So $\mathcal{H}$ is at most $2k-$conformal.   □

Example 3. The hypergraph $\mathcal{H} = \{123, 124, 34\}$ can be decomposed in 2 conformal hypergraphs. So, according to Proposition 3, $\mathcal{H}$ is at most 4-conformal. In this case $\mathcal{H}$ is 3-conformal since it's keys are $\Sigma(\mathcal{H}) = \{134, 234\}$.

We have no idea on the difficulty of the problem MCD, but it seems to be intractable. Nevertheless we will give a naive algorithm to compute a decomposition without guarantee of minimality. Theorem 3 gives a characterization of conformal hypergraphs.

Theorem 3. *[14] An hypergraph $\mathcal{H}$ on a set V is conformal if and only if for all $E_1, E_2, E_3 \in \mathcal{H}$ we have $E \in \mathcal{H}$ such that:*

$$(E_1 \cap E_2) \cup (E_1 \cap E_3) \cup (E_2 \cap E_3) \subseteq E$$

Theorem 3 tells us that any hypergraph with at most two hyperedges is conformal. Thus the maximum size of a conformal decomposition is at most $\lceil \frac{|\mathcal{H}|}{2} \rceil$. Note also that Theorem 3 implies a polynomial-time algorithm to test if a given hypergraph is conformal. We can deduce from this the following algorithm which compute a conformal decomposition of a hypergraph.

Algorithm 2. Computing a conformal decomposition

    Data : A hypergraph $\mathcal{H} = \{E_1, ..., E_m\}$.

    Result: $\mathbb{H} = \{\mathcal{H}_1, ..., \mathcal{H}_k\}$ a conformal decomposition of $\mathcal{H}$.

(1) $k = 1$;
    $\mathcal{H}_1 = \{E_1, E_2\}$.;
    $\mathbb{H} = \{\mathcal{H}_1\}$;

(2) Let $3 \leq i \leq m$. If exists $1 \leq j \leq k$ such that $\mathcal{H}_j \cup E_i$ is conformal then add E_i to $\mathcal{H}_j$. Otherwise $k = k + 1$, $\mathcal{H}_k = \{E_i\}$ and add $\mathcal{H}_k$ to $\mathbb{H}$.

Example 4. Consider the hypergraph $\mathcal{H} = \{123, 234, 34, 45\}$ defined on the set $V = \{1, 2, 3, 4, 5\}$. Algorithm 2 produces two conformals hypergraphs $\mathbb{H} = \{\mathcal{H}_1 = \{123, 234\}, \mathcal{H}_2 = \{34, 45\}\}$.

Note that Algorithm 2 does not guaranty a minimal conformal decomposition. Its complexity depends on the cost of testing condition of Theorem 3.

Remark 2. Ordering hyperedges according their size (largest first) seems to be an interesting heuristic

Proposition 4. *Algorithm 2 computes a conformal decomposition using $O(k.|V|.|\mathcal{H}|^4)$ time complexity.*

Proof. The correctness of Algorithm 2 comes directly from Theorem 3. The time complexity depends on the complexity of step (2). In step (2), for each hyperedge E_i, $3 \leq i \leq m$ and each component $\mathcal{H}_j$ in $\mathbb{H}$, we check if $\mathcal{H}_j \cup E_i$ is conformal. Knowing that $\mathcal{H}_j$ is a conformal hypergraph, it remains to test the condition of Theorem 3 on (E, E', E_i) with $E, E' \in \mathcal{H}_j$. This can be done in $O(|V|.|\mathcal{H}|^3)$ for each component. Thus the total complexity of step (2) can by bounded by $O(k.|V|.|\mathcal{H}|^4)$. □

The following proposition shows how to compute keys for a conformal hypergraph.

Proposition 5. *Let $\mathcal{H} \subseteq 2^V$ be a nonempty conformal hypergraph and $G(\mathcal{H}) = (X, E)$ its 2-section graph. We have*

$$\Sigma(\mathcal{H}) = X \setminus V \cup \{xy \in X^2 |\ xy \notin E\}$$

Proof. According Corollary 1, the size of keys of a conformal hypergraph is at most 2. Clearly, vertices in $V \setminus X$ satisfy definition of a key since it can't be contained in a hyperedge. According the definition of 2−section graph, every pair $xy \in E$ is included in a hyperedge of $\mathcal{H}$. Thus other keys are exactly non edges in $G(\mathcal{H})$. □

Theorem 4. *Let $\mathcal{H} \subseteq 2^V$ be a hypergraph and $\mathbb{H} = \{\mathcal{H}_1, ..., \mathcal{H}_k\}$ a conformal decomposition of $\mathcal{H}$. Then Algorithm 1 computes $\Sigma(\mathcal{H})$ using $O(|V|^{4k})$ time complexity.*

Proof. Correctness follows directly from Propositions 2 and 5. It remains to show time complexity. In Step (2), each graph can have at most $|V|^2$ pairs. The size of the product is at most $|V|^{2k}$. Thus computing minimal sets of the product can be done in $O(|V|^{4k})$ time complexity. □

A Complete Running Example

Consider the following formal context.

$\mathbb{K}$	1	2	3	4	5
a	1	1	1	0	0
b	1	1	0	1	0
c	0	0	1	1	0
d	0	1	0	0	1
e	0	0	0	1	1
f	1	0	1	0	1

The associated hypergraph is $\mathcal{H}_{\mathbb{K}} = \{123, 124, 34, 25, 45, 135\}$ defined on the set $V = \{1, 2, 3, 4, 5\}$. Algorithm 2 produces a collection of 3 conformal hypergraphs : $\mathcal{H}_1 = \{123, 124, 25\}$, $\mathcal{H}_2 = \{34, 45\}$ and $\mathcal{H}_3 = \{135\}$. Their 2-section graphs are shown in Figure 2.

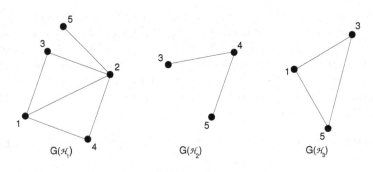

Fig. 2.

Step (2) of Algorithm 1 produces keys of each component in this collection: $\Sigma(\mathcal{H}_1) = \{15, 34, 35, 45\}$, $\Sigma(\mathcal{H}_2) = \{1, 2, 35\}$ and $\Sigma(\mathcal{H}_3) = \{2, 4\}$. The minimal sets of the product $\Sigma(\mathcal{H}_1) \vee \Sigma(\mathcal{H}_2) \vee \Sigma(\mathcal{H}_3)$ are $\Sigma(\mathcal{H}_{\mathbb{K}}) = \{125, 145, 134, 235, 234, 245, 345\}$ which correspond to keys of $\mathbb{K}$.

5   Conclusion

In this paper we have established a link between the FCA and the hypergraphs domain. In particular, we have pointed out equivalence between $k-$conformal hypergraphs and the largest size of a key in a formal context. We have also defined a decomposition into conformal hypergraphs which generalize the Berge algorithm to generate keys of a formal context (or a hypergraph).

Acknowledgment

We are grateful to the anonymous reviewers for their useful remarks and suggestions.

References

1. Eiter, T., Gottlob, G.: Identifying the minimal transversals of a hypergraph and related problems. SIAM J. Comput. 24(6), 1278–1304 (1995)
2. Eiter, T., Gottlob, G.: Hypergraph transversal computation and related problems in logic and ai. In: JELIA, pp. 549–564 (2002)
3. Khardon, R.: Translating between horn representations and their characteristic models. J. Artif. Intell. Res (JAIR) 3, 349–372 (1995)

4. Fredman, M.L., Khachiyan, L.: On the complexity of dualization of monotone disjunctive normal forms. J. Algorithms 21(3), 618–628 (1996)
5. Boros, E., et al.: Generating maximal independent sets for hypergraphs with bounded edge-intersections. In: Farach-Colton, M. (ed.) LATIN 2004. LNCS, vol. 2976, pp. 488–498. Springer, Heidelberg (2004)
6. Boros, E., et al.: An efficient incremental algorithm for generating all maximal independent sets in hypergraphs of bounded dimension. Parallel Processing Letters 10(4), 253–266 (2000)
7. Boros, E., et al.: Computing Many Maximal. Independent Sets for Sparse. Hypergraphs in Parallel. Technical report, RUTCOR (October 2004)
8. Gunopulos, D., et al.: Data mining, hypergraph transversals, and machine learning. In: Proceedings of the Sixteenth ACM SIGACT-SIGMOD-SIGART Symposium on Principles of Database Systems, Tucson, Arizona, May 12-14, 1997, pp. 209–216. ACM Press, New York (1997)
9. Khachiyan, L., et al.: A new algorithm for the hypergraph transversal problem. In: Wang, L. (ed.) COCOON 2005. LNCS, vol. 3595, pp. 767–776. Springer, Heidelberg (2005)
10. Ganter, B., Wille, R.: Formal Concept Analysis. In: Mathematical Foundation, Springer, Heidelberg (1999)
11. Berge, C.: Hypergraphes, Combinatoires des ensembles finis. Bordas (1987), Number ISBN: 5-04-016906-7
12. Demetrovics, J., Thi, V.D.: Describing candidate keys by hypergraphs. Computers and artificial intelligence 18(2), 191–207 (1999)
13. Edmonds, J., Fulkerson, D.: Bottleneck Extrema. Journal of Combinatorial Theory 8, 299–306 (1970)
14. Gilmore, P.: Families of sets with faithful graph representation. IBM Research Note N.C. 184, Thomas J. Watson Research Center, YorkTown Heights, New York (1962)
15. Garey, M.R., Johnson, D.S.: Computers and Intractability, A Guide to the Theory of NP-Completeness. W.H. Freeman and Company, New York (1979)

An Algebraization of
Linear Continuum Structures

Rudolf Wille

Technische Universität Darmstadt, Fachbereich Mathematik,
Schloßgartenstr. 7, D–64289 Darmstadt,
wille@mathematik.tu-darmstadt.de

Abstract. This paper continuous the approach of developing an order-theoretic structure theory of *one-dimensional continuum structures* as elaborated in [Wi07] (see also [Wi83],[Wi03]). The aim is to extend the order-theoretic structure theory by a meaningful algebraization; for this, we concentrate on the real linear continuum structure with its derived concept lattice which gives rise to the so-called *"real half-numbers"*. The algebraization approaches an ordered algebraic structure on the set of all real half-numbers to make the continuum structure of the reals more transparent and tractable.

Contents

1   Introduction

It is still an open question whether a *continuum* should consist of points or not. This paper is based on the view that, on the phenomenological level, a continuum does not contain points, but, on the logical level, it might be useful to construct points as limits of convergence processes. Aristotle made this already clear by analysing *Zenon's paradox of the flying arrow*: it says that the flying arrow is in each moment in some place, hence does not change the place in any moment, and therefore does not move. Aristotle has convincingly analysed this paradox by making clear "that the movement is not performed in a 'now', but in some time; time however does not consist of nows, but of durations" ([We72], p.431). Aristotle understood time and durations as continua which means according to his continuum definition that they "are unlimitedly divisible into smaller parts" ([We72], p.431). Therefore, for Aristotle, durations do not consist of time points, but *time points are only limits of durations*. Aristotle's conception of the time and space continuum yields in general that a continuum does not consist of

R. Medina and S. Obiedkov (Eds.): ICFCA 2008, LNAI 4933, pp. 150–157, 2008.

points, but has as parts only continua again whose nature is to be extensive. In contrast to that, points are in principle of different nature: they are not extensive and can only be understood as limits of extensives.

2   Linear Continuum Structures

First we recall the mathematical analysis of the one-dimensional continuum of an *unlimited straight line* which is based by the following definition cited from [Wi07]:

Definition 1. *A* linear continuum structure *is defined as an ordered set* $\underline{C} :=$ $(C, \leq)$ *satisfying the following conditions:*
(1) $\underline{C}$ *is a* $\bigvee$*-semilattice with greatest element 1 and without smallest element;*
(2) *the* $\wedge$*-irreducible elements of* $\underline{C}$ *form two disjoint dense chains* $C^{\dashv}$ *and* $C^{\vdash}$
without greatest and smallest element, where $c^{\dashv} \vee c^{\vdash} = 1$ *for all* $c^{\dashv} \in C^{\dashv}$
and $c^{\vdash} \in C^{\vdash}$*;*
(3) $c_1 \wedge c_2 = d_1 \wedge d_2$ *implies* $c_1 = d_1$, $c_2 = d_2$ *for all* $c_1, d_1 \in C^{\dashv}$, $c_2, d_2 \in C^{\vdash}$*;*
(4) *there exists an antiisomorphism* $c^{\dashv} \mapsto c^{\vdash}$ *from* $C^{\dashv}$ *onto* $C^{\vdash}$ *such that*
$$C = \{1\} \cup C^{\dashv} \cup C^{\vdash} \cup \{c^{\dashv} \wedge d \mid c^{\dashv} \in C^{\dashv} \text{ and } d \in C^{\vdash} \text{ with } c^{\vdash} < d\}.$$
The elements of $\underline{C}$ *are called* (linear) continua, *and the pairs* $(c^{\dashv}, c^{\vdash})$ *of corresponding elements* $c^{\dashv} \in C^{\dashv}$ *and* $c^{\vdash} \in C^{\vdash}$ *are called the* cuts *of* $\underline{C}$*. Notice that no pair consisting of* $c^{\dashv}$ *and* $c^{\vdash}$ *has a lower bound in the ordered set* $\underline{C}$*.*   ◊

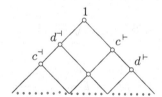

Fig. 1. A linear continuum represented by an ordered set

Fig. 1 visualizes a linear continuum structure $\underline{C} := (C, \leq)$. In this figure the phenemonological conception of a linear continuum is indicated by the horizontal sequence of dots on the bottom. Each element of $\underline{C}$ stands for a subcontinuum of the phenemonological continuum; for instance, the element $d^{\dashv} \wedge c$ stands for the subcontinuum indicated by the dots below the two line segments drawn downwards from the circle representing $d^{\dashv} \wedge c$. The elements $c^{\dashv}$ and $c^{\vdash}$ stand for two subcontinua which are disjoint, but cover together the total continuum coded by 1; hence the pair $(c^{\dashv}, c^{\vdash})$ indicates a division of a one-dimensional phenomenological continuum into two subcontinua.

In this paper we concentrate on the *real linear continuum structure* introduced in [Wi07] as follows: The ordered field $\mathbb{R}$ of real numbers gives rise to the *linear continuum structure* $\underline{C_{\mathbb{R}}} := (C_{\mathbb{R}}, \subseteq)$ for which the set $C_{\mathbb{R}}$ consists of the open intervals $]r, s[:= \{x \in \mathbb{R} \mid r < x < s\}$ with $r \in \mathbb{R} \cup \{-\infty\}$ and $s \in \mathbb{R} \cup \{+\infty\}$.

In this continuum structure the convex hull of the set-theoretic union is the supremum and $]-\infty, +\infty[$ is the greatest element; the $\wedge$-irreducible elements form the dense chains $C^{\dashv} := \{]-\infty, s[\mid s \in \mathbb{R}\}$ and $C^{\vdash} := \{]r, +\infty[\mid r \in \mathbb{R}\}$; the accompanying antiisomorphism between $C^{\dashv}$ and $C^{\vdash}$ can be described by $]-\infty, r[\mapsto]r, +\infty[$. The cuts of $\underline{C}_{\mathbb{R}}$ are therefore the pairs $(]-\infty, r[,]r, +\infty[)$ $(r \in \mathbb{R})$.

3   Concept Lattices Derived from Ordered Sets

Formal Concept Analysis [GW99] can be activated to make mathematically explicit that points, according to Aristotle [Ar95], can be understood as limits of continua; such limits cannot be parts of continua. The extension by points will be performed by using a general method of Formal Concept Analysis which mathematically establishes the transfer from ideas to concepts in the sense of the *structure-genetic psychology* of Jean Piaget [Pi59]. The mathematization of this transfer is grounded on ordered sets $\underline{C} := (C, \leq)$ of preconceptual 'ideas' (cf. [SW86]). For analysing the general method we have to refer to the Basic Theorem of Concept Lattices which therefore shall be recalled here (for the proof see [GW99]):

Basic Theorem on Concept Lattices. [Wi82] *Let* $\mathbb{K} := (G, M, I)$ *be a formal context. Then* $\underline{\mathfrak{B}}(\mathbb{K})$ *is a complete lattice, called the* concept lattice *of* $\mathbb{K}$, *whose infima and suprema can be described as follows:*

$$\bigwedge_{t \in T}(A_t, B_t) = (\bigcap_{t \in T} A_t, (\bigcup_{t \in T} B_t)^{II}), \quad \bigvee_{t \in T}(A_t, B_t) = ((\bigcup_{t \in T} A_t)^{II}, \bigcap_{t \in T} B_t).$$

In general a complete lattice L *is isomorphic to* $\underline{\mathfrak{B}}(\mathbb{K})$ *if and only if there exist mappings* $\tilde{\gamma} : G \to L$ *and* $\tilde{\mu} : M \to L$ *such that* $\tilde{\gamma}G$ *is* $\bigvee$-*dense in* L *(i.e.* $L = \{\bigvee X \mid X \subseteq \tilde{\gamma}G\}$), $\tilde{\mu}M$ *is* $\bigwedge$-*dense in* L *(i.e.* $L = \{\bigwedge X \mid X \subseteq \tilde{\mu}M\}$), *and* $gIm \iff \tilde{\gamma}g \leq \tilde{\mu}m$ *for* $g \in G$ *and* $m \in M$; *in particular,* $L \cong \underline{\mathfrak{B}}(L, L, \leq)$ *and, if the set* $J(L)$ *of all* $\bigvee$-*irreducible elements is* $\bigvee$-*dense in* L *and the set* $M(L)$ *of all* $\bigwedge$-*irreducible elements is* $\bigwedge$-*dense, then we have that* $L \cong \underline{\mathfrak{B}}(J(L), M(L), \leq)$.

The process of concept building models formal objects by filters of an arbitrary ordered set $\underline{C}$ and formal attributes by ideals of $\underline{C}$ (cf. [SW86]). A *filter of* $\underline{C}$ is a non-empty subset F of C, for which $a \in F$ and $a \leq b$ imply $b \in F$ and $a, c \in F$ guarantees the existence of some $d \in F$ with $d \leq a, c$; an *ideal of* $\underline{C}$ is dually defined to the filter[1]. This modelling leads to the derived context $\mathbb{K}(\underline{C}) := (\mathfrak{F}(\underline{C}), \mathfrak{I}(\underline{C}), \Delta)$ for which $\mathfrak{F}(\underline{C})$ is the set of all non-empty filters F of $\underline{C}$ and $\mathfrak{I}(\underline{C})$ is the set of all non-empty ideals I of $\underline{C}$ with $F \Delta I : \iff F \cap I \neq \emptyset$; hence a filter as 'object' has an ideal as 'attribute' if and only if filter and ideal have at least one idea in common. Important are the *ideal-maximal filters* F in $\mathfrak{F}(\underline{C})$ for which an ideal I exists in $\mathfrak{I}(\underline{C})$ so that F is a maximal filter having the

---

[1] Filters and ideals represent dual processes of convergence of ordered ideas.

property $F \cap I = \emptyset$; F is named an *I-maximal filter* and, furthermore, if I is a maximal ideal with $F \cap I = \emptyset$ then I is called an *F-opposite*. As dual notions we have *filter-maximal ideals*, *F-maximal ideals*, and *I-opposites*. The set of all ideal-maximal filters is denoted by $\mathfrak{F}_0(\underline{C})$ and the set of all filter-maximal ideals is denoted by $\mathfrak{I}_0(\underline{C})$. The following theorem informs about meaningful structural properties of the concept lattice of $\mathbb{K}(\underline{C})$ (for the proof see [Wi07]):

Theorem 1. The ordered set $\underline{C}$ of ideas is naturally embedded by the map $\iota : x \mapsto (\{F \in \mathfrak{F}(\underline{C}) \mid x \in F\}, \{I \in \mathfrak{I}(\underline{C}) \mid x \in I\})$ into the concept lattice of the derived context $\mathbb{K}(\underline{C})$ where $\iota(x \wedge y) = \iota(x) \wedge \iota(y)$ resp. $\iota(x \vee y) = \iota(x) \vee \iota(y)$ if $x \wedge y$ resp. $x \vee y$ exists in $\underline{C}$; in $\mathfrak{B}(\mathbb{K}(\underline{C}))$, the set of all $\bigvee$-irreducibles $J(\mathfrak{B}(\mathbb{K}(\underline{C})))(= \gamma \mathfrak{F}_0(\underline{C}))$ is $\bigvee$-dense and the set of all $\bigwedge$-irreducibles $M(\mathfrak{B}(\mathbb{K}(\underline{C})))(= \mu \mathfrak{I}_0(\underline{C}))$ is $\bigwedge$-dense, i.e., $\mathfrak{B}(\mathbb{K}(\underline{C})) \cong \mathfrak{B}(\mathfrak{F}_0(\underline{C}), \mathfrak{I}_0(\underline{C}), \Delta)$.

Theorem 1 yields a *general method* to derive from an ordered set of preconceptual ideas a concept lattice in which every formal concept is the supremum of $\bigvee$-irreducible concepts and the infimum of $\bigwedge$-irreducible concepts. For applying Theorem 1 to *linear continuum structures*, the ideal-maximal filters and filter-maximal ideals of those continuum structures are determined by the following lemma (for the proof see [Wi07]).

Lemma 1. Let $\underline{C}$ be a linear continuum structure. In the ordered set $\underline{C}$, $F^{\dashv} := C^{\dashv} \cup \{1\}$ and $F^{\vdash} := C^{\vdash} \cup \{1\}$ are the 'extreme' ideal-maximal filters and $I^{\dashv} := \{x \in C \mid x \leq c \text{ for some } c \in C^{\dashv}\}$ and $I^{\vdash} := \{x \in C \mid x \leq c \text{ for some } c \in C^{\vdash}\}$ are the 'extreme' filter-maximal ideals. The cuts $(c^{\dashv}, c^{\vdash})$ of $\underline{C}$ supply the other ideal-maximal filters of $\underline{C}$ by

$$F_{c^{\dashv}} := \{x \in C \mid x \geq c^{\dashv} \wedge d \text{ for some } d \in C^{\vdash} \text{ with } c^{\vdash} < d\},$$
$$F_{c^{\vdash}} := \{y \in C \mid y \geq c^{\vdash} \wedge d \text{ for some } d \in C^{\dashv} \text{ with } c^{\dashv} < d\},$$

and the other filter-maximal ideals of $\underline{C}$ by

$$I_{(c^{\dashv})} := \{x \in C \mid x \leq \bar{x} < c^{\dashv} \text{ for some } \bar{x} \in C^{\dashv}\} \quad \text{and} \quad I_{(c^{\vdash}]} := \{y \in C \mid y \leq c^{\vdash}\},$$
$$I_{(c^{\vdash})} := \{y \in C \mid y \leq \bar{y} < c^{\vdash} \text{ for some } \bar{y} \in C^{\vdash}\} \quad \text{and} \quad I_{(c^{\dashv}]} := \{x \in C \mid x \leq c^{\dashv}\}.$$

Theorem 2. In the concept lattice of the formal context $\mathbb{K}(\underline{C}) := (\mathfrak{F}(\underline{C}), \mathfrak{I}(\underline{C}), \Delta)$ of a linear continuum structure $\underline{C}$,
(1) $\iota(1)(= \gamma F^{\dashv} \vee \gamma F^{\vdash})$ is the greatest element of $\mathbb{K}(\underline{C})$,
(2) $\gamma \mathfrak{F}_0(\underline{C})$ is the set of all atoms and is the disjoint union of the sets
$A_1 := \{\gamma F^{\dashv}\} \cup \{\gamma F_{c^{\vdash}} \mid c^{\vdash} \in C^{\vdash}\}$ and $A_2 := \{\gamma F^{\vdash}\} \cup \{\gamma F_{c^{\dashv}} \mid c^{\dashv} \in C^{\dashv}\}$,
(3) $\mu \mathfrak{I}_0(\underline{C})(= \{\gamma F^{\dashv} \vee \gamma F \mid F \in A_1\} \cup \{\gamma F^{\vdash} \vee \gamma F \mid F \in A_2\})$ is the set of all $\bigwedge$-irreducible elements and is the disjoint union of the convex chains $[\gamma F^{\dashv}, \iota(1)[$ and $[\gamma F^{\vdash}, \iota(1)[$,
(4) for each cut $(c^{\dashv}, c^{\vdash})$, we have $\gamma F^{\dashv} \vee \gamma F_{c^{\dashv}} = \mu I_{(c^{\dashv}]}$ and $\gamma F^{\vdash} \vee \gamma F_{c^{\vdash}} = \mu I_{(c^{\vdash}]}$, $\mu I_{(c^{\dashv}]}$ is a lower neighbour of $\mu I_{(c^{\dashv}]} \vee \gamma F_{c^{\vdash}}$ and an upper neighbour of $\mu I_{(c^{\dashv})}$, $\mu I_{(c^{\vdash}]}$ is a lower neighbour of $\mu I_{(c^{\vdash}]} \vee \gamma F_{c^{\dashv}}$ and an upper neighbour of $\mu I_{(c^{\vdash})}$,
(5) for $x = d^{\dashv} \wedge c^{\vdash}$, we have $\iota(x) := (\{F \in \mathfrak{F}(\underline{C}) \mid x \in F\}, \{I \in \mathfrak{I}(\underline{C}) \mid x \in I\})$
$= \gamma F_{c^{\vdash}} \vee \gamma F_{d^{\dashv}} = \mu I_{(d^{\dashv}]} \wedge \mu I_{(c^{\vdash}]}$.

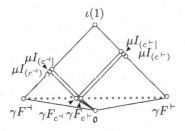

Fig. 2. A concept lattice derived from a linear continuum structure

Since the linear continuum structure $\underline{C}$ is *embeddable* into the concept lattice $\mathfrak{B}(\mathbb{K}(\underline{C}))$ by Theorem 1, that concept lattice yields for the linear continuum structure an extended conceptual coherence which is made explicit in several aspects by Theorem 2. Most important for the theme of this paper is that the atoms of $\mathfrak{B}(\mathbb{K}(\underline{C}))$ represent '*point concepts*', which have every other formal concept as supremum by (2). By (4), each cut of the linear continuum structure gives rise to two cut-limiting point concepts the supremum of which may be viewed as a point in the common sense. (5) desribes how point concepts can be represented as limits of continua. These hints shall suffice to demonstrate the fruitfulness of the concept-analytic method and support Aristotle's conception of continua. The concept-analytic results of Theorem 2 shall be concretized by the real linear continuum:

The *real linear continuum structure* $\underline{C}_{\mathbb{R}}$ described on page 2 can be embedded by the mapping ι into the concept lattice of the formal context $\mathbb{K}(\underline{C}_{\mathbb{R}})$ by Theorem 1. This can be illustrated by a linear ordered set $(\check{\mathbb{R}}, \leq)$ extending $(\mathbb{R}, \leq)$. $(\check{\mathbb{R}}, \leq)$ is defined by

$$\check{\mathbb{R}} := \mathbb{R} \cup \overline{\mathbb{R}} \cup \{\underline{-\infty}, \overline{+\infty}\} \text{ with } \underline{\mathbb{R}} := \{\, \underline{r} \mid r \in \mathbb{R}\}, \overline{\mathbb{R}} := \{\overline{r} \mid r \in \mathbb{R}\}, \text{ and}$$
$$\underline{r} \leq \underline{s} :\Leftrightarrow r \leq s, \quad \overline{r} \leq \overline{s} :\Leftrightarrow r \leq s, \quad \underline{r} \leq \overline{s} :\Leftrightarrow r \leq s, \ \overline{r} < \underline{s} :\Leftrightarrow r < s, \text{ and}$$
$$\underline{-\infty} < \underline{r} < \overline{+\infty} \text{ and } \underline{-\infty} < \overline{r} < \overline{+\infty} \text{ for all } r \in \mathbb{R}.$$

The linear ordered set $(\check{\mathbb{R}} \setminus \{\underline{-\infty}, \overline{+\infty}\}, \leq)$ clearly evolves out of $(\mathbb{R}, \leq)$ by dividing each real number r into the two elements $\underline{r} < \overline{r}$. $\check{\mathbb{R}}$ is bijectively mapped onto the set of all atoms of $\mathfrak{B}(\mathbb{K}(\underline{C}_{\mathbb{R}}))$ by the mapping α with $\alpha(\underline{-\infty}) := \gamma F^{\dashv}$, $\alpha(\underline{r}) := \gamma F_{]-\infty,r[}$, $\alpha(\overline{r}) := \gamma F_{]r,+\infty[}$, and $\alpha(\overline{+\infty}) := \gamma F^{\vdash}$. Simplifying definitions are $-\infty := \gamma F^{\dashv}$, $\underline{r} := \gamma F_{]-\infty,r[}$, $\overline{r} := \gamma F_{]r,+\infty[}$ and $+\infty := \gamma F^{\vdash}$. The linear order of $(\check{\mathbb{R}}, \leq)$ is transferred onto the set of all atoms of $\mathfrak{B}(\mathbb{K}(\underline{C}_{\mathbb{R}}))$; according to this linear order $\leq$ on $\mathfrak{B}(\mathbb{K}(\underline{C}_{\mathbb{R}}))$, $-\infty$ is the smallest atom, $+\infty$ is the greatest atom, and $\underline{r} < \overline{r} < \underline{s} < \overline{s}$ if $r < s$ in $\mathbb{R}$. The continua of the real linear continuum structure $\underline{C}_{\mathbb{R}}$ are represented in the concept lattice by the formal concepts $\iota(]r, s[)$, respectively. By Theorem 2(5), we have $\iota(]r, s[) = \overline{r} \vee \underline{s}$ which shows that the atoms below $\iota(]r, s[)$ are exactly the atoms $\mathfrak{a}$ with $\overline{r} \leq \mathfrak{a} \leq \underline{s}$; therefore it is meaningful to say that the point concepts $\overline{r}$ and $\underline{s}$ are the limits of the continuum concept $\iota(]r, s[)$. The cuts of the real linear continuum structure are represented in the concept lattice by the pairs $(\underline{r}, \overline{r})$; in this conceptual connection $\underline{r}$ and $\overline{r}$ are standing for the two irreducible subpoints of the reducible real point which is represented by the formal concept $\underline{r} \vee \overline{r}$.

4   An Algebraization by Real Half-Numbers

It is, of course, interesting to find answers to the question whether the irreducible subpoints of the real linear continuum structure may carry a meaningful algebraic structure. For approaching such an answer, the irreducible subpoints, which have been created by the maximal filters of the real linear continuum structure, shall be understood as the *"real half-numbers"*. They divide into three sets: the set $\underline{\mathbb{R}}$ of all *"real lower half-numbers"*, the set $\overline{\mathbb{R}}$ of all *"real upper half-numbers"*, and the set $\{-\infty, +\infty\}$ of the *lower* and *upper infinite numbers*, respectively, which bound the real half-numbers from below and above.

Now, an algebraization shall be proposed on the set $\overline{\underline{\mathbb{R}}} := \underline{\mathbb{R}} \cup \overline{\mathbb{R}}$ of all real half-numbers where the real lower half-numbers of $\underline{\mathbb{R}}$ are generally denoted by $\underline{x}$ and the real upper half-numbers of $\overline{\mathbb{R}}$ are generally denoted by $\overline{y}$. First a *binary addition* is defined as follows:

$$\underline{r} + \underline{s} = \underline{r+s}, \quad \overline{r} + \overline{s} = \overline{r+s}, \quad \underline{r} + \overline{s} = \overline{r+s}, \quad \text{and} \quad \overline{r} + \underline{s} = \overline{r+s};$$

furthermore, a unary operation is formed by two *inversions* $\underline{\overline{\rho}}_0$ which are defined by

$$\underline{\rho}_0(\underline{r}) := \overline{-r} \quad \text{and} \quad \overline{\rho}_0(\overline{r}) := \underline{-r}.$$

The mapping $\pi : (\overline{\underline{\mathbb{R}}}, +) \longrightarrow (\mathbb{R}, +)$ with $\pi(\underline{r}) = r$ and $\pi(\overline{r}) = r$ for all $r \in \mathbb{R}$ is a surjective homomorphism. Conversely, we have two injective homomorphisms $\underline{\iota} : (\mathbb{R}, +) \longrightarrow (\overline{\underline{\mathbb{R}}}, +)$ with $\underline{\iota}(\underline{r}) = \underline{r}$ and $\overline{\iota} : (\mathbb{R}, +) \longrightarrow (\overline{\underline{\mathbb{R}}}, +)$ with $\overline{\iota}(\overline{r}) = \overline{r}$. Of course, $(\underline{\mathbb{R}}, +)$ and $(\overline{\mathbb{R}}, +)$ are isomorphic to $(\mathbb{R}, +)$ which becomes explicit by the isomorphisms $\pi\underline{\iota}$ and $\pi\overline{\iota}$. The isomorphisms $\underline{\pi} : (\underline{\mathbb{R}}, +) \longrightarrow (\mathbb{R}, +)$ with $\underline{\pi}(\underline{r}) = r$ and $\overline{\pi} : (\overline{\mathbb{R}}, +) \longrightarrow (\mathbb{R}, +)$ with $\overline{\pi}(\underline{r}) = r$ are called *"lower projection"* and *"upper projection"*, respectively.

The inversion $\underline{\rho}_0 : (\overline{\underline{\mathbb{R}}}, +, \underline{\rho}_0, \leq) \longrightarrow (\overline{\underline{\mathbb{R}}}, +, \overline{\rho}_0, \leq)$ with $\underline{\rho}_0(\underline{r}) = \overline{-r}$ and the inversion $\overline{\rho}_0 : (\overline{\underline{\mathbb{R}}}, +, \overline{\rho}_0, \leq) \longrightarrow (\overline{\underline{\mathbb{R}}}, +, \underline{\rho}_0, \leq)$ with $\overline{\rho}_0(\overline{r}) = \underline{-r}$ combine to an antiisomorphism with respect to the order $\leq$. An analogous antiautomorphism can be defined for every fixed real number x:

$$\underline{\rho}_x(\underline{r}) := \overline{2x - r} \quad \text{and} \quad \overline{\rho}_x(\overline{r}) := \underline{2x - r}$$

Now, a meaningful algebraization of the real linear continuum structures can be established. First we define the binary addition of real linear continuum substructures which are bounded by the pairs of real half-numbers $\underline{r1} < \overline{r2}$ and $\underline{s1} < \overline{s2}$, respectively:

$$[\underline{r1}, \overline{r2}] + [\underline{s1}, \overline{r2}] := [\underline{r1+s1}, \overline{r2+s2}]$$

Notice, if $\underline{r1} = \underline{0} = \underline{s1}$, the addition yields just the sum of the positive lengths of the two continua and, if $\overline{r2} = \overline{0} = \overline{s2}$, the addition yields just the sum of the negative length of the two continua. Another special case is when $r2 = (\overline{r2} \vee \underline{s1}) = s1$ for the real numbers $r2$ and $s1$, then the sum of the lengths of the two continua equals the length of the subcontinuum $[\underline{r1}, \overline{s2}]$. If $\overline{r2} < \underline{s1}$, then the length of the subcontinuum $[\underline{r1}, \overline{s2}]$ is the sum of the lengths of the three disjoint continua $[\underline{r1}, \overline{r2}]$, $[\underline{r2}, \overline{s1}]$, and $[\underline{s1}, \overline{r2}]$.

Secondly, the unary inversions ρ_x and $\overline{\rho}_x$ can be applied to the real linear continuum structures as follows:

$$[\rho_x(\underline{r1}), \overline{\rho_x}(\overline{r2})] = [\underline{2x - r2}, \overline{2x - r1}]$$

The inversions ρ_x and $\overline{\rho}_x$ reflect the real half-numbers on the real number x where lower half-numbers are mapped onto upper half-numbers and upper half numbers are mapped onto lower half-numbers.

The started algebraization of real linear continuum structures shall be interrupted here, but should be continued more deeply in the near future. The aim of this paper was to show how an algebraization of continuum structures could be approached.

5   Further Research

The natural continuation of the work on continuum structures would be to elaborate first an order-theoretic and then an algebraic approach to mathematize higher dimensional continuum structures. The basic task of developing such an approach is to find adequate order-theoretic and algebraic models in higher dimensions. Already for the 2-dimensional case those continuum structures are not known yet, which indicates that new ideas are necessary. There is some hope that the results about the one-dimensional continuum structures can be used and generalized for building higher dimensional order-theoretic and algebraic models. A promising idea is to inductively construct those models top-down by starting with a network of infinitely many one-dimensional continuum structures.

References

[Ar95]   Aristoteles Werke in deutscher Übersetzung. Bd. 11: Physikvorlesung. 5. Aufl. Akademie Verlag, Berlin (1995)

[GW99]   Ganter, B., Wille, R.: Formal Concept Analysis: Mathematical Foundations. Springer, Heidelberg (1999)

[Pi59]   Piaget, J.: La formation du symbole chez l'enfant-imitation, jeu et rêve - Image et représentation. Delachaux et Niestlé S.A., Neuchâtel (1959)

[SW86]   Stahl, J., Wille, R.: Preconcepts and set representations of contexts. In: Gaul, W., Schader, M. (eds.) Classification as a tool of research, pp. 431–438. North-Holland, Amsterdam (1986)

[We72]   von Weizsäcker, C.F.: Möglichkeit und Bewegung. Eine Notiz zur aristotelischen Physik. In: von Weizsäcker, C.F., Die Einheit der Natur. 3. Aufl., pp. 428–440. Hanser, München (1972)

[Wi82]   Wille, R.: Restructuring lattice theory: an approach based on hierarchies of concepts. In: I. Rival (ed.) Ordered sets, pp. 445–470 Reidel, Dordrecht-Boston (1982)

[Wi83]   Wille, R.: Zur Ordnung von Zeit und Raum - eine Untersuchung im Rahmen der formalen Begriffsanalyse. Vortrag auf der Jahrestagung der Deutschen Mathematikervereinigung, Köln (1983)

[Wi03]   Wille, R.: Sind unsere Vorstellungen von Raum und Zeit richtig? oder:
          Besteht ein Kontinuum aus Punkten? In: Hefendehl-Hebeker, L., Hußmann,
          S., (Hrsg.): Mathematikdidaktik: Zwischen Fachorientierung und Empirie.
          Franzbecker Verlag, Hildesheim, pp. 266–279 (2003)
[Wi07]   Wille, R.: Formal Concept Analysis of one-dimensional continuum struc-
          tures. Algebra Universalis (to appear)

On the Complexity of Computing
Generators of Closed Sets

Miki Hermann[1] and Barış Sertkaya[2]

[1] LIX (CNRS, UMR 7161), École Polytechnique, 91128 Palaiseau, France
hermann@lix.polytechnique.fr
[2] Institut für Theoretische Informatik, TU Dresden, Germany
sertkaya@tcs.inf.tu-dresden.de

Abstract. We investigate the computational complexity of some decision and counting problems related to generators of closed sets fundamental in Formal Concept Analysis. We recall results from the literature about the problem of checking the existence of a generator with a specified cardinality, and about the problem of determining the number of minimal generators. Moreover, we show that the problem of counting minimum cardinality generators is #·coNP-complete. We also present an incremental-polynomial time algorithm from relational database theory that can be used for computing all minimal generators of an implication-closed set.

1 Introduction

Closed sets and pseudo-closed sets play an important rôle in Formal Concept Analysis (FCA) [5]. For instance, the sets closed under implications are fundamental to the attribute exploration algorithm [4]. In addition, pseudo-closed sets form the left-hand sides of the implications in the canonical implication base called the Duquenne-Guigues Base [7] of a formal context. As a result, many problems related to closed and pseudo-closed sets have been by now well investigated in the FCA community. For instance, there exist several polynomial-delay algorithms[1] that generate all concept intents of a formal context. Other computational problems related to pseudo-closed sets have been analyzed in [11,12,14].

Beside closed and pseudo-closed sets, generators of closed sets also play an important rôle in FCA. Inspite of this, as mentioned in [17], they have been paid little attention in the FCA community, especially computational problems related to them have not been well investigated. Different aspects of minimal generators have been investigated in the literature [23,17,3]. Valtchev et al. presented in [23] an efficient method for maintaining the set of minimal generators of all intents of a formal context upon increases in the object set of the underlying context. Nehmé et al. investigated in [17] the same problem in the dual setting. They presented a method for maintaining the set of minimal generators upon increases in the attribute set of the context. They characterized how the

---

[1] See [13] for a comprehensive list and a detailed comparison of these algorithms.

R. Medina and S. Obiedkov (Eds.): ICFCA 2008, LNAI 4933, pp. 158–168, 2008.
© Springer-Verlag Berlin Heidelberg 2008

set of minimal generators changes when a new attribute is added to the context. Using this characterization they developed an efficient incremental algorithm for generating concept intents. Frambourg et al. worked in [3] on evolution of the the set of minimal generators during lattice assembly.

The present paper aims to given an overview of the computational complexity of some decision and counting problems on generators of closed sets. In particular we consider the two types of closed sets that are fundamental in FCA, namely concept intents and sets closed under a set of implications. Throughout the text, for the latter type of sets, we use the term implication-closed set. We recall results from the literature about the problem of checking the existence of a generator with a specified cardinality, and about the problem of determining the number of minimal generators. Moreover, we define a new problem about the second type of closed sets, namely the problem of determining the number of minimum cardinality generators, and show that this problem is #·coNP-complete, i.e., it is even more difficult than determining the number of minimal generators. We also point out that an incremental-polynomial time algorithm from relational database theory can be used for computing all minimal generators of an implication-closed set.

Our motivation for analyzing these problems is not only theoretical, but also practical. A good analysis of these problems can help to develop methods that support the expert during attribute exploration by making the implication questions "simpler". We know that the attribute exploration algorithm asks the smallest number of questions to the expert, i.e., none of the questions it asks is redundant. However, it might still be possible to shorten an implication question by removing redundant attributes from its premise and conclusion. Moreover, a good analysis of the problems related to generators of concept intents can help to develop efficient lattice construction and merge algorithms.

2   Counting Complexity

We assume that the reader has a basic knowledge of complexity theory. Additional information can be found in the book [19].

A *counting problem* is presented using a suitable *witness* function which for every input x returns a set of *witnesses* for x. Formally, a *witness* function is a function $A\colon \Sigma^* \to \mathcal{P}^{<\omega}(\Gamma^*)$, where Σ and Γ are two alphabets, and $\mathcal{P}^{<\omega}(\Gamma^*)$ is the collections of all finite subsets of Γ^*. Every such witness function gives rise to the following *counting problem*: given a string $x \in \Sigma^*$, find the cardinality $|A(x)|$ of the *witness* set $A(x)$.

Complexity of counting problems was first investigated by Valiant in [21,22]. For a systematical study and classification of counting problems he introduced the counting complexity class #P, defined as the class of functions counting the number of accepting paths of nondeterministic polynomial-time Turing machines. A typical member is the problem #SAT, counting the number of satisfying assignments to a propositional formula in conjunctive normal form. Valiant showed in [21,22] that #SAT and many other problems are #P-complete.

Hemaspaandra and Vollmer introduced in [9] a predicate-based approach for defining higher counting complexity classes. In this approach, the counting complexity classes are denoted by $\#{\cdot}\mathcal{C}$.

Definition 1. $\#{\cdot}\mathcal{C}$ *is the class of all counting problems whose witness function A satisfies the following conditions:*
(i) *There is a polynomial $p(n)$ such that every $x \in \Sigma^*$ and every $y \in A(x)$ satisfy the relation $|y| \leq p(|x|)$;*
(ii) *The decision problem "given x and y, does y belong to $A(x)$?" is in $\mathcal{C}$.*

Completeness of the problems in $\#\mathrm{P}$ is often proved by using *parsimonious reductions*, which are polynomial-time reductions preserving the number of solutions by establishing a bijection between the solution sets of the problems. There are, however, two shortcoming of parsimonious reductions. First, they are not powerful enough, since they represent a particular case of many-one reductions, whereas Valiant was obliged to use Turing reductions in [21, 22] to be able to prove $\#\mathrm{P}$-completeness of several problems like $\#\mathrm{PERMANENT}$ or $\#\mathrm{PERFECT\ MATCHINGS}$. Second, even if the many-one reduction is powerful enough for proving completeness, there does not need to exist a one-to-one correspondence between the solutions of the reduced problems. On the other hand, Turing reductions turned out to be too powerful, since as it was proved in [20], they collapse all counting classes $\#{\cdot}\Sigma_k\mathrm{P}$ and $\#{\cdot}\Pi_k\mathrm{P}$ to $\#\mathrm{P}$.

In order to overcome this problem, Durand et al. introduced in [2] a new kind of reductions called *subtractive reduction*, under which $\#\mathrm{P}$ and the higher classes $\#{\cdot}\Pi_k\mathrm{P}$ for each $k \in \mathbb{N}$ are closed. A subtractive reduction between counting problems first overcounts the number of solutions and then carefully subtracts any surplus. It is formally defined as follows.

Definition 2. *Let Σ, Γ be two alphabets and let $\#{\cdot}A$ and $\#{\cdot}B$ be two counting problems determined by the binary relations A and B between strings from Σ to Γ. We say that $\#{\cdot}A$ reduces to $\#{\cdot}B$ via a strong subtractive reduction if there exist two polynomial-time computable functions f and g such that for every string $x \in \Sigma^*$ the following conditions hold.*
1. $B(f(x)) \subseteq B(g(x))$;
2. $|A(x)| = |B(g(x))| - |B(f(x))|$.
A subtractive reduction is a transitive closure of strong subtractive reductions.

Parsimonious reductions constitute a special case of subtractive reductions with $B(f(x)) = \emptyset$. In [2] it was pointed out that subtractive reductions are well-suited tools to study the higher counting complexity classes $\#{\cdot}\Sigma_k\mathrm{P}$ and $\#{\cdot}\Pi_k\mathrm{P}$.

3   Generators of Concept Intents

We assume that the reader is familiar with the theory of FCA. We briefly mention the necessary basic notions and refer the reader to the standard textbook [5] for additional information. In the present section we shortly recall the notion of generators of a concept intent, and some well-known computational problems about them.

Definition 3. *Let $\mathbb{K} = (G, M, I)$ be a formal context and $C \subseteq M$ be a concept intent, i.e., $C'' = C$. The subset $D \subseteq C$ is a minimal generator of C under $(\cdot)''$ if $D'' = C$ holds and D is subset-minimal, i.e., for all $E \subsetneq D$ we have $E'' \subsetneq C$.*

We first recall the computational complexity of checking whether a concept intent has a generator of cardinality less than or equal to a specified size. It is well-known that the following problem is NP-complete.

Problem: INTENT GENERATOR
Input: A formal context $\mathbb{K} = (G, M, I)$, the intent D of a formal concept (C, D) from $\mathbb{K}$, and a positive integer $m \leq |A|$.
Question: Is there a subset $Q \subseteq D$ of cardinality less than or equal to m that generates D, i.e., is there a $Q \subseteq D$ such that $Q'' = D$ and $|Q| \leq m$?

Frambourg et al. mentioned in [3] that the number of minimal generators of an intent can be exponential in the size of the context. Apart from this exponential bound, it is common folklore that the following problem is #P-complete.

Problem: #MINIMAL INTENT GENERATOR
Input: A formal context $\mathbb{K} = (G, M, I)$ and the intent D of a formal concept (C, D) in $\mathbb{K}$.
Output: Number of all subset-minimal intent generators of D with respect to the closure operator $(\cdot)''$, i.e., $|\{Q \subseteq D \mid Q'' = D \wedge \forall P \subsetneq Q, \ P'' \neq D\}|$.

4   Generators of Implication-Closed Sets

In the present section we first shortly recall the notion of minimal generators of an implication-closed set, and some well-known computational problems about minimal generators. Later we define a new problem about minimal generators, and work its computational complexity.

Definition 4. *Let $\mathcal{L}$ be a set of implications on a finite attribute set A and $P \subseteq A$ be closed with respect to $\mathcal{L}$, i.e., $\mathcal{L}(P) = P$. The subset $Q \subseteq P$ is a minimal generator of P under $\mathcal{L}$ if $\mathcal{L}(Q) = P$ holds and Q is subset-minimal, i.e., for all $R \subsetneq Q$ we have $\mathcal{L}(R) \subsetneq P$.*

Minimal generators appear in the literature under different names in various fields. For instance, in relational databases they are called minimal keys, and various properties of them have been considered in the literature. In order to make this connection clear, let us briefly recall some basic notions of relational databases.

4.1   Connection to Relational Databases

Functional dependencies are a way of expressing constraints on data in relational databases [16]. Informally, a functional dependency occurs when the values of a tuple on one set of attributes uniquely determine the values on another set of attributes. Formally, given a relation R and a set of attribute names A, a *functional dependency* is a pair of sets $X, Y \subseteq A$ written as $X \rightarrow Y$. The relation R *satisfies* the functional dependency $X \rightarrow Y$ if the tuples with equal X-values also have equal Y-values. In this case we say that the set of attributes X *functionally determine* the set of attributes Y.

Another important concept in relational databases is the notion of a key. Given a relation R on the attribute set A, a set $K \subseteq A$ is called a *key* of R if K functionally determines A. It is called a *minimal key* if no proper subset of it is a key. Alternatively, given a set of functional dependencies F that are satisfied by R, a set $K \subseteq A$ is called a key of the *relational system* $\langle A, F \rangle$ if $K \rightarrow A$ can be inferred from F by using Armstrong's axioms [1]. In practical applications, it is important to find "small" keys of a given relation. Lucchesi and Osborn analyzed in [15] how difficult it is to check whether a given relation has a key of cardinality bounded by a specified size. This problem is known as the MINIMUM CARDINALITY KEY problem (see problem [SR26] in [6]).

Problem: MINIMUM CARDINALITY KEY
Input: A set A of attribute names, a collection F of functional dependencies, and a positive integer $m \in \mathbb{N}$.
Question: Is there a key of cardinality m or less for the relational system $\langle A, F \rangle$?

Lucchesi and Osborn proved in [15] that MINIMUM CARDINALITY KEY is NP-complete. It is well-known that minimal generators of a closed set are the minimal keys of the subrelation defined by this closed set. Based on this observation, it is easy to see that the following problem is also NP-complete.

Problem: MINIMUM CARDINALITY GENERATOR
Input: A set A of attribute names, a set $\mathcal{L}$ of implications on A, an $\mathcal{L}$-closed subset P of A, and a positive integer $m \leq |A|$.
Question: Is there a subset $Q \subseteq P$ of cardinality $|Q| \leq m$ that generates P under $\mathcal{L}$, i.e., is there a $Q \subseteq P$ such that $\mathcal{L}(Q) = P$ and $|Q| \leq m$?

4.2   Counting Minimal Generators

Osborn showed in [18] that the number of minimal keys for a relational system $\langle A, F \rangle$ can be exponential in $|A|$. Moreover, Gunopulos et al. proved in [8] that the problem of determining the number of minimal keys of a relational system is #P-complete. Due to the correspondence between minimal keys and minimal generators of a closed set, it is also well-known that the number of minimal generators can be exponential in the size of the attribute set, and that the following counting problem is #P-complete.

Problem: #MINIMAL GENERATOR
Input: A set A of attribute names, a set $\mathcal{L}$ of implications on A, and an $\mathcal{L}$-closed subset P of A.
Output: Number of all subset-minimal generators of P under $\mathcal{L}$.

Algorithm 1. Minimal generator

Input: Implications $\mathcal{L}$ on the attribute set A and a subset $P \subseteq A$ such that $\mathcal{L}(P) = P$
Output: A minimal generator Q of P
1: $Q \leftarrow P$
2: **for all** $m \in P$ **do**
3:    **if** $\mathcal{L}(Q \smallsetminus \{m\}) = P$ **then**
4:       $Q \leftarrow Q \smallsetminus \{m\}$
5:    **end if**
6: **end for**

4.3   Finding All Minimal Generators

In some cases, it might not be enough to find only one minimal generator of an implication-closed set. For instance during attribute exploration it might be useful to show the expert different minimal generators of the premise and conclusion of the implication question for better understandability. The expert might want to browse among them to find a shortened version of the question which is most comprehensible to him. In the sequel we are going to investigate the problem of determining all minimal generators of a closed set.

Lucchesi and Osborn presented in [15] an algorithm to determine all minimal keys of a given relation. Given a set of attributes R and a set of functional dependencies F, the algorithm returns the set of all minimal keys for the relational system $\langle R, F \rangle$. Below we present an adaptation of this algorithm to find all minimal generators of a given implication-closed set. The algorithm is based on the following property shown in [15]. Here we formulate the property in terms of implications and minimal generators, and leave out its proof.

Lemma 5. *Let $\mathcal{L}$ be a set of implications on the attributes A and $\mathcal{G}$ be a nonempty set of minimal generators for a given $P \subseteq A$ under $\mathcal{L}$. The complement set $2^P \smallsetminus \mathcal{G}$ contains a minimal generator if and only if $\mathcal{G}$ contains a minimal generator G and $\mathcal{L}$ contains an implication $L \to R$, such that $L \cup R \cup G \subseteq P$ holds and $L \cup (G \smallsetminus R)$ does not include any minimal generator from $\mathcal{G}$.*

Lemma 5 assumes the existence of a nonempty set of minimal generators, thus the algorithm following from the lemma needs one minimal generator before it can proceed to find all other minimal generators. It is not difficult to find one minimal generator of a given implication-closed set P. We can start with P, iterate over all elements of P, and remove an element if the remaining set still generates P. Algorithm 1 implements this idea. It determines a minimal generator of a given set of attributes P closed under a given set of implications $\mathcal{L}$. Algorithm 1 terminates since P is finite. Upon termination, Q is a minimal generator of P since it does not contain any redundant attributes. For checking whether $Q \smallsetminus \{m\}$ generates P we can use the well-known implicational closure algorithm LINCLOSURE from [16]. The LINCLOSURE algorithm runs in time $O(|\mathcal{L}||A|)$. Algorithm 1 makes at most $|A|$ iterations of LINCLOSURE and therefore it runs in time $O(|\mathcal{L}||A|^2)$.

Algorithm 2. All minimal generators

Input: Set of implications $\mathcal{L}$ on the attribute set A and an $\mathcal{L}$-closed set $P \subseteq A$
Output: All minimal generators $\mathcal{G}$ of P
 1: $\mathcal{G} \leftarrow \{MinGen(P, \mathcal{L})\}$ {Initial set of minimal generators}
 2: **for all** $G \in \mathcal{G}$ **do**
 3:    **for all** $L \rightarrow R \in \mathcal{L}$ such that $L \cup R \cup G \subseteq P$ **do**
 4:       $S \leftarrow L \cup (G \setminus R)$
 5:       $flag \leftarrow true$
 6:       **for all** $H \in \mathcal{G}$ **do**
 7:          **if** $H \subseteq S$ **then**
 8:             $flag \leftarrow false$
 9:          **end if**
10:       **end for**
11:       **if** $flag$ **then**
12:          $\mathcal{G} \leftarrow \mathcal{G} \cup \{MinGen(S, \mathcal{L})\}$
13:       **end if**
14:    **end for**
15: **end for**

Now that we have an algorithm to determine one minimal generator, we can proceed with the algorithm determining the set of all minimal generators of an implication-closed set.

Algorithm 2 terminates, since $\mathcal{G}$ and $\mathcal{L}$ are both finite. Following Lemma 5, upon termination of the algorithm the set $\mathcal{G}$ contains all minimal generators of the given set of attributes P under $\mathcal{L}$. Let $|\mathcal{L}| = \ell$, $|\mathcal{G}| = g$, and $|P| = p$ be the cardinalities of the corresponding sets. The algorithm runs in time $O(\ell g(p + gp)) + O(gm)$, where m is the complexity of Algorithm 1. Hence Algorithm 2 has time complexity $O(\ell gp(g + p))$. Note that the algorithm finds minimal generators in *incremental polynomial time*, which is a notion introduced in [10] for analyzing the performance of algorithms that generate all solutions of a problem. An algorithm is said to run in incremental polynomial time if given an input and a prefix of the set of solutions (say, a closed set and a collection of the first k minimal generators), it finds another solution, or determines that none exists, in time polynomial in the combined sizes of the input and the given prefix. For finding a minimal generator, Algorithm 2 needs to perform at most $g\ell p(g + p)$ operations, which is polynomial in the size of the input, i.e., in the size of $\mathcal{L}$ and P, as well as polynomial in the size of the already found minimal generators $\mathcal{G}$.

Another notion introduced in [10] for analyzing algorithms that enumerate solutions is polynomial delay. An algorithm is said to run with *polynomial delay* if the delay until the first solution is written, as well as thereafter the delay between any two consecutive solutions, is bounded by a polynomial in the size of the input. Polynomial delay is a stronger notion than incremental polynomial time, i.e., if an algorithm runs with polynomial delay it is also runs in incremental polynomial time. To the best of our knowledge, there is no polynomial delay algorithm that finds all minimal keys of a relation, which is equivalent to finding all minimal generators of an attribute set closed under a set of implications.

4.4   Counting Minimum Cardinality Generators

In this section we consider a modified version of the #MINIMAL GENERATOR. For this problem, we slightly change the notion of "generates" as follows. For a given set $\mathcal{L}$ of implications on an attribute set A, and an $\mathcal{L}$-closed set $P \subseteq A$, we say that a $Q \subseteq A$ is a *minimum cardinality generator* of P if $\mathcal{L}(Q) \setminus Q = P$ holds and no subset of A with smaller cardinality satisfies this property. In other words, we require that P should be the "new consequences" of closing Q under $\mathcal{L}$ and that no set with smaller cardinality can have this property. It turns out that the problem of counting such sets is #·coNP-complete, which means that it is even harder than the #MINIMAL GENERATOR problem.

Problem: #MINIMUM CARDINALITY GENERATOR
Input: A set A of attribute names, a set $\mathcal{L}$ of implications on A, an $\mathcal{L}$-closed subset P of A.
Question: Number of all minimum cardinality generators of P under $\mathcal{L}$, i.e., number of the subsets $Q \subseteq A$ such that $\mathcal{L}(Q) \setminus Q = P$ and no other subset $R \subseteq A$ with $|R| < |Q|$ satisfies the condition $\mathcal{L}(R) \setminus R = P$.

Theorem 6. #MINIMUM CARDINALITY GENERATOR *is* #·coNP-*complete.*

Proof. The problem is clearly in #·coNP what can be shown as follows. Given a set of attributes Q, we have to check (i) whether Q generates P, and if so (ii) whether there is another generator R with $|R| < |Q|$. The first test can be done in polynomial time using a closure algorithm based on the reachability algorithm for graphs. The second test, which dominates the overall complexity, can be done by a coNP-algorithm. Indeed, checking whether Q is *not* a minimum cardinality generator can be done by the following NP-algorithm: Guess a subset of attributes $R \subseteq A$ such that $|R| < |Q|$ and check if R generates P. Again, checking if R generates P can be done in polynomial time, thus checking whether Q is a minimum cardinality generator can be done in coNP and counting such sets can be done in #·coNP.

We show the #·coNP-hardness by a strong subtractive reduction from the problem #Π_1SAT. #Π_1SAT is #·coNP-complete according to [2]. Consider an instance of the #Π_1SAT problem given by a formula $\varphi(X) = \forall Y \ \psi(X, Y)$ where $X = \{x_1, \ldots, x_k\}$ and $Y = \{y_1, \ldots, y_l\}$ are disjoint sets of variables. Without loss of generality we can assume that $\psi(X, Y)$ is in 3DNF, i.e., it is of the form $C_1 \vee \cdots \vee C_n$ where each C_i is of the form $C_i = l_{i1} \wedge l_{i2} \wedge l_{i3}$, and the l_{ij}'s are propositional literals over $X \cup Y$.

Let $x'_1, \ldots, x'_k, q_1, \ldots, q_k, y'_1, \ldots, y'_l, r_1, \ldots, r_l, g_1, \ldots, g_n, u$ denote fresh pairwise distinct variables and let us regroup them in the sets $X_= \{x'_1, \ldots, x'_k\}$, $Y_= \{y'_1, \ldots, y'_l\}$, $Q_1 = \{q_1, \ldots, q_k\}$, $R_1 = \{r_1, \ldots, r_l\}$, and $G = \{g_1, \ldots, g_n\}$. We define two instances of the minimum cardinality generator problem. The first problem $\mathbb{P}_1$ is defined as follows:

$$A_1 = A = X \cup X_1 \cup Y \cup Y_1 \cup Q_1 \cup R_1 \cup G \cup \{u\}$$
$$P_1 = Q_1 \cup R_1 \cup G$$
$$\mathcal{L}_1 = \{\{x_i, x_i'\} \rightarrow A, \ x_i \rightarrow q_i, \ x_i' \rightarrow q_i \mid 1 \leq i \leq k\} \ \cup$$
$$\{\{y_i, y_i'\} \rightarrow A, \ y_i \rightarrow r_i, \ y_i' \rightarrow r_i \mid 1 \leq i \leq l\} \ \cup$$
$$\{z_{ij} \rightarrow g_i \mid 1 \leq i \leq n \ \text{ and } \ 1 \leq j \leq 3\}$$

where, for $1 \leq s \leq k$ and $1 \leq t \leq l$, z_{ij} is in one of the forms $x_s, x_s', y_t,$ or, y_t' depending on whether the literal l_{ij} in C_i is in one of the forms $\neg x_s, x_s, \neg y_t,$ or y_t, respectively. In other words, z_{ij} encodes the negation of l_{ij}. Now we define the second problem $\mathbb{P}_2$.

$$A_2 = A, \quad P_2 = P_1, \quad \mathcal{L}_2 = \mathcal{L}_1 \cup \{\{y_1, \ldots, y_l\} \rightarrow g_i \mid 1 \leq i \leq n\}.$$

Now let $\mathcal{A}(\varphi)$ denote the set of all satisfying truth assignments of a $\#\Pi_1\text{SAT-}$ formula φ and let $\mathcal{B}(\mathbb{P})$ denote the set of all solutions of a minimum cardinality generator problem $\mathbb{P}$. We claim that the following holds:

$$\mathcal{B}(\mathbb{P}_1) \subseteq \mathcal{B}(\mathbb{P}_2) \quad \text{and} \quad |\mathcal{A}(\varphi)| = |\mathcal{B}(\mathbb{P}_2)| - |\mathcal{B}(\mathbb{P}_1)|.$$

Consider the problem $\mathbb{P}_1$. Solutions of $\mathbb{P}_1$, i.e., minimum cardinality generators of P_1 satisfy the following 3 conditions: (1) An attribute q_i can be generated only in two ways, by the implication $x_i \rightarrow q_i$ or by the implication $x_i' \rightarrow q_i$. So a solution of $\mathbb{P}_1$ contains one of x_i and x_i'. Moreover, it cannot contain both of them due to the implication $\{x_i, x_i'\} \rightarrow A$, since this implication would also generate the attribute u, and u is not contained in P_1. This means, for each $1 \leq i \leq k$ a solution of $\mathbb{P}_1$ contains either x_i or x_i' in order to be able to generate the q_i's. (2) Similarly, it also contains either y_i or y_i' for each $1 \leq i \leq l$ in order to be able to generate the r_i's. (3) In addition to these, in order to be able to generate an attribute g_i, a solution contains at least one attribute that encodes the negation of a literal occurring in the implicant C_i. In order be able to generate all g_i's, a solution contains at least one such attribute for each implicant C_i. Subsets of A that satisfy these 3 conditions are solutions of $\mathbb{P}_1$. Each such subset has exactly the size $|X| + |Y| = k + l$. Moreover, they are the only solutions of $\mathbb{P}_1$, since any subset of A that has cardinality less than $k + l$ fails to generate at least one attribute in P_1. Conditions (1) and (2) enforce a solution to be a truth assignment over $X \cup Y$. Condition (3) enforces this truth assignment to contain the negation of at least one literal in every implicant, i.e., it enforces this truth assignment to falsify the formula $\psi(X, Y)$.

Consider now the problem $\mathbb{P}_2$. Each solution of $\mathbb{P}_1$ is also a solution of $\mathbb{P}_2$ since $P_2 = P_1$ and $\mathcal{L}_2$ contains all implications from $\mathcal{L}_1$. In addition to the implications from $\mathcal{L}_1$, $\mathcal{L}_2$ also contains implications of the form $\{y_1, \ldots, y_l\} \rightarrow g_i$ for each $1 \leq i \leq n$. These new implications give rise to the following new solutions. Like the solutions of $\mathbb{P}_1$, in order to be able to generate the q_i's and r_i's, they satisfy the conditions (1) and (2) mentioned above. In order to be able to generate the g_i's, they contain every y_i for each $1 \leq i \leq l$. In other words, these new solutions are truth assignments over $X \cup Y$ that set every $y_1, \ldots, y_l$ to *true*.

Based on the above descriptions, $\mathcal{B}(\mathbb{P}_1)$ is the set of truth assignments that *falsify* $\psi(X,Y)$ and $\mathcal{B}(\mathbb{P}_2)$ is the set of truth assignments that *falsify* $\psi(X,Y)$, plus the set of truth assignments that set every $y_1, \ldots, y_l$ to *true*. Obviously, the claim $\mathcal{B}(\mathbb{P}_1) \subseteq \mathcal{B}(\mathbb{P}_2)$ is satisfied. Moreover, the difference $\mathcal{B}(\mathbb{P}_1) \setminus \mathcal{B}(\mathbb{P}_2)$ is the set of truth assignments that set every $y_1, \ldots, y_l$ to *true* and at the same time *satisfy* $\psi(X,Y)$ (since by taking the set difference from $\mathcal{B}(\mathbb{P}_1)$ we remove the truth assignments that falsify $\psi(X,Y)$). In other words, this set contains the models of $\psi(X,Y)$ such that all Y values are fixed by setting them to *true*. This set has exactly the same cardinality as the set of models of $\varphi(X) = \forall Y \; \psi(X,Y)$, thus the other claim $|\mathcal{A}(\varphi)| = |\mathcal{B}(\mathbb{P}_2)| - |\mathcal{B}(\mathbb{P}_1)|$ holds.   □

5 Concluding Remarks

We analyzed some decision and counting problems related to generators of closed sets fundamental in FCA, namely concept intents and implication-closed sets. We have recalled results from the literature on the problem of checking the existence of a generator with cardinality less than a specified size, and on the problem of determining the number of minimal generators. Moreover, we have defined a new problem, which is determining the number of minimum cardinality generators, and shown that this problem is #·coNP-complete, i.e., it is even more difficult than counting minimal generators. We have also given an incremental-polynomial time algorithm from relational databases that can be used for computing all minimal generators of an implication-closed set.

It is not surprising to see that the mentioned problems about generators of concept intents and generators of implication-closed sets are of the same complexity. In fact, the closure operator induced by a formal context and the closure operator induced by the set of implications that are valid in this formal context coincide. That is, one can easily transfer these results from one case to the other.

References

1. Armstrong, W.W.: Dependency structures of data base relationships. In: Rosenfeld, J.L. (ed.) Proceedings 6th Information Processing Conference (IFIP 1974), Stockholm, Sweden, pp. 580–583. North-Holland, Amsterdam (1974)
2. Durand, A., Hermann, M., Kolaitis, P.G.: Subtractive reductions and complete problems for counting complexity classes. Theoretical Computer Science 340(3), 496–513 (2005)
3. Frambourg, C., Valtchev, P., Godin, R.: Merge-based computation of minimal generators. In: Dau, F., Mugnier, M.-L., Stumme, G. (eds.) ICCS 2005. LNCS (LNAI), vol. 3596, pp. 181–194. Springer, Heidelberg (2005)
4. Ganter, B.: Two basic algorithms in concept analysis. Technical Report Preprint-Nr. 831, Technische Hochschule Darmstadt, Germany (1984)
5. Ganter, B., Wille, R.: Formal Concept Analysis: Mathematical Foundations. Springer, Heidelberg (1999)
6. Garey, M.R., Johnson, D.S.: Computers and intractability: A guide to the theory of NP-completeness. W.H. Freeman, New York (1979)

7. Guigues, J.-L., Duquenne, V.: Familles minimales d'implications informatives resultant d'un tableau de données binaries. Mathématiques, Informatique et Sciences Humaines 95, 5–18 (1986)
8. Gunopulos, D., et al.: Discovering all most specific sentences. ACM Transactions on Database Systems 28(2), 140–174 (2003)
9. Hemaspaandra, L.A., Vollmer, H.: The satanic notations: Counting classes beyond #P and other definitional adventures. SIGACT News, Complexity Theory Column 8 26(1), 2–13 (1995)
10. Johnson, D.S., Yannakakis, M., Papadimitriou, C.H.: On generating all maximal independent sets. Information Processing Letters 27(3), 119–123 (1988)
11. Kuznetsov, S.O.: On computing the size of a lattice and related decision problems. Order 18(4), 313–321 (2001)
12. Kuznetsov, S.O.: On the intractability of computing the Duquenne-Guigues base. Journal of Universal Computer Science 10(8), 927–933 (2004)
13. Kuznetsov, S.O., Obiedkov, S.A.: Comparing performance of algorithms for generating concept lattices. Journal of Experimental and Theoretical Artificial Intelligence 14(2-3), 189–216 (2002)
14. Kuznetsov, S.O., Obiedkov, S.O.: Counting pseudo-intents and #P-completeness. In: Missaoui, R., Schmidt, J. (eds.) Formal Concept Analysis. LNCS (LNAI), vol. 3874, pp. 306–308. Springer, Heidelberg (2006)
15. Lucchesi, C.L., Osborn, S.L.: Candidate keys for relations. Journal of Computer and System Science 17(2), 270–279 (1978)
16. Maier, D.: The Theory of Relational Databases. Computer Science Press (1983)
17. Nehmé, K., et al.: On computing the minimal generator family for concept lattices and icebergs. In: Ganter, B., Godin, R. (eds.) ICFCA 2005. LNCS (LNAI), vol. 3403, pp. 192–207. Springer, Heidelberg (2005)
18. Osborn, S.L.: Normal Forms for Relational Data Bases. PhD thesis, University of Waterloo, Canada (1977)
19. Papadimitriou, C.H.: Computational complexity. Addison-Wesley, Reading (1994)
20. Toda, S., Watanabe, O.: Polynomial-time 1-Turing reductions from #PH to #P. Theoretical Computer Science 100(1), 205–221 (1992)
21. Valiant, L.G.: The complexity of computing the permanent. Theoretical Computer Science 8(2), 189–201 (1979)
22. Valiant, L.G.: The complexity of enumeration and reliability problems. SIAM Journal on Computing 8(3), 410–421 (1979)
23. Valtchev, P., Missaoui, R., Godin, R.: Formal concept analysis for knowledge discovery and data mining: The new challenges. In: Eklund, P.W. (ed.) ICFCA 2004. LNCS (LNAI), vol. 2961, pp. 352–371. Springer, Heidelberg (2004)

Generating Positive and Negative Exact Rules Using Formal Concept Analysis: Problems and Solutions

Rokia Missaoui[1], Lhouari Nourine[2], and Yoan Renaud[2]

[1] Département d'informatique et d'ingénierie
Université du Québec en Outaouais
C.P. 1250, succursale B, Gatineau (Québec) Canada, J8X 3X7
[2] LIMOS - CNRS UMR 6158
Université Blaise Pascal, Clermont-Ferrand
Rokia.Missaoui@uqo.ca, {nourine,renaud}@isima.fr

Abstract. The objective of this article is to investigate the problem of generating both positive and negative exact association rules when a formal context K of (positive) attributes is provided. A straightforward solution to this problem consists of conducting an apposition of the initial context K with its complementary context $\tilde{K}$, construct the concept lattice $\mathfrak{B}(K|\tilde{K})$ of apposed contexts and then extract rules. A more challenging problem consists of exploiting rules generated from each one of the contexts K and $\tilde{K}$ to get the whole set of rules for the context $K|\tilde{K}$.

In this paper, we analyze a set of identified situations based on distinct types of input, and come out with a set of properties. Obviously, the global set of (positive and negative) rules is a superset of purely positive rules (i.e., rules with positive attributes only) and purely negative ones since it generally contains mixed rules (i.e., rules in which at least a positive attribute and a negative attribute coexist).

The paper presents also a set of inference rules to generate a subset of all mixed rules from positive, negative and mixed ones. Finally, two key conclusions can be drawn from our analysis: (i) the generic basis containing negative rules, $\Sigma_{\tilde{K}}$, cannot be completely and directly inferred from the set Σ_K of positive rules or from the concept lattice $\mathfrak{B}(K)$, and (ii) the whole set of mixed rules may not be completely generated from Σ_K alone, $\Sigma_K \cup \Sigma_{\tilde{K}}$ alone, or $\mathfrak{B}(K)$ alone.

1 Introduction

Association rule mining [1] is an extensively studied problem in data mining and consists of extracting a set of association rules from data (e.g., a set of transactions describing a collection of items bought together). An association rule r is an implication of the form $Y \rightarrow Z$ $[sup, conf]$, where Y and Z are subsets of attributes (called *itemsets*), $Y \cap Z = \emptyset$, and sup and $conf$ represent the support and the confidence of the rule, respectively. The support of a rule is defined as $Prob(Y \cup Z)$ (i.e., the probability that a set of objects have $Y \cup Z$)

R. Medina and S. Obiedkov (Eds.): ICFCA 2008, LNAI 4933, pp. 169–181, 2008.
© Springer-Verlag Berlin Heidelberg 2008

while the confidence is computed as the conditional probability $Prob(Z/Y)$. One interesting problem is the efficient generation of negative rules like the following one: *customers who buy smoked salmon buy also caviar but not Coke drink*. Such rules may exhibit unexpected patterns such as exceptions (e.g., *ostrich is a bird that exceptionally does not fly*).

In this paper we investigate the problem of generating both positive and negative (exact) association rules in formal concept analysis when a formal context K of (positive) attributes is initially provided. A straightforward but not efficient solution to this problem consists of conducting an apposition (see Section 2) of the initial context K with its complementary context $\tilde{K}$ to get the concept lattice $\mathfrak{B}(K|\tilde{K})$ and then extract the generic basis out of that lattice. However, data collections in many real-life applications tend to be very sparse and hence the corresponding complementary contexts are dense, and generate a very likely huge set of candidate itemsets and a tremendous set of uninteresting rules (mainly purely negative and mixed ones). To handle our general problem, we first identify a set of scenarios based on the sorts of the provided input and discuss their tractability.

Obviously, the set of all rules, called *complete* set in the rest of the paper, is a superset of purely positive rules (i.e., rules with positive attributes only) and purely negative ones since it generally contains mixed rules (i.e., rules in which at least a negative attribute and a positive attribute coexist). In this paper, we propose a set of inference rules to deduce mixed rules from positive, negative and even other mixed ones. However, completeness of the inference rules is not guaranteed.

The paper is organized as follows. Section 2 provides a background on formal concept analysis and association rule mining. Section 3 gives a brief overview about the generation of association rules with negation. Identified problems and their analysis are described in Section 4 while the generation of positive, negative and mixed rules is studied in Section 5. Finally, a conclusion and further work are given in Section 6.

2   Background

2.1   Formal Concept Analysis

Formal concept analysis (FCA) is a branch of applied mathematics, which is based on a formalization of concept and concept hierarchy. It has been successfully used for conceptual clustering and rule generation [9]. Let $K = (G, M, I)$ be a formal context, where G, M and I are a set of objects, a set of attributes or properties, and a binary relation between G and M respectively. Two functions, f and g, summarize the links between subsets of objects and subsets of attributes induced by I. Function f maps a set of objects into a set of its common attributes, whereas g is the dual for attribute sets:

- $f : \mathcal{P}(G) \to \mathcal{P}(M)$, $f(X) := X' =: \{a \in M \mid \forall o \in X, oIa\}$,
- $g : \mathcal{P}(M) \to \mathcal{P}(G)$, $g(Y) := Y' := \{o \in G \mid \forall a \in Y, oIa\}$.

The operators $g \circ f$ and $f \circ g$ (denoted by $''$) are *closure* operators over $\mathcal{P}(G)$ and $\mathcal{P}(M)$ respectively.

A formal concept C is a pair of sets (X, Y) where $X \in \mathcal{P}(G)$, $Y \in \mathcal{P}(M)$, $X = Y'$ and $Y = X'$. The closed subset X is called the extent of C, and Y its intent. In the association rule mining problem [1,15,23,20], X and Y correspond to the notion of closed *tidset* (e.g., a set of transactions or objects) and closed *itemset* (e.g., a set of bought items) respectively.

The set $\mathfrak{B}(K)$ of all concepts extracted from the context K, partially ordered by:

$$(X_1, Y_1) \le (X_2, Y_2) \Leftrightarrow X_1 \subseteq X_2, Y_2 \subseteq Y_1.$$

forms a complete lattice, called a concept lattice.

Object (resp. attribute) set reduction of a context $K = (G, M, I)$ consists of discarding from the set G (resp. M) all objects (resp. attributes) that may be obtained through the intersection of some other objects (resp. attributes). The concept lattice of a reduced context is isomorphic to the concept lattice of the initial one.

The apposition $K = K_1|K_2$ of two contexts $K_1 = (G, M_1, I_1)$ and $K_2 = (G, M_2, I_2)$ is the horizontal concatenation of contexts sharing the same set G of objects [9]. It represents the context $K = (G, M_1 \dot\cup M_2, I_1 \dot\cup I_2)$ whose corresponding lattice is a substructure of the direct product of $\mathfrak{B}(K_1)$ and $\mathfrak{B}(K_2)$ [19].

In the rest of the paper and unless otherwise indicated, we will use uppercase letters (e.g., B, Y), lower-case letters and letters with tilde to mean sets of attributes (itemsets), atomic attributes and elements with negation respectively. For example, $\tilde{a}$ stands for the negation of attribute a and means that object o belongs to the extent of $\tilde{a}$ iff o does not belong to the extent of a, and $\tilde{A}$ represents the set $\{\tilde{a} \mid a \in A\}$ [8].

2.2   Association Rule Mining

One interesting trend in association rule mining is to use formal concept analysis to extract frequent closed itemsets (*i.e.*, closed itemsets having a support greater than or equal to a predefined value) and compute a reduced set of association rules. A set of studies in FCA were conducted on the generation of concise representations (i.e., minimal bases) of rules [12] such as informative rules (i.e., with minimized premise and maximized consequence), Guigues-Duquenne basis [9,10], generic basis [15], Luxenburger basis, and so on.

A *generic basis* [15] associated with a context K, denoted by Σ_K, is a concise representation of exact rules (implications) of the form $r\colon Y \to Y''\backslash Y$ $[sup, 1]$ such that Y is a generator for Y''. The (minimal) generator Y [16] of a closed itemset Z is a minimal subset of Z such that $Y'' = Z$. The support of the rule r is $|Y'|/|G|$.

Since we are using the generic basis as a container for the extracted rules, any rule in Σ_K, $\Sigma_{\tilde{K}}$ and $\Sigma_{K|\tilde{K}}$ will be further represented by: $Y \to Z$ $[sup]$ because the confidence is always equal to 1.

2.3   Illustrative Example

To further illustrate notions and propositions, let us take the following example in which a context $K = (G, M, I)$ is given, with $G = \{1, 2, 3, 4\}$ and $M = \{a, b, c, d\}$. The corresponding concept lattice[1] is illustrated in Figure 1.

Example 1

The following table provides the complete set of rules generated from Example 1.

Table 1. A context K, its complementary context $\tilde{K}$ and the apposition $K|\tilde{K}$ of the two contexts

K	a	b	c	d
1	1	0	0	1
2	0	1	0	1
3	1	1	0	0
4	0	1	1	1

$\tilde{K}$	$\tilde{a}$	$\tilde{b}$	$\tilde{c}$	$\tilde{d}$
1	0	1	1	0
2	1	0	1	0
3	0	0	1	1
4	1	0	0	0

$K|\tilde{K}$	a	b	c	d	$\tilde{a}$	$\tilde{b}$	$\tilde{c}$	$\tilde{d}$
1	1	0	0	1	0	1	1	0
2	0	1	0	1	1	0	1	0
3	1	1	0	0	0	0	1	1
4	0	1	1	1	1	0	0	0

A first glance at Table 2 indicates that rules in Σ_K (i) do not convey interesting information about the absence of some items, and (ii) rules with a null support seem useless. However, the set $\Sigma_{K|\tilde{K}}$ brings additional associations about the absence of items, and we will see later that rules with a null support (either in Σ_K or $\Sigma_{\tilde{K}}$) can be exploited to generate mixed rules.

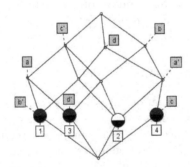

Fig. 1. The concept lattice $\mathcal{B}(K|\tilde{K})$ corresponding to Example 1

3   Related Work

In the classical problem of association rule mining, only attributes (items) present in data are recorded and positive rules are extracted. This class of rules is a subclass of the larger and more general set of boolean association rules (i.e., rules with negation, conjunction and disjunction) [13].

---

[1] The lattice is constructed with reduced labelling using the SourceForge project called Concept Explorer.

Table 2. Positive, negative and complete rule set of Example 1

Positive rules Σ_K	Negative rules $\Sigma_{\tilde{K}}$	Complete set of rules $\Sigma_{K|\tilde{K}}$	
$c \to bd[0.25]$	$\tilde{b} \to \tilde{c}[0.25]$	$a \to \tilde{c}[0.5]$	$abd \to \tilde{a}\tilde{b}c\tilde{c}\tilde{d}[0]$
$ac \to bd[0]$	$\tilde{d} \to \tilde{c}[0.25]$	$bd \to \tilde{a}[0.5]$	$ac \to \tilde{a}\tilde{b}\tilde{b}\tilde{c}d\tilde{d}[0]$
$abd \to c[0]$	$\tilde{a}\tilde{b} \to \tilde{c}\tilde{d}[0]$	$ad \to \tilde{b}\tilde{c}[0.25]$	$\tilde{a}\tilde{d} \to ab\tilde{b}c\tilde{c}d[0]$
	$\tilde{a}\tilde{d} \to \tilde{b}\tilde{c}[0]$	$c \to \tilde{a}bd[0.25]$	$\tilde{b}\tilde{d} \to a\tilde{a}b\tilde{c}d[0]$
	$\tilde{b}\tilde{d} \to \tilde{a}\tilde{c}[0]$	$ab \to \tilde{c}\tilde{d}[0.25]$	$\tilde{a}\tilde{b} \to abc\tilde{c}d\tilde{d}[0]$
		$\tilde{a} \to bd[0.5]$	$a\tilde{a} \to \tilde{b}bc\tilde{c}d\tilde{d}[0]$
		$\tilde{b} \to ad\tilde{c}[0.25]$	$b\tilde{b} \to a\tilde{a}c\tilde{c}d\tilde{d}[0]$
		$\tilde{d} \to ab\tilde{c}[0.25]$	$c\tilde{c} \to a\tilde{a}\tilde{b}bd\tilde{d}[0]$
		$\tilde{a}\tilde{c} \to bd[0.25]$	$d\tilde{d} \to a\tilde{a}\tilde{b}bc\tilde{c}[0]$
		$b\tilde{c}d \to \tilde{a}[0.25]$	$\tilde{b}c \to a\tilde{a}b\tilde{c}d\tilde{d}[0]$
			$c\tilde{d} \to a\tilde{a}\tilde{b}b\tilde{c}d[0]$

Association rules with negation could be in one of the following forms:

- $B \wedge \tilde{C} \to D$ indicating that if an object owns all the items in B and does not have the items in C (i.e., has all the items in $\tilde{C}$), then it owns all the items in D.
- $B \to C \wedge \tilde{D}$ indicating that if an object owns all the items in B, then it owns all the items in C but does not have anyone of the items in D.
- $\tilde{B} \to \tilde{C}$ indicating that if an object does not own the items in B, then it does not have the items in C.

The expressions *mixed* rules and *purely negative* rules will be used to refer to the first two forms and the last form, respectively.

In the area of data mining, the notion of negative associations (relationships) between itemsets was initially discussed by Brin and Motwani [5] who proposed a procedure that exploits the Chi-square test to search for a border between correlated and uncorrelated elements in the itemset lattice. Many studies recognize that mining rules with negation (i.e., rules that contain negative items) is a very challenging problem [4] and there is an urgent need to define interestingness measures (other than confidence) and pruning procedures to generate negative association rules in an efficient and correct way [3,5,21]. For example, Wu *et al.* [21] define a new algorithm for negative association rule generation as well as a new quality measure for an efficient pruning of generated frequent itemsets. In [17], positive frequent itemsets are combined with background knowledge to mine negative association rules, while in [2] a new technique based on Kullback-Leibler divergence is defined.

The notion of negative rules has different meanings. In [17], it represents rules of the form $Y \nrightarrow Z$ whose actual support deviates at least $MinRI \times MinSup$ from its expected support (based on the support of items in closed itemsets and the taxonomy on attributes). $MinRI$ and $MinSup$ correspond to the minimal value of an interest measure RI and the support, respectively. In [4], the rule $Y \rightarrow \tilde{Z}$ means that the presence of items in Y implies the absence of all items in Z.

Mining association rules with negation cannot be studied by a simple adaptation of work on discovering positive association rules. Indeed, the former raises new problems such as dealing with non frequent itemsets, assigning meaningful interpretation to negation and computing appropriate interestingness measures that reflect the negative associations among itemsets. For database miners, negative rules may help identify unexpected (surprising) patterns and suggest item substitution [22].

4   Problem Statement

In this section we identify a set of problems related to the generation of both positive and negative association rules from a finite formal context. To conduct our analysis, we consider the production of the generic basis with a support greater than or equal to 0 and we take a set of distinct situations whose analysis is summarized in Table 3. Both positive and negative rules with a null support will be used to infer some (not necessarily all) mixed rules as will be explained later.

There are several situations that we have identified when analyzing the issue of generating rules with negation. In the following, we describe each situation/problem and discuss its complexity.

Problem 1 : Positive Rule Generation (PRG)
Instance: A formal context $K = (G, M, I)$.
Question: Compute Σ_K.

As stated, the PRG problem aims to compute the set of positive rules when a formal context K is given. This is a classical problem in FCA and association rule mining that has been extensively studied and lead to a set of algorithms for minimal rule set generation and efficient computation [10,9,23,15,12,20]. The complexity of this problem is still open for some specific cases [7] and has been discussed in many studies (see e.g., [6,11]).

Problem 2 : Negative Rule Generation (NRG)
Instance : A formal context $K = (G, M, I)$.
Question : Compute $\Sigma_{\tilde{K}}$.

This problem consists of generating a set of purely negative rules given a formal context K. It is almost equivalent to the previous one except that a pre-computation of the complementary context $\tilde{K}$ is handled before the generation of rules. However, since most real-life formal contexts are sparse, complementary contexts are therefore dense and lead to an exponential number of concepts (and closed itemsets) and possibly large sets of purely negative rules. In a transaction database, for example, a customer's basket contains a reduced set of items compared to the large set of items that could be provided in the market. Therefore, the corresponding context is sparse since there are more absent items than present ones in individual transactions.

Problem 3 : Complete Rule Generation (CRG)
Instance: A formal context $K = (G, M, I)$.
Question: Compute $\Sigma_{K|\tilde{K}}$.

Starting from a given context K, we first compute its complementary context $\tilde{K}$ and the apposition $K|\tilde{K}$ of context K and $\tilde{K}$, and then compute from $\mathcal{B}(K|\tilde{K})$ the complete set of rules that includes all positive, negative and mixed ones. The complexity of this problem is similar to the previous ones. Moreover, the observation about the sparsity/density of contexts holds for this case too.

Problem 4 : Dual Rule Generation (DRG)
Instance: A formal context $K = (G, M, I)$ and Σ_K.
Question: Compute $\Sigma_{\tilde{K}}$.

This problem takes both the context and the set of positive rules as input to attempt to compute the set $\Sigma_{\tilde{K}}$ of purely negative rules. The size of the output may be exponential in the size of the input, and hence a polynomial algorithm is impossible. The following example illustrates this observation.

Example 2. Consider the context $K = (G, M, I)$ with $G = \{1, ..., n-1\}$, $M = \{a_1, a_2, ..., a_n\}$ and $(i, a_{n-i+1}) \in I$. This means in particular that no object has property a_1. The generic bases Σ_K and $\Sigma_{\tilde{K}}$ are given by:

- $\Sigma_K = \{a_1 \rightarrow M \backslash \{a_1\} \ [0] \cup \{a_i a_j \rightarrow a_1 \ [0] \ | \ i, j \in \{2, ..., n\}, i \neq j\}$
- $\Sigma_{\tilde{K}} = \{\tilde{A} \rightarrow \tilde{a_1} [sup] \ | \ \emptyset \subset \tilde{A} \subseteq \tilde{M} \backslash \{\tilde{a_1}\}$ where sup=0 when $\tilde{A} = \tilde{M} \backslash \{\tilde{a_1}\}$.

  Thus, $|\Sigma_K| = 1 + (n-1)(n-2)/2$ and $|\Sigma_{\tilde{K}}| = 2^{n-1} - 1$.

It is important to note that the set Σ_K cannot be used alone to generate $\Sigma_{\tilde{K}}$ since it does not give information about the binary relation that holds between objects and attributes. The following example illustrates this fact by showing that for a given Σ_K, there may exist more than one context and hence more than one $\Sigma_{\tilde{K}}$.

Example 3. Consider the following contexts where K_1 is a reduced context of K_2 with respect to objects. Indeed, K_2 is not reduced since $f(7) = f(2) \cap f(3)$. Obviously, Σ_{K_1} and Σ_{K_2} have the same set of rules (where matching rules have distinct non null supports) since $\mathcal{B}(K_1)$ is isomorphic to $\mathcal{B}(K_2)$. The set of positive rules is given by: $\{ad \rightarrow bc \ [0], bd \rightarrow ac \ [0]\}$. The computation of purely negative rules leads to two different sets:
$\Sigma_{\tilde{K_1}} = \{\tilde{a}\tilde{c} \rightarrow \tilde{b} \ [0.16], \tilde{b}\tilde{c} \rightarrow \tilde{a} \ [0.16], \tilde{a}\tilde{c}\tilde{d} \rightarrow \tilde{b} \ [0], \tilde{a}\tilde{b}\tilde{d} \rightarrow \tilde{c} \ [0], \tilde{b}\tilde{c}\tilde{d} \rightarrow \tilde{a} \ [0]\}$
$\Sigma_{\tilde{K_2}} = \{\tilde{a}\tilde{c} \rightarrow \tilde{b} \ [0.14], \tilde{a}\tilde{c}\tilde{d} \rightarrow \tilde{b} \ [0], \tilde{a}\tilde{b}\tilde{d} \rightarrow \tilde{c} \ [0]\}$.

Problem 5 : Negative Concept Generation (NCG)
Instance: A formal context $K = (G, M, I)$ and $\mathcal{B}(K)$ the lattice of K.
Question: Compute $\mathcal{B}(\tilde{K})$.

K_1	a	b	c	d	e
1	1	1	0	0	0
2	1	0	0	0	1
3	1	1	1	0	1
4	1	1	0	1	1
5	0	1	0	0	1

K_2	a	b	c	d	e
1	1	1	0	0	0
2	1	0	0	0	1
3	1	1	1	0	1
4	1	1	0	1	1
5	0	1	0	0	1
6	1	1	0	0	1

$$K_1 \qquad\qquad K_2$$

Fig. 2. K_1 is a reduced context of K_2

In this case, the size of the output may be exponential in the size of the input (see Example 4 below). However, concepts in $\mathcal{B}(\tilde{K})$ may be generated in a polynomial total time from K [14].

Example 4. Consider the context $K = (G, M, I)$ with $G = \{1, ..., n\}$, $M = \{a_1, a_2, ..., a_n\}$ and $(i, a_i) \in I$. This context K has $n + 2$ concepts, while the complementary context $\tilde{K}$ has 2^n concepts.

Problem 6 : Complete Rule Generation from Concepts (CRGC)
Instance: A formal context $K = (G, M, I)$ as well as the lattices $\mathcal{B}(K)$ and $\mathcal{B}(\tilde{K})$ produced from K and $\tilde{K}$, respectively.
Question: Compute $\Sigma_{K|\tilde{K}}$.

This problem can be solved by first performing the assembly of $\mathcal{B}(K)$ and $\mathcal{B}(\tilde{K})$ as a substructure of the direct product of these lattices [19], and then computing the set of rules. The size of the resulting lattice $\mathcal{B}(K|\tilde{K})$ is generally less than the size of the product of the two initial lattices. However, the drawback of this straightforward solution is that if one of the two contexts is dense and large, it will generate an impressive number of concepts and lead to an important execution time for concept and rule generation.

Corollary 1. *CRGC is polynomial.*

Problem 7 : Complete Rule Generation Twice (CRGT)
Instance : Σ_K and $\Sigma_{\tilde{K}}$.
Question : Compute $\Sigma_{K|\tilde{K}}$.

If the input elements Σ_K and $\Sigma_{\tilde{K}}$ are not already computed, their calculation is done in a polynomial time with respect to the size of concept lattices $\mathcal{B}(K)$ and $\mathcal{B}(\tilde{K})$.

This problem is open and will be discussed in the next section where we illustrate through an example that $\Sigma_{K|\tilde{K}}$ is not unique when Σ_K and $\Sigma_{\tilde{K}}$ are given.

The following table summarizes the identified problems in terms of the input, output, complexity and size of the output.

Table 3. List of problems and their related complexity

Problem	Input	Output	Class of Complexity	Size of the output
1. PRG	K	Σ_K	open	exponential
2. NRG	K	$\Sigma_{\tilde{K}}$	open	exponential
3. CRG	K	$\Sigma_{K|\tilde{K}}$	open	exponential
4. DRG	K, Σ_K	$\Sigma_{\tilde{K}}$	open	exponential
5. NCG	$K, \mathcal{B}(K)$	$\mathcal{B}(\tilde{K})$	polynomial	exponential
6. CRGC	$K, \mathcal{B}(K), \mathcal{B}(\tilde{K})$	$\Sigma_{K|\tilde{K}}$	polynomial	polynomial
7. CRGT	$\Sigma_K, \Sigma_{\tilde{K}}$	$\Sigma_{K|\tilde{K}}$	open	open

5 Complete Set of Association Rules

As indicated earlier, the complete set of association rules includes purely positive ones, purely negative ones as well as mixed rules. While the first two sets are relatively easy to compute, the last group raises some challenging issues if we attempt to generate mixed rules based only on the information about purely positive and purely negative rules. In this section, we propose a set of inference rules to deduce a partial set of mixed association rules, using either positive, negative or even mixed association rules. Then, we show that in general cases, it is impossible to compute all mixed rules from the sets of positive and negative association rules.

It is important to note that the first four propositions making reference to context K and its corresponding generic basis Σ_K can be adapted to hold for $\tilde{K}$ and its related generic basis $\Sigma_{\tilde{K}}$.

5.1 Properties

Let $K = (G, M, I)$ be a formal context, Σ_K the set of positive association rules and $\Sigma_{\tilde{K}}$ the set of negative association rules. For every attribute x, a concept of $K|\tilde{K}$ cannot contain x and $\tilde{x}$ unless its extent is an empty set.

Property 1. For all $x \in M$, we have $x\tilde{x} \rightarrow M\tilde{M}\backslash\{x\tilde{x}\}$ $[0] \in \Sigma_{K|\tilde{K}}$.

Proposition 1. *Let K be a context, and $A \rightarrow B$ [sup] $\in \Sigma_K$ an association rule with sup > 0. Then, $A \rightarrow B$ is derivable from rules in $\Sigma_{K|\tilde{K}}$ by Armstrong's inference axioms[2] [18].*

Proof. If $A \rightarrow B$ [sup] $\in \Sigma_K$, then there are two concepts: $C = (X, A \cup B)$ in $\mathcal{B}(K)$ such that A is a generator of the intent $A \cup B$, and $C' = (X, T)$ in $\mathcal{B}(K|\tilde{K})$ such that $A \cup B \subseteq T$. In particular, $A \rightarrow B$ [sup] holds with the same set X of supporting objects in $\Sigma_{K|\tilde{K}}$. □

---

[2] The inference system for functional dependencies includes reflexivity, augmentation, and transitivity axioms. Additional axioms are: union, decomposition and pseudo-transitivity.

Based on our previous remark, the above proposition also holds for any negative rule $\tilde{C} \to \tilde{D}$. For example (see Table 2), $\tilde{b} \to \tilde{c}$ [0.25] $\in \Sigma_{\tilde{K}}$ can be inferred from $\tilde{b} \to ad\tilde{c}$ [0.25] $\in \Sigma_{K|\tilde{K}}$.

Given an initial context K, the following proposition shows that extracted positive association rules with a non null support help generate some mixed association rules of the context $K|\tilde{K}$.

Proposition 2. *Let K be a context and $Ax \to y$ [sup] $\in \Sigma_K$ an association rule with sup > 0. Then $A\tilde{y} \to \tilde{x}$ [sup₁] is an association rule in $K|\tilde{K}$, with sup₁ possibly different from sup.*

Proof. We consider two distinct cases:

- Y is a closed itemset (intent) of $K|\tilde{K}$ such that $A\tilde{y} \subseteq Y$. Since $Ax \to y \in \Sigma_{K|\tilde{K}}$ (by Proposition 1), the attribute x cannot be in Y because otherwise Y contains y. Therefore, $A\tilde{y} \to \tilde{x} \in \Sigma_{K|\tilde{K}}$.
- Y is not a closed itemset of $K|\tilde{K}$ such that $A\tilde{y} \subseteq Y$. Then, $A\tilde{y} \to M \cup \tilde{M}$[0], and subsequently $A\tilde{y} \to \tilde{x}$.   □

For example (see Table 2), from $c \to bd$ [0.25] in Σ_K, one can infer $\tilde{b} \to \tilde{c}d$ [0.25] and $\tilde{d} \to \tilde{c}b$ [0.25].

In a similar way, we show that we can deduce some mixed association rules using positive association rules with a null support.

Proposition 3. *Whenever the association rule $Ax \to M\backslash\{Ax\}$ [0] holds in Σ_K, then the rule $A \to \tilde{x}$ [sup] holds in $\Sigma_{K|\tilde{K}}$ (with sup > 0), and vice versa.*

Proof. If $Ax \to M\backslash\{Ax\} \in \Sigma_K$ has a null support, then there is no closed set other than $M \cup \tilde{M}$ in $K|\tilde{K}$ which contains Ax. Thus, any closed set of $K|\tilde{K}$ which contains A, does not contain x and therefore contains $\tilde{x}$. Moreover, since a generic basis uses the notion of generator to extract exact rules, the closure of A is a proper subset of $M \cup \tilde{M}$. Therefore, there exists at least one object in G which owns the attributes in A, and hence sup > 0. Conversely, if $A \to \tilde{x}$ [sup] holds in $K|\tilde{K}$, then there does not exist a rule in Σ_K with a premise Ax and a non null support. Consequently, $Ax \to M\backslash\{Ax\}$ [0] holds in Σ_K.   □

Example 5. From rule $abd \to c$ [0] in Σ_K (see Example 1), one can infer $ab \to \tilde{d}$ [0.25], $ad \to \tilde{b}$ [0.25], and $bd \to \tilde{a}$ [0.5] in $\Sigma_{K|\tilde{K}}$. Conversely, from $ad \to \tilde{b}$ [0.25], one can infer that $abd \to M\backslash\{a\}$ [0] holds in Σ_K.

The following proposition states that when a context K has at least one object with all the attributes in M, then there does not exist a mixed rule of the form $A \to \tilde{x}$.

Proposition 4. *Let $K = (G, M, I)$ be a formal context. If the infimum of the lattice has a non empty extent (i.e., at least one object in K has all the attributes in M), then $A \to \tilde{x}$ [sup] $\notin \Sigma_{K|\tilde{K}}$, with sup not null and for any $A \subseteq M$ and $\tilde{x} \in \tilde{M}$.*

Proof. Let $A \subseteq M$ and o be an object in G which owns all attributes in M. Clearly, the extent corresponding to A must contain o and the closure of A in $K|\tilde{K}$ does not include any $\tilde{x} \in \tilde{M}$ since the complementary tuple of o in $\tilde{K}$ contains zero values only. □

The following example illustrates this proposition.

Example 6. Object 5 owns all attributes and any subset A in M (e.g., bd, bcd) can not imply a negative item (with a non null support).

$K|\tilde{K}$	a	b	c	d	$\tilde{a}$	$\tilde{b}$	$\tilde{c}$	$\tilde{d}$
1	1	0	0	1	0	1	1	0
2	0	1	0	1	1	0	1	0
3	1	1	0	0	0	0	1	1
4	0	1	1	1	1	0	0	0
5	1	1	1	1	0	0	0	0

Proposition 5. *The rule $A \to \tilde{x}$ [sup] $\in \Sigma_{K|\tilde{K}}$ for $A \subseteq M \cup \tilde{M}$ is equivalent to $Ax \to M\tilde{M}\backslash\{Ax\}$ [0] $\in \Sigma_{K|\tilde{K}}$.*

This proposition is a generalization of Proposition 3 to the case of $A \subseteq M \cup \tilde{M}$, and helps infer new mixed rules from existing ones. For example, $adb \to M\tilde{M}\backslash\{adb\}$ [0] is inferred from $ad \to \tilde{b}$ [0.25].

Proof. This statement can be proved by conducting a reasoning similar to the one provided for Proposition 3. □

5.2   Computation of Mixed Rules from Positive and Negative Rules

In this part, we show that the sets of purely positive and purely negative rules are not enough to infer the whole set of mixed rules. The reason is that different formal contexts can be associated with a same couple of sets Σ_K and $\Sigma_{\tilde{K}}$.

Example 7. Consider K_1 as a reduced context of K_2 where $f(6) = f(3) \cap f(4)$ in K_2. The sets Σ_{K_1} and Σ_{K_2} have the same collection of rules but matching rules have distinct (non null) supports because the number of objects in K_1 is smaller than the object set in K_2. In this special case, $\Sigma_{\tilde{K}_1}$ and $\Sigma_{\tilde{K}_2}$ have also the same set of rules. However, $\Sigma_{K_1|\tilde{K}_1}$ is different from $\Sigma_{K_2|\tilde{K}_2}$. For example, $ab\tilde{c}\tilde{d} \to \tilde{e}$ [0.2] belongs to $\Sigma_{K_1|\tilde{K}_1}$ but does not appear in $\Sigma_{K_2|\tilde{K}_2}$.

   This example illustrates the fact that if we consider positive and negative association rules without key information about the corresponding context, it is not possible to generate in a unique way the whole set of mixed association rules.

Proposition 6. *Given a set Σ_K of purely positive rules and a set $\Sigma_{\tilde{K}}$ of purely negative rules extracted from a context K, then rules in $\Sigma_{K|\tilde{K}}$ may not be completely derivable from Σ_K and $\Sigma_{\tilde{K}}$ using the proposed inference rules.*

Proof. Since for a same set of (positive or negative) rules, there may exist more than one context, the apposition $K|\tilde{K}$ will lead to different contexts and hence different sets $\Sigma_{K|\tilde{K}}$.

K_1	a	b	c	d	e
1	1	1	0	0	0
2	1	0	0	0	1
3	1	1	1	0	1
4	1	1	0	1	1
5	0	1	0	0	1

K_2	a	b	c	d	e
1	1	1	0	0	0
2	1	0	0	0	1
3	1	1	1	0	1
4	1	1	0	1	1
5	0	1	0	0	1
6	1	1	0	0	1

K_1 K_2

Fig. 3. K_1 is a reduced context of K_2

6 Conclusion

This paper studies the problem of computing the generic basis of positive, negative and mixed rules from a given input. To that end, a set of situations are proposed based on the type of available input (e.g., the formal context $K = (G, M, I)$, the set Σ_K of positive rules, the concept lattice $\mathfrak{B}(K)$) and the sort of output to produce (e.g., $\mathfrak{B}(K)$, the set of negative rules $\Sigma_{\tilde{K}}$, the whole set of rules $\Sigma_{K|\tilde{K}}$).

We also propose a set of inference rules to deduce a partial set of mixed association rules, using either positive, negative or mixed rules. Then, we illustrate through an example the fact that the sets of positive and negative association rules are not enough to generate the whole set of mixed rules.

We are currently working on the characterization of mixed rules that are missed by our inference rules and the efficient computation of the complete set of rules $\Sigma_{K|\tilde{K}}$.

Since generators are needed for generic basis computation, we plan to study the case when the generators associated with concepts in contexts K and $\tilde{K}$ are given, and design an efficient algorithm for computing concepts, generators and rules for context $K|\tilde{K}$.

Acknowledgment

The first author acknowledges the financial support of the Natural Sciences and Engineering Research Council of Canada (NSERC). All the authors would like to thank anonymous referees for providing constructive feedback.

References

1. Agrawal, R., Srikant, R.: Fast algorithms for mining association rules. pp. 487–499 (September 1994)
2. Alachaher, L.N., Guillaume, S.: Mining negative and positive influence rules using kullback-leibler divergence. ICCGI 00, 25 (2007)
3. Antonie, M.-L., Zaïane, O.R.: Mining positive and negative association rules: An approach for confined rules. In: Boulicaut, J.-F., et al. (eds.) PKDD 2004. LNCS (LNAI), vol. 3202, pp. 27–38. Springer, Heidelberg (2004)

4. Boulicaut, J.-F., Bykowski, A., Jeudy, B.: Towards the tractable discovery of association rules with negations. In: FQAS, pp. 425–434 (2000)
5. Brin, S., Motwani, R., Silverstein, C.: Beyond market baskets: generalizing association rules to correlations. In: SIGMOD '97: Proceedings of the 1997 ACM SIGMOD international conference on Management of data, pp. 265–276. ACM Press, New York (1997)
6. Eiter, T., Gottlob, G.: Hypergraph transversal computation and related problems in logic and ai. In: Flesca, S., et al. (eds.) JELIA 2002. LNCS (LNAI), vol. 2424, pp. 549–564. Springer, Heidelberg (2002)
7. Fredman, M.L., Khachiyan, L.: On the complexity of dualization of monotone disjunctive normal forms. J. Algorithms 21(3), 618–628 (1996)
8. Ganter, B., Wille, R.: Contextual attribute logic. In: Tepfenhart, W.M. (ed.) ICCS 1999. LNCS, vol. 1640, pp. 377–388. Springer, Heidelberg (1999)
9. Ganter, B., Wille, R.: Formal Concept Analysis: Mathematical Foundations. Springer, New York, Translator-C. Franzke (1999)
10. Guigues, J.L., Duquenne, V.: Familles minimales d'implications informatives résultant d'un tableau de données binaires. Mathématiques et Sciences Humaines 95(1), 5–18 (1986)
11. Khachiyan, L., et al.: A global parallel algorithm for the hypergraph transversal problem. Inf. Process. Lett. 101(4), 148–155 (2007)
12. Kryszkiewicz, M., Gajek, M.: Concise representation of frequent patterns based on generalized disjunction-free generators. In: Chen, M.-S., Yu, P.S., Liu, B. (eds.) PAKDD 2002. LNCS (LNAI), vol. 2336, pp. 159–171. Springer, Heidelberg (2002)
13. Mannila, H., Toivonen, H.: Multiple uses of frequent sets and condensed representations (extended abstract). In: KDD, pp. 189–194 (1996)
14. Nourine, L., Raynaud, O.: A fast incremental algorithm for building lattices. J. Exp. Theor. Artif. Intell. 14(2-3), 217–227 (2002)
15. Pasquier, N., et al.: Efficient Mining of Association Rules Using Closed Itemset Lattices. Information Systems 24(1), 25–46 (1999)
16. Pfaltz, J., Taylor, C.: Scientific discovery through iterative transformations of concept lattices. In: Proceedings of the 1st International Workshop on Discrete Mathematics and Data Mining, April 2002, pp. 65–74 (2002)
17. Savasere, A., Omiecinski, E., Navathe, S.B.: Mining for strong negative associations in a large database of customer transactions. In: ICDE, pp. 494–502 (1998)
18. Ullman, J.D., Widom, J.: A First Course in Database Systems. Prentice-Hall, Englewood Cliffs (1997)
19. Valtchev, P., Missaoui, R., Lebrun, P.: A partition-based approach towards constructing galois (concept) lattices. Discrete Math. 256(3), 801–829 (2002)
20. Valtchev, P., Missaoui, R., Godin, R.: Formal concept analysis for knowledge discovery and data mining: The new challenges. In: Eklund, P.W. (ed.) ICFCA 2004. LNCS (LNAI), vol. 2961, pp. 352–371. Springer, Heidelberg (2004)
21. Wu, X., Zhang, C., Zhang, S.: Efficient mining of both positive and negative association rules. ACM Trans. Inf. Syst. 22(3), 381–405 (2004)
22. Yuan, X., et al.: Mining negative association rules. In: ISCC 2002: Proceedings of the Seventh International Symposium on Computers and Communications, Washington, DC, USA, p. 623. IEEE Computer Society, Los Alamitos (2002)
23. Mohammed Javeed Zaki and Ching-Jiu Hsiao. Charm: An efficient algorithm for closed itemset mining. In: Proceedings of the Second SIAM International Conference on Data Mining, April 11-13, 2002, Arlington, VA, USA (2002)

On the Merge of Factor Canonical Bases

Petko Valtchev[1] and Vincent Duquenne[2]

[1] Département d'informatique, UQÀM, C.P. 8888, Succ. CV, Montréal (Qc), Canada
[2] CNRS - UMR 7090 - ECP6, 175, rue Chevaleret, 75013 Paris, France

Abstract. *Formal concept analysis* (FCA) has a significant appeal as a formal framework for knowledge discovery not least because of the mathematical tools it provides for a range of data manipulations such as splits and merges. We study the computation of the *canonical* basis of a context starting from the bases of two apposed subcontexts, called *factors*. Improving on a previous method of ours, we provide here a deeper insight into its pivotal implication family and show it represents a *relative basis*. Further structural results allow for more efficient computation of the global basis, in particular, the relative one admits, once added to factor bases, an inexpensive reduction. A method implementing the approach as well as a set of further combinatorial optimizations is shown to outperform *NextClosure* on at least one dataset.

1 Introduction

Since the early eighties, formal concept analysis (FCA) [10] has developed as a series of technics for representing and structuring qualitative data. Albeit rooted in lattice theory [3], it has produced a strong impact in data analysis and later on in data mining, generating a large body of work on closed-itemset representations [16, 25, 17] and association rule bases [19, 24], within the association rule mining (ARM) discipline [1]. This impact is a direct consequence of the double concern within FCA with both the concept systems (formalized as lattices) and the implication families stemming out of a tableau-shaped binary relation (formal context).

The duality between lattices (equivalently in the finite case semilattices/closure operators, etc.) and implications (also called rules/dependencies) can be simply spelled: in extracting a semilattice out of a Boolean lattice, the more elements one removes, the more implications are generated and *vice versa*. Hence semilattices and implicational systems represent two sides of a same reality indicating, roughly speaking, what is existing/missing in the data. The relationship between lattices and implications has always been a key concern in FCA [23]. Noteworthily, this duality is known and used in the theory of relational databases [5] and in artificial intelligence [13, 12]. For instance, in database normalization theory, the computation of a minimal cover of a family of functional dependencies is a major task whose resolution makes extensive use of that duality [14, 15].

Usually, when dealing with combinatorial objects, a first concern is to put them in some "canonical" form. For a lattice such a form is the reduced standard context that only comprises irreducible intents from the semi-lattice. The implicational system can in turn be reduced to a basis of minimal size whereby the canonical basis [11] (called

R. Medina and S. Obiedkov (Eds.): ICFCA 2008, LNAI 4933, pp. 182–198, 2008.

Duquenne-Guigues in FCA's folklore [10], section 2.3) enjoys a particular status among all possible bases since uniquely defined. It is defined on top of the family of *pseudo-closed* of a binary relation which are still much of a challenge since their inherently recursive definition prevents an easy computation.

A second concern arises with the combinatorial generation of lattices and canonical bases. Both problems are hard ones as the respective target structures may grow exponentially in the size of the input context.

The overall goal of our study is to improve the construction of the canonical basis by revisiting a product/merge-based approach. The latter has proven efficient on lattice construction [21], by embedding the lattice into a semilattice product of its factors. The motivation behind our study is three-fold:

1. From a practical viewpoint, in cases where the two factor lattices/bases are already computed, reusing them to speed-up the computation the global canonical basis is a matter of principle. This will be useful for data merging purposes or, alternatively, for revision/maintenance of analysis results in time. Even if the factor results are not given beforehand, in many situations it might still be advantageous to operate the problem reduction with the split-process-merge schema.
2. On the algorithmic axis, we look for a new compromise between time/memory consumption in basis computation. While the classic NEXTCLOSURE algorithm [9, 10] requires minimal memory allocation, our proposal achieves a significant search space reduction through the factor result reuse which is expected to save significant amount of time. In fact, although the direct product of two semi-lattices might be huge, it will usually remain much smaller than the powerset of the attributes.
3. From a methodological viewpoint, we are intrigued by the potential gains from mixing the use of merge/apposition of contexts/product of semilattices with the simultaneous computation of both the lattice and the canonical basis. In this respect, summarizing the gap between a merged lattice and the full product of its factors – at last providing tools to decipher the structure of a nested line diagram – can be most useful as a measure of what could be called algebraic independence between two factors.

Yet the pseudo-closed sets of a context, unlike closed ones, do not project on pseudo-closed of the apposed factors, whereas their strongly recursive nature of pseudo-closed sets imposes ordering constraints on their generation. Consequently, a direct product-based generate-and-test approach as in [21] will not work.

In [20] we presented a first method for computing the canonical basis from factor bases and intent semilattices. It exploits what we called the *hybrid* implications, i.e., whose premises mix attributes from both factor contexts, and defines a basis thereof. The hybrid basis is computed step-wise along a traversal of the direct product of the factor semilattices. As hybrid and factor canonical bases jointly form a cover of the global basis, the latter is yielded through a classical reduction [15].

Here we propose a further insight into the relationships between the global and the factor pseudo-closed families that underlies an alternative basis definition which also exploits the notion of *relative* pseudo-closedness as defined in [18]. Additional structural results are provided that enabled the design of a more parsimonious reduction method for the cover of the canonical basis.

In what follows, we first recall the theory behind concept lattices, lattice assembly and the canonical basis (section 2). Then we summarize the original approach for merging factor bases (section 3). Next, the new basis definition and the corresponding algorithmic design are presented (section 4) followed by the results on reduction and the method description (section 5). Finally, details about current implementation and some performance evidence are provided (section 6).

2 Background on Relevant FCA Notions

Formal concept analysis (FCA) [10] studies the various combinatorial structures induced by a *Galois connection* [6] between two power-set lattices and approaches these from both theoretical and applied, i.e., data analysis, standpoints.

2.1 Contexts, Concepts and Lattices

We use the standard FCA constructs and terminology (see [10]) except for a small set of notations. Thus, we consider a dataset made of *objects* (O) described by *attributes* (A) which are introduced by a two-way data table $\mathcal{K}$, the (formal) *context*, expressing the incidence relation I between objects and attributes. Two *derivation* operators $'$ expand I: the object one maps sets of objects to the maximal sets of attributes that are shared by the objects while the attribute one works dually. Both $'$ operators form a *Galois connection* between $\wp(O)$ and $\wp(A)$ whose closed subset families, denoted hereafter $\mathcal{C}_{\mathcal{K}}^{o}$ and $\mathcal{C}_{\mathcal{K}}^{a}$, respectively, once provided with set-theoretic inclusion form two semi-lattices that are dually isomorphic. The pairs of mutually corresponding closed sets, i.e., the *concepts* ((X, Y) where $X = Y'$ and $Y = X'$), constitute a complete lattice, the *concept lattice* of $\mathcal{K}$ (*aka* the *Galois* lattice of I). Fig. 1 shows a sample context (left) and its lattice (right).

2.2 Implications

An *implication* (aka functional dependency) $X \to Y$ ($X, Y \subseteq A$) is an implicit assertion that "any object having X also has Y". $X \to Y$ is *valid* in a context whenever no

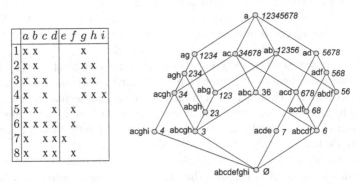

Fig. 1. A sample context (left) and its corresponding concept lattice (right). Both adapted from [10].

object violates it, which basically amounts to $Y \subseteq X''$ (Prop. 19 in [10]). Given $\mathcal{K}$, its family of valid implications, $\Sigma_{\mathcal{K}}$, is usually prohibitively large whereas many implications are *redundant*. Redundancy reduces to *derivation* ($\models$) between implication sets: $\Sigma_1 \models \Sigma_2$ means, in FCA-terminology, that contexts validating Σ_1 necessarily validate Σ_2 too (Σ_1 is called a *cover* of Σ_2). $X \rightarrow Y$ is redundant within $\Sigma \subseteq \wp(A)^2$ if it can be derived from the the the rest of Σ, i.e., $\Sigma - \{X \rightarrow Y\} \models X \rightarrow Y$. Armstrong axioms [2] automate derivation yet the Σ-*closure* operator [15] (denoted $\_^{\Sigma}$) is more practical. Given Y, Y^{Σ} is the infinite member X^{∞} of the sequence X^i below:

$$X^0 = Y; \quad X^{k+1} = X^k \cup \bigcup_{Z \rightarrow T \in \Sigma; Z \subseteq X^k} T$$

It is readily shown that for any $\mathcal{K}$ both closure operators are identical, i.e., for all $Y \subseteq A$, $Y'' = Y^{\Sigma_{\mathcal{K}}}$. Now redundancy of $X \rightarrow Y$ in Σ means its removal does not change the closure operator, i.e., $Y \subseteq X^{\Sigma - \{X \rightarrow Y\}}$.

A non-redundant cover of $\Sigma_{\mathcal{K}}$, i.e., one free of redundant implications, whose cardinality is minimal, is a *minimal cover* [14]. Similar to reduced contexts for lattices minimal covers of $\Sigma_{\mathcal{K}}$ represent it in the most compact way. Among all possible minimal covers, the *canonical* basis, also known as *Duquenne-Guigues* [11] or *stem* basis [10], is the only one admitting a unique definition. A generalization thereof is used here which comprises a set of *a priori* implications (due to Stumme [18]). Thus, given $\mathcal{K} = (O, A, I)$ and $\Omega \subseteq \Sigma_{\mathcal{K}}$, the Ω-basis of $\mathcal{K}$, $\mathcal{B}_{\Omega}$ is composed of all implications $Y \rightarrow Y''$ where Y is Ω-*pseudo-closed* as defined below:

Definition 1. *Given a context $\mathcal{K} = (O, A, I)$ and a set of valid implications $\Omega \subseteq \Sigma_{\mathcal{K}}$, a set $Y \subseteq A$ is a pseudo-closed relative to Ω (or Ω-pseudo-closed) if:*

- *$Y \neq Y''$,*
- *$Y = Y^{\Omega}$ (closed for Ω, but not for $\Sigma_{\mathcal{K}}$),*
- *for each Ω-pseudo-closed Z, $Z \subset Y$ entails $Z'' \subseteq Y$.*

$\mathcal{B}_{\Omega}$ behaves like a minimal cover relative to Ω, i.e., (i) $\mathcal{B}_{\Omega} \cup \Omega \models \Sigma_{\mathcal{K}}$, (ii) it is non-redundant modulo Ω ($\forall X \rightarrow Y \in \mathcal{B}_{\Omega}$, $\Omega \cup \mathcal{B}_{\Omega} - \{X \rightarrow Y\} \not\models X \rightarrow Y$), and (ii) has minimal cardinality. To get the standard canonical basis $\mathcal{B}_{\mathcal{K}}$, Ω is set to $\emptyset$ in Definition 1 which yields the pseudo-closure family $\mathcal{PC}_{\mathcal{K}}$. $\mathcal{B}_{\mathcal{K}}$ of the running example is:

$$adg \rightarrow bcefhi \quad abcghi \rightarrow def \quad acg \rightarrow h \quad ah \rightarrow g \quad \rightarrow a$$
$$acdef \rightarrow bghi \quad ai \rightarrow cgh \quad abd \rightarrow f \quad ae \rightarrow cd \quad af \rightarrow d$$

Furthermore, $\mathcal{PC}_{\mathcal{K}} \cup \mathcal{C}_{\mathcal{K}}^a$ defines a richer closure semi-lattice whose closure operator is a refinement of $\_^{\Sigma_{\mathcal{K}}}$ (which, trivially, equals $\_^{\mathcal{B}_{\mathcal{K}}}$). Thus, following Prop. 26 in [10], the operator $\_^{\mathcal{B}_{\mathcal{K}}^{\neq}}$ on $Y \subseteq A$ is defined as the infinite member X^{∞} in:

$$X^0 = Y; \quad X^{k+1} = X^k \cup \bigcup_{Z \rightarrow T \in \mathcal{B}_{\mathcal{K}}; Z \subsetneq X^k} T$$

Observe that testing X for pseudo-closedness only requires the portion of $\mathcal{B}_{\mathcal{K}}$ made of implications $Y \rightarrow Z$ where $Y \subset X$: X fails the test iff an invalidating implication is found ($Y \subset X$ and $Z \not\subseteq X$). Hence the $\mathcal{B}_{\mathcal{K}}$ could be computed gradually provided the pseudo-closures are generated in an order compatible with set inclusion (see NEXTCLOSURE in Section 2.4). The otherwise recursive generation is grounded on the layer of minimal pseudo-closures which correspond to the minimal non-closed sets $(\min(\mathcal{PC}_{\mathcal{K}}) = \min(\wp(A) - \mathcal{C}_{\mathcal{K}}^a))$.

In a further restriction of the admitted implications from a cover $\Theta \models \Sigma_{\mathcal{K}}$, one bans all rules whose premises have the same $''$-closure as the argument set Y. This yields the *saturation* operator [8], denoted $\_^{\Theta \otimes 1}$. The $\_^{\Theta \otimes}$ closures are called *quasi-closed* sets. Their family, $\mathcal{QC}_{\mathcal{K}}$, comprises, but is not limited to, both $''$ closures and pseudo-closures: $\mathcal{PC}_{\mathcal{K}} \cup \mathcal{C}_{\mathcal{K}}^a \subseteq \mathcal{QC}_{\mathcal{K}}$. Within $\mathcal{QC}_{\mathcal{K}}$, pseudo-closures are uniquely determined as both non $\Sigma_{\mathcal{K}}$-closed and minimal for their respective $\Sigma_{\mathcal{K}}$-closures.

2.3 Context Splits and Lattice Products

Two contexts sharing their objects are said to be *apposed* if they agree on all common attributes (see Chap.3 in [10]). Thus, given $\mathcal{K}_1 = (O, A_1, I_1)$ and $\mathcal{K}_2 = (O, A_2, I_2)$, the context $\mathcal{K}_3 = (O, A_1 \dot\cup A_2, I_1 \dot\cup I_2)$ is called the *apposition* of $\mathcal{K}_1$ and $\mathcal{K}_2$. Apposition models in a natural way useful manipulations of relational database such as vertical table splits (in a distributed database) or projections upon complementary views. For example, the context in Fig. 1, can be split into two sub-contexts by dividing A into $A_1 = \{a, b, c, d\}$ and $A_2 = \{e, f, g, h, i\}$. The corresponding lattices, further referred to as *factor* ones, $\mathcal{L}_1$ and $\mathcal{L}_2$, are drawn on the left-hand-side of Fig. 2. The canonical bases $\mathcal{B}_1$ and $\mathcal{B}_2$ are given below:

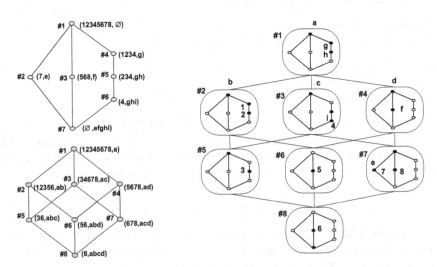

Fig. 2. Left: Factor lattices $\mathcal{L}_1$ and $\mathcal{L}_2$ of the apposed contexts in Fig. 1; **Right:** The NLD thereof.

$\mathcal{B}_1$	$\mathcal{B}_2$		
$\to a$	$i \to gh$	$h \to g$	$fg \to ehi$
	$eg \to fhi$	$ef \to ghi$	

---

[1] Summarizes the *direct determination* relation between sets as in [14].

A natural question is how $\mathcal{L}_3$ relates to factors $\mathcal{L}_1$ and $\mathcal{L}_2$. In fact, $\mathcal{L}_1$ and $\mathcal{L}_2$ only represent real factors, i.e., constructed by means of a congruence relation, whenever reduced to their (intent) *join*-semi-lattices [10]. In the extreme case, $\mathcal{L}_3$ will be isomorphic to the direct product of the factors, $\mathcal{L}_1 \times \mathcal{L}_2$ (denoted shortly $\mathcal{L}_{1,2}$ hereafter), while in general it will be merely isomorphic to a *join* sub-semi-lattice of $\mathcal{L}_{1,2}$. Spelled in more intuitive terms, the global intents in $\mathcal{C}_3^a$ are unions of two factor intents, one from each $\mathcal{C}_i^a$ ($i = 1, 2$). However, some unions in the resulting family $\mathcal{C}_{1,2}^a = \{Y_1 \cup Y_2 \mid Y_i \in \mathcal{C}_i^a, \ i = 1, 2\}$ may be non closed in $\mathcal{K}_3$.

Apposition underlies the *nested line diagrams* (NLD) visualization mode for lattices [10] which typically presents $\mathcal{L}_3$ as embedded into $\mathcal{L}_{1,2}$. The embedding of the example global lattice into the factor product is illustrated on the right-hand-side of Fig. 2 where filled nodes correspond to concepts (global intents in $\mathcal{C}_3^a$) and unfilled ones to non-closed factor intent unions from $\mathcal{C}_{1,2}^a$.

2.4   Computing the Canonical Basis

The interplay between implications and intents, on the one hand, and between exhaustive structures and minimal representations, on the other hand, generates a number of challenging problems, many of whom are known to be hard[2].

The computation of the canonical basis and of the pseudo-closed family has been explored under various settings yet few practical algorithms have been published. Among them, NEXTCLOSURE [9] is the reference. In fact, NEXTCLOSURE is a generic algorithm for computing the closed sets of any closure operator provided as parameter. Thus, for computing $\mathcal{C}_{\mathcal{K}}^a$, the $''$ operator is enough while $\mathcal{P}\mathcal{C}_{\mathcal{K}}$ requires also $\_^{\mathcal{B}_{\mathcal{K}}^{\neq}}$ (jointly emulated by $\_^{\mathcal{B}_{\mathcal{K}}}$ and the traversal order).

Technically speaking, it behaves as a typical listing procedure, i.e., one using a canonical representation for concept intents to generate them according to a fixed total order (called *lectic*). The search space, i.e., the Boolean lattice $2^{\mathbf{A}}$, is traversed in an ascending lectic order while computing the closure of the current element. Though a closure may be computed several times, it is only validated whenever canonically generated, i.e., by a minimal generating prefix thereof. The perceived strength of NEXTCLOSURE is in its minimal memory requirements – a single cell is used to store the current closure – whereas a perceived weakness thereof is the high number of computed closures (targeted with various optimization techniques).

In [8], $\mathcal{P}\mathcal{C}_{\mathcal{K}}$ is targeted with saturation. The starting point is the set of all minimal generators, i.e., X s.t., $\forall Y \subset X, Y'' \subset X''$, which are not closed whereas the saturation exploits the respective implications $X \to X''$.

$\mathcal{B}_{\mathcal{K}}$ can also be computed as a minimal cover of $\Sigma_{\mathcal{K}}$ by starting with an arbitrary cover thereof, say Θ, and applying a reduction procedure to Θ as described above. Classical reduction merely removes the redundant rules $X \to Y$ one by one while examining Θ in an arbitrary order. In fact, the removed rules fall in two categories: (i) rules for which there is no pseudo-closure in the order interval $[X, X^{\Theta}]$, and (ii) rules capturing a pseudo-closure for which another rule exists in the yet unreduced part of Θ. A polynomial algorithm for the task, LINCLOSURE (see Chap. 4 in [15]) jointly applies

---

[2] M. Wild proposes an exhaustive and insightful presentation of these problems in [22].

indexing and counters to avoid re-scaning of the entire implication set at each step of the iterative computation of $X^{\Sigma-\{X\to Y\}}$. Reduction is readily reorganized to output $\mathcal{B}_{\mathcal{K}}$ by replacing all non redundant premises X by the closure $X^{\Sigma-\{X\to Y\}}$ available at the end of the reduction task. Two variations of that algorithm aimed at the canonical basis can be found in [4].

3   Merge-Based Computation of the Canonical Basis

The relationships between the two factor lattices and the global one have been explored from a computational viewpoint, for constructing $\mathcal{L}_3$ and, more recently, $\mathcal{B}_3$.

In [21], a method for extracting $\mathcal{L}_3$ from $\mathcal{L}_{1,2}$ was proposed whose motivation could be summarized as follows. Provided $\mathcal{L}_1$ from $\mathcal{L}_2$ are available, the direct product $\mathcal{L}_{1,2}$ constitutes a (usually) much smaller search space for global intents (and concepts) than the entire Boolean lattice $2^{\mathbf{A}_3}$ (as most of the time $|\mathcal{L}_1| * |\mathcal{L}_2| \ll 2^{|A_3|}$). The proposed approach amounts to a bottom-up traversal of $\mathcal{L}_{1,2}$ which filters the closed intent unions in $\mathcal{C}^a_{1,2}$ and upgrades those to full-scale global concepts. By applying the technique to factors, i.e., in a *divide-and-conquer* manner, a complete lattice construction method is devised which has been shown to perform well on specific classes of datasets.

The equivalent for implications would be to combine $\mathcal{B}_1$ and $\mathcal{B}_2$ into a product-based search space, which motivated the study in [20]. As a direct merge-based approach for the simultaneous computation of both $\mathcal{C}^a_3$ and $\mathcal{B}_3$ from factor counterparts ($\mathcal{C}^a_i$ and $\mathcal{B}_i$, $(i = 1, 2)$) hit some hurdles (see Section 4.1), we moved to a less ambitious two-stage method. In short, the approach exploits the availability of $\mathcal{C}^a_{1,2}$ (hence the benefit of the direct product and the reduced search space it represents) to compute an approximation of $\mathcal{B}_3$, i.e., a structurally close cover thereof, which is then reduced to a minimal size and set in canonical form. The cover is made of both factor bases $\mathcal{B}_i$, $(i = 1, 2)$ plus *some* of the non-factor implications from Σ_3, called *hybrid*.

Technically speaking, the premise of a hybrid rule is a non 3-closed union from $\mathcal{C}^a_{1,2}$ and the conclusion its respective 3-closure[3]:

$$\mathcal{I}_{3/1,2} = \{Y \to Y^{33} \mid Y \in \mathcal{C}^a_{1,2} - \mathcal{C}^a_3\}.$$

It is readily shown that $\mathcal{I}_{3/1,2} \cup \mathcal{B}_1 \cup \mathcal{B}_2$ is a cover of of Σ_3, hence of $\mathcal{B}_3$. Intuitively, one shows that any $X \to Y \in \Sigma_3$ is derivable from $\mathcal{I}_{3/1,2} \cup \mathcal{B}_1 \cup \mathcal{B}_2$, by establishing $X^{\mathcal{I}_{3/1,2} \cup \mathcal{B}_1 \cup \mathcal{B}_2} = X^{33}$ This is done in two steps: (i) $X^{\mathcal{B}_1 \cup \mathcal{B}_2}$ is a member of $\mathcal{C}^a_{1,2}$, and (ii) provided the latter is not 3-closed (if it is, we are done), there is a rule in $\mathcal{I}_{3/1,2}$ where it is the premise so the rule "pushes" the final result up to the 3-closure of X.

Yet the goal is not to work with the entire $\mathcal{I}_{3/1,2}$ which can be huge but rather to reduce it to a smaller cover that would allow for reasonable reduction cost. Hence we defined a subset thereof, $\mathcal{B}_{3/1,2}$, that mimics (yet it is not equal to) the canonical basis, hence it name, the *hybrid basis*. Thus, we defined the family $\mathcal{PC}_{3/1,2}$:

Definition 1. *A set $Y \in \mathcal{C}^a_{1,2} - \mathcal{C}^a_3$ is in $\mathcal{PC}_{3/1,2}$ whenever for each $Z \in \mathcal{PC}_{3/1,2}$, $Z \subset Y$ implies $Z^{33} \subset Y$.*

---

[3] To avoid confusion, we denote hereafter $'$ in $\mathcal{K}_i$ by $\_^i$ and speak about i-(pseudo-)closed.

Now $\mathcal{B}_{3/1,2}$ is made of hybrid rules with premises in $\mathcal{PC}_{3/1,2}$. In our example, it is:

$$adg \rightarrow bcefhi \quad abcghi \rightarrow def \quad acg \rightarrow h \quad af \rightarrow d$$
$$aghi \rightarrow c \quad\quad abd \rightarrow f \quad\quad ae \rightarrow cd$$

Definition 1 insures a good trade-off between compact size and easy computation from the input structures. Indeed, as the example suggests, the two bases $\mathcal{B}_{3/1,2}$ and $\mathcal{B}_3$ have a large intersection.

The computation of $\mathcal{B}_{3/1,2}$ (see Algorithm 1) is a gradual process unfolding along a traversal of $\mathcal{C}^a_{1,2}$ that follows a linear extension of the set-inclusion order (e.g., lectic). To that end, the predicate associated with $\mathcal{B}^{\neq}$ as presented above is evaluated: A candidate Y fails the $\mathcal{B}^{\neq}_{3/1,2}$ test if an *invalidating rule* can be found in the already constructed part of $\mathcal{B}_{3/1,2}$, i.e., $Z \rightarrow T$ s.t., $Z \subseteq Y$ and $T \not\subseteq Y$.

A dedicated variant of the $\mathcal{B}^{\neq}$ predicate is used: ground hybrid pseudo-closures are all minimal non 3-closed elements from $\mathcal{C}^a_{1,2}$ while the members of $\mathcal{PC}_{3/1,2}$ at higher set-inclusion levels are exactly those who pass the $\mathcal{B}^{\neq}_{3/1,2}$ test. Once the hybrid basis is computed, a plain reduction step shrinks $\mathcal{B}_{3/1,2} \cup \mathcal{B}_1 \cup \mathcal{B}_2$ to $\mathcal{B}_3$.

1: **procedure** MERGE-BIS(**In:** $\mathcal{C}^a_1, \mathcal{C}^a_2$ set families; $\mathcal{B}_1, \mathcal{B}_2$ implication sets; **Out:** $\mathcal{C}^a_3$ set family; $\mathcal{B}_3$ an implication set)

2:

3: $\mathcal{C}^a_3 \leftarrow \emptyset$

4: $\mathcal{B}_w \leftarrow \emptyset$ {The already discovered part of $\mathcal{B}_{3/1,2}$}

5: SORT($\mathcal{C}^a_1$); SORT($\mathcal{C}^a_2$)   {Decreasing order; if needed}

6: **for all** (Y_1, Y_2) **in** $\mathcal{C}^a_1 \times \mathcal{C}^a_2$ **do**

7:   $Y \leftarrow Y_1 \cup Y_2$

8:   $r \leftarrow$ FINDINVALIDATINGRULE($\mathcal{B}_w, Y$)   {The $\mathcal{B}^{\neq}_{3/1,2}$ test}

9:   **if** $r =$ NULL **then**

10:     $Y_c \leftarrow$ CLOSURE(Y)   {``Survived'', test for closure}

11:     **if** $Y = Y_c$ **then**

12:       $\mathcal{C}^a_3 \leftarrow \mathcal{C}^a_3 \cup \{Y\}$ {Global closure detected}

13:     **else**

14:       $r_n \leftarrow$ NEWRULE(Y, Y_c)

15:       $\mathcal{B}_w \leftarrow \mathcal{B}_w \cup \{r_n\}$ {Global pseudo-closure, a new rule in $\mathcal{B}_w$}

16: $\mathcal{B}_3 \leftarrow$ REDUCE($\mathcal{B}_w, \mathcal{B}_1, \mathcal{B}_2$)

Algorithm 1. Straightforward merge of factor intent families and canonical bases

4 An Optimized Merge of Factor Bases

$\mathcal{B}_{3/1,2}$ happens to possess further properties enabling an improved algorithmic design.

4.1 Do We Need an Intermediate Cover?

As our initial motivation was to design a structure emulating $\mathcal{C}^a_{1,2}$ for $\mathcal{PC}_3$ it is worth examining the real necessity of using the hybrid basis. In fact, given the available mathematical and algorithmic tools there are four hypothesis about the possible search space to yield pseudo-closures by a product-like transformation. Spelled differently, global pseudo-closures may systematically split along the factor attribute sets into factor:

H1: pseudo-closures, i.e., $\mathcal{PC}_3 \subseteq \{Y_1 \cup Y_2 \mid Y_i \in \mathcal{PC}_i, \ i = 1, 2\}$.
H2: closures, i.e., $\mathcal{PC}_3 \subseteq \{Y_1 \cup Y_2 \mid Y_i \in \mathcal{C}_i^a, \ i = 1, 2\}$.
H3: pseudo-closures or closures, i.e., $\mathcal{PC}_3 \subseteq \{Y_1 \cup Y_2 \mid Y_i \in \mathcal{C}_i^a \cup \mathcal{PC}_i, \ i = 1, 2\}$.
H4: quasi-closures.

Obviously, **H1** and **H2** are mutually exclusive and also stronger than **H3** (i.e., imply it) whereas **H3** implies **H4**. So far, **H1** and **H2** have been invalidated by immediate counter-examples (e.g., in the above context acg, $ai \in \mathcal{PC}_3$ but $ac = acg \cap A_1$ is not in $\mathcal{PC}_1$ while $i = ai \cap A_2$ is not in $\mathcal{C}_2^a$). Moreover, **H4** trivially holds.

Proposition 1. *Given contexts* $\mathcal{K}_i = (O, A_i, I_i)$ $(i = 1, 2, 3)$ *such that* $\mathcal{K}_3 = \mathcal{K}_1 | \mathcal{K}_2$, *for any* $Y \in \mathcal{PC}_3$ *and* $Y_j = Y \cap A_j$ $(j = 1, 2)$ *holds* $Y_j \in \mathcal{QC}_j$

In summary, the only open question remains whether mixing factor closures and pseudo-closures would be enough (**H3**).

From a computational viewpoint, however, Proposition 1 is of little practical use as quasi-closed sets may outnumber by far pseudo-closed plus closed. Thence there is no point in computing and maintaining them which further means that pending a proof of **H3** the only reasonable alternative is to use an intermediate cover of $\mathcal{B}_3$ such as $\mathcal{B}_{3/1,2}$.

4.2   What $\mathcal{B}_{3/1,2}$ Represents and Reveals about the Factors

$\mathcal{B}_{3/1,2}$ happens to be a true canonical basis, yet a *relative* one, as defined in [18]. Indeed, it is uniquely defined, and, as we show below, covers Σ_3 together with $\mathcal{B}_{1,2} = \mathcal{B}_1 \cup \mathcal{B}_2$ while being of minimal size. In a sense, $\mathcal{B}_{3/1,2}$ represents the minimal set of implications that, whenever applied to $\mathcal{C}_{1,2}^a$ transform it into $\mathcal{C}_3^a$.

To fix the terminology, we shall say that $\mathcal{A} \subseteq \wp(A)$ *respects* ([10], § 2.3) an implication family Θ whenever all members of Θ are valid within $\mathcal{A}$ (e.g., $\mathcal{C}_{1,2}^a$ respects $\mathcal{B}_{1,2}$ but not $\mathcal{B}_{1,2} \cup \mathcal{I}_{3/1,2}$). In our setting, we say that a set of implications Θ *extracts* $\mathcal{C}_3^a$ from $\mathcal{C}_{1,2}^a$ if $\mathcal{C}_3^a$ respects Θ but none of the elements of $\mathcal{C}_{1,2}^a - \mathcal{C}_3^a$ respects Θ. Such Θ will be said *non-redundant* if no proper subset thereof extracts $\mathcal{C}_3^a$ from $\mathcal{C}_{1,2}^a$.

Theorem 1. $\mathcal{B}_{3/1,2}$ *extracts* $\mathcal{C}_3^a$ *from* $\mathcal{C}_{1,2}^a$ *and is non-redundant.*

Proof. Clearly $\mathcal{C}_3^a$ respects $\mathcal{B}_{3/1,2}$ (subset of $\mathcal{I}_{3/1,2}$). Now let $Y \in \mathcal{C}_{1,2}^a - \mathcal{C}_3^a$, so that $Y \subset Y^{33}$ holds, and assume Y respects $\mathcal{B}_{3/1,2}$. This means Y satisfies the conditions for being a $3/1, 2$-pseudo-closed, hence there should be an implication $Y \to Y^{33}$ in $\mathcal{B}_{3/1,2}$. Yet this contradicts the hypothesis that Y respects $\mathcal{B}_{3/1,2}$. Hence no such Y exists and, consequently, $\mathcal{B}_{3/1,2}$ extracts $\mathcal{C}_3^a$ from $\mathcal{C}_{1,2}^a$.

To prove $\mathcal{B}_{3/1,2}$ is minimal, it is enough to show that for any $Y \to Y^{33} \in \mathcal{B}_{3/1,2}$, Y respects $\mathcal{B}_{3/1,2} - \{Y \to Y^{33}\}$ (i.e., $\mathcal{B}_{3/1,2} - \{Y \to Y^{33}\} \not\models Y \to Y^{33}$). Yet Y is $3/1, 2$-pseudo-closed, hence for all $Z \to Z^{33} \in \mathcal{B}_{3/1,2} - \{Y \to Y^{33}\}$, $Z \subset Y$ entails $Z^{33} \subset Y$. Hence Y respects $\mathcal{B}_{3/1,2} - \{Y \to Y^{33}\}$.

Theorem 1 shows that $\mathcal{B}_{3/1,2}$ is a *relative canonical basis* expressing how $\mathcal{C}_3^a$ is embedded into $\mathcal{C}_{1,2}^a$. It is directly adapted from the standard theorem for arbitrary closure operators (Th. 8 in [10]) and can be rephrased in terms of background knowledge and implications [18].

In this second setting, we keep the same definition for $3/1, 2$-pseudo-closed and let the role of $C_{1,2}^a$ be played by $\mathcal{B}_{1,2}$ (which is the canonical basis of the concept lattice whose intents coincide with $C_{1,2}^a$). Assume now $\mathcal{B}_{1,2}$ is given as background knowledge. We say that a set of implications Θ is $3/1, 2$-complete if $\Theta \cup \mathcal{B}_{1,2} \models \Sigma_3$. Again, Θ is non-redundant if no proper subset thereof is $3/1, 2$-complete.

Theorem 2. $\mathcal{B}_{3/1,2}$ is $3/1, 2$-complete and non redundant.

Proof. The demonstration is analogous to the previous one.

The above twin theorems have their respective interests. The first one formalizes the Algorithms 1 and 2 both of whom compute $\mathcal{B}_{3/1,2}$ by scanning through $C_{1,2}^a$ without referring to $\mathcal{B}_1$ and $\mathcal{B}_2$. In this sense, it provides the *correctness* of both algorithms.

It will be useful for evaluating the gap between C_3^a and $C_{1,2}^a$, hence the gap to direct product $\mathcal{L}_1 \times \mathcal{L}_2$ and global independence. More precisely, small-size hybrid bases hint at independent factors (few interactions) with the extreme case being $\mathcal{B}_{3/1,2} = \emptyset$. Void $\mathcal{B}_{3/1,2}$ entails $\mathcal{B}_3 = \mathcal{B}_1 \cup \mathcal{B}_2$ and therefore $C_3^a = C_{1,2}^a$. Conversely, large hybrid bases mean lots of non-closed combinations in $C_{1,2}^a$ and therefore few global closures that combine two non-empty factor ones.

The second theorem focuses on the fact that $\mathcal{B}_{3/1,2} \cup \mathcal{B}_1 \cup \mathcal{B}_2$ is a cover of Σ_3, hence it can be transformed into $\mathcal{B}_3$ by standard reduction [15, 4].

4.3 Optimized Computational Strategy for $\mathcal{B}_{3/1,2}$

A key concern in the approach is the quickest elimination of $C_{1,2}^a$ elements which are not recognized by $\mathcal{B}_{3/1,2}^{\neq}$. To address this concern, we examine some small-scale optimizations that help speed-up the detection of invalidating implications for unsuccessful candidates. Intuitively, we must insure that a potential invalidating, or "killer", implications will lay somewhere *close to the head* of the implication list $\mathcal{B}_w$ (see Algorithm 1). To that end, some inexpensive rearrangement of the list could be carried out, e.g., adding new applications always at the head and pushing forward previous successful "killers". However, there will still be too many cases in which a significant portion of $\mathcal{B}_w$ is tested before the invalidating implication pops up. Hence further invalidation tools are needed.

A first idea is to structure the search space in a way that eases transfer of successful invalidating implications downwards in the lattice (inheritance of invalidating implications). Recall that for the $\mathcal{B}_{3/1,2}^{\neq}$ test to work, the traversal of $C_{1,2}^a$ must be done with respect to a linear extension of the product order. A step further is to require this order to preserve the highest degree of "continuity", i.e., the neighbor nodes in the order to share a large number of attributes.

This would allow the buffering of recent successful "killers" for preferential examination, i.e., before the rest of the implications. The assumption here is that the closer the composition of the neighbors of a non recognized set Y, the greater the chances for those to be invalidated, or "killed", by the same implication that discarded Y. The underlying implication list, anecdotically called "smoking gun", is gradually updated along the traversal and may have variable length.

A further tool aims at a partial remedy for the situations of *continuity disruption*, i.e., a sharp structural change between two neighbour nodes in the order. In this case, the chances of having the smoking gun list work are low while those of spending large efforts on lookup of killing implication within $\mathcal{B}_w$ are much higher. A way out could be to shift immediately to closure computation from the context and avoid potential bottleneck in searching the $\mathcal{B}_{3/1,2}$ while capitalizing on the cost to pay in the subsequent steps. Indeed, the new candidate Y and its 3-closure could form a local smoking gun although $Y \rightarrow Y^{33}$ may end up discarded once having become irrelevant in the local search. However, its chances to succeed in killing locally the successors of Y are high. Therefore, keeping and applying these *local* implications constitutes another fine-grained mechanism for rapid elimination of unsuccessful candidates.

The above mechanisms, together with the lookup for a invalidating implication in $\mathcal{B}_w$ and the closure computation from $\mathcal{K}_3$ bring the number of algorithmic tools for canonicity check to four. Therefore, their appropriate composition becomes an issue.

1: **procedure** MERGE-TERS(**In:** $\mathcal{C}_1^a$, $\mathcal{C}_2^a$ intent families; $\mathcal{B}_1$, $\mathcal{B}_2$ implication sets; **Out:** $\mathcal{C}_3^a$ an intent family; $\mathcal{B}_3$ an implication set)
2:
3: $\mathcal{C}_3^a \leftarrow \emptyset$
4: $\mathcal{B}_{3/1,2} \leftarrow \emptyset$
5: $Z_l \leftarrow \emptyset$; $T_l \leftarrow \emptyset$ {The **local implication**, $Z_l \rightarrow T_l$, set to a trivial value}
6: $r_s = (Z_s \rightarrow T_s) \leftarrow first(\mathcal{B}_1)$ {The **smoking gun** is any valid implication}
7: **for all** (Y_1, Y_2) in $\mathcal{C}_1^a \times \mathcal{C}_2^a$ **do**
8:    $Y \leftarrow Y_1 \cup Y_2$
9:    **if** $Z_l \not\subseteq Y$ **or** $Z_l = T_l$ **then**
10:      $Z_l \leftarrow Y$; $T_l \leftarrow$ CLOSE($\mathcal{K}_3$, Y)   {Reinstall the local implication}
11:    **if** $Y \subseteq T_l$ **then**
12:      **if** $Z_s \not\subseteq Y$ **or** $Y \subseteq T_s$ **then**
13:        **if** $Y = T_l$ **then**
14:          $\mathcal{C}_3^a \leftarrow \mathcal{C}_3^a \cup \{Y\}$ {Survived all guns, closed}
15:          $Z_l \leftarrow T_l$ {Signal need for changing the local implication}
16:        **else**
17:          $r \leftarrow$ FINDINVALIDATINGRULE($\mathcal{B}_{3/1,2}$,Y) {Survived all guns, non closed}
18:          **if** $r =$ NULL **then**
19:            $r_n \leftarrow$ NEWRULE(Y,CLOSE($\mathcal{K}_3$,Y)) {Survived all guns, pseudo-closed}
20:            $\mathcal{B}_{3/1,2} \leftarrow \mathcal{B}_{3/1,2} \cup \{r_n\}$ {Add to the basis}
21:          **else**
22:            $r_{sg} \leftarrow r$ {Reinstall the smoking gun}
23: $\mathcal{B}_3 \leftarrow$ REDUCE($\mathcal{B}_{3/1,2}$, $\mathcal{B}_1$, $\mathcal{B}_2$)

Algorithm 2. Enhanced merge of factor lattices and canonical implication bases

---

4.4   Algorithmic Design

The order between tools reflects their relative cost. For instance, the original tools are clearly heavy-weight ones, whereas the mechanisms introduced in Section 4.3 are rather light-weight (single implication or few of these). Therefore, the former must be applied more sparingly whereas the latter could work as checkpoints for candidates barring them the way to more costly examinations. In other terms, the external layers in a nested control structure should relie on smoking gun and local implications. Only candidates which "survive" those checks and thus have higher chances of becoming 3-closed or $3/1, 2$-pseudo-closed will be admitted to the inner layers of the control where their status can be thoroughly examined.

Algorithmic Scheme. The following algorithmic code is a product of a preliminary performance study that helped establish an order based on the relative contribution of each tool in the global discarding effort.

Thus, the outer most level performs the local check on every new candidate before any other processing. The local rule is systematically re-computed whenever irrelevant, i.e., its premise is no more a subset of the candidate set (lines 9-10). Once this preliminary relevance test fixed, the implication is applied to only let in candidates that represent subsets of its conclusion (line 11).

Next, the smoking gun is fired (line 12), and the survivors are then let into the heavy-weight tool area (lines 13-22). Here candidates are further dispatched on their closeness status. Closed sets are easily detected (line 13), while the local implication is set to a tautological value to force a change at the following step (line 15).

Non closed survivors to all other tools come to the basic invalidation mechanism as described in Algorithm 2 (lines 17-22). The only difference is that rules having discarded a candidate are kept as the next "smoking gun" implication (line 22).

The processing ends by the reduction of $\mathcal{B}_{3/1,2} \cup \mathcal{B}_1 \cup \mathcal{B}_2$.

Correctness of the algorithm. The overall correctness of Algorithm 2 follows from the Theorem 1. Below, some clarifications about the additional mechanisms are provided.

Concerning local implication, only candidates that are subsets of its conclusion (a 3-closed set) are effectively processed. In fact, the "killed" candidates could be either incomparable with the conclusion or supersets thereof. While the first cannot be closed for $\_^{\mathcal{B}_{3/1,2}^{\neq}}$, the second case actually never occurs.

Proposition 2. *Let the candidate set be Y and the local premise at the time Y is examined Z_l. Then $Z_i^{33} \not\subset Y$.*

Moreover, the way for 3-closed candidates to quit the **for** loop is through a test for equality with the local implication conclusion. In fact, as every candidate comprises the local premise, its closure necessarily equals the local conclusion.

5   Parsimonious Reduction Strategy

The reduction process can be expensive as computing $Y^{\mathcal{B}_{3/1,2} \cup \mathcal{B}_1 \cup \mathcal{B}_2 - (Y \rightarrow Z)}$ may take a time linear in the total number of attributes in the initial rule set. Thus, the complete reduction could cost as much as a square function of the number of implicatons. Yet for large parts of $\mathcal{B}_{3/1,2} \cup \mathcal{B}_1 \cup \mathcal{B}_2$ a much smaller reduction effort will suffice.

5.1 Composition of the Intermediate Cover

First, some $Y \rightarrow Z$ from $\mathcal{B}_1 \cup \mathcal{B}_2$ will be in canonical form (Y in $\mathcal{PC}_3$, Z in $\mathcal{C}_3^a$).

Proposition 3. *For all $Y \rightarrow Y^{ii}$ from $\mathcal{B}_i$ ($i = 1, 2$), if $Y^{ii} = Y^{33}$ then $Y \in \mathcal{PC}_3$.*

Indeed, part of the factor closures remain closed in $\mathcal{K}_3$ – these form an upper-set in the respective factor intent semi-lattice, as well as in the global one. Thus, there is no interference with the implications from the opposite factor in the corresponding sub-order of the Boolean lattice. Consequently, nothing challenges the status of global pseudo-closures for the premises of the rules corresponding to each factor closure upper-set. In the canonical basis of the example, the only stable factor implication is $\rightarrow a$.

Next, any $Y \rightarrow Z$ from $(\mathcal{B}_1 \cup \mathcal{B}_2) - \mathcal{B}_3$ is valid in $\mathcal{K}_3$ yet incomplete in a sense since neither of the conditions $Y \in \mathcal{PC}_3$ and $Z = Y^{33}$ need to hold. However, only a part of these implications will be inherently redundant, i.e., capture no 3-pseudo-closed between Y and Y^{33}. For example, the 2-pseudo-closed fg is no more a pseudo-closed in $\mathcal{K}_3$. More dramatically, fg is not even a part of a 3-pseudo-closed set.

Recall that even if a 3-pseudo-closed lays in that interval, $Y \rightarrow Z$ may still be eliminated since coming before the last implication generating that pseudo-closed.

Regarding $Y \rightarrow Z$ in $\mathcal{B}_{3/1,2}$, there is no point in trying to expand Y using other rules from $\mathcal{B}_{3/1,2}$ because of its pseudo-closure status. Neither is it reasonable to try with factor rules given that $Y \in \mathcal{C}_{1,2}^a$, whence $Y = Y^{\mathcal{B}_1 \cup \mathcal{B}_2}$. Hence any effective expansion can only come from factor rules that have had their parts completed, in particular, conclusions. Therefore completed version of each factor basis, B_i^{33} ($i = 1, 2$), is considered where each original rule $X \rightarrow T$ is replaced by $X \rightarrow X^{33}$. This reasoning only holds for the first expansion of Y, which opens the space for all the remaining rules in the cover. Luckily enough, there is no need to follow with a full-scale expansion since whenever a relative premise "violates" an implication from B_i^{33}, the underlying hybride rule can be discarded as redundant.

Proposition 4. *For a given $Y \rightarrow Y^{33}$ from $\mathcal{B}_{3/1,2}$, if there exists $Z \rightarrow Z^{33}$ in B_i^{33} with $Z \subseteq Y$ and $Z^{33} \not\subseteq Y$, then $Z^{33} = Y^{33}$.*

Further to Proposition 4, as there is a less constrained implication $Z \rightarrow Z^{33}$ which nevertheless will produce $Y^{\mathcal{B}_{3/1,2}^{\neq}}$ (whether pseudo-closed or closed) in the global set, $Y \rightarrow Y^{33}$ can be discarded. Moreover, in scanning $\mathcal{B}_{3/1,2}$, one needs to test only factor implications from $B_i^{33} - B_i$ ($i = 1, 2$) since the remaining ones are expansion-neutral.

To sum up, among the rules in $\mathcal{B}_{3/1,2} \cup \mathcal{B}_1 \cup \mathcal{B}_2$, only those from $\mathcal{B}_1 \cup \mathcal{B}_2$ require full-scale redundancy tests and these tests involve the rest of the global rule set. Before doing that, the conclusions must be set to their effective 3-closures. Redundant $\mathcal{B}_{3/1,2}$ rules are much easier to detect: a sufficient condition is the existance of at least one invalidating rule from B_i^{33} ($i = 1, 2$).

5.2 Reduction Procedure

The reduction process is an application of the results presented in Section 5.1. It is described in Algorithm 3. As a first step, the conclusions of the rules in $\mathcal{B}_i$ ($i = 1, 2$) are set to the respective 3-closures (lines 4-8). For each rule in $\mathcal{B}_{3/1,2}$ possible invalidations by rules that have their conclusions changed (in the $\mathcal{B}_w$ list) are examined (lines 9-11). Finally, the premises of modified rules in $\mathcal{B}_i^{33}$ are expanded in order to establish the potential redundancy of each rule (lines 12-16).

```
 1: procedure REDUCE(In: B₃/₁,₂, B₁, B₂ implication sets; Out: B₃ an implication set)
 2:
 3:   Bw ← ∅ ; B₃ ← ∅
 4:   for all Y → Z in B₁ ∪ B₂ do
 5:     if Z ≠ Y³³ then
 6:         Bw ← Bw ∪ {Y → Y³³}
 7:     else
 8:         B₃ ← B₃ ∪ {Y → Z}
 9:   for all Y → Z in B₃/₁,₂ do
10:     if FIND-INVALIDATING-RULE(Bw, Y) = NULL then
11:         B₃ ← B₃ ∪ {Y → Z}
12:   for all Y → Z in Bw do
13:       Bw ← Bw − {Y → Z}
14:       Yᶜ ← LINCLOSURE(Bw ∪ B₃, Y)
15:       if Z ⊄ Yᶜ then
16:           B₃ ← B₃ ∪ {Yᶜ → Z}
```

Algorithm 3. Reduction of the global implication set $\mathcal{B}_{3/1,2} \cup \mathcal{B}_1 \cup \mathcal{B}_2$

6 Implementation and Experimental Results

We have implemented a number of variants of the basic algorithmic scheme described above. Each of them features a specific combination of speed-up heuristics. These have been compared to the standard NEXTCLOSURE algorithm. The latter has been chosen for its highly valuable properties of reduced additional space requirements and relatively high efficiency. Moreover, both schemes benefit from the same implicit tree structures that "organize" the memory of past computations in a way that helps carry out the current test more rapidly. The comparison has been made on several synthetic datasets. An average case of the still on-going study is described as follows. The dataset is made out of 267 objects described by 23 attributes. The contexts generates 21550 concepts, with 2169 pseudo-closed and 2069 3/1, 2-pseudo-closed.

The current implementations of the various merge methods use the environment provided by the GLAD system. Thus, the implementation language is FORTRAN. The tests have been run on a Windows platform, with a Pentium I, 166 MHz. Moreover, in all merge algorithms the factor lattices and implication bases are "prepared" by recursively executing NEXTCLOSURE on each factor.

Among the versions of highest level of optimization, the one that produces only concepts takes 0.82s while the most rapid method for both concepts and implications, with the computation of the basis $\mathcal{B}_3$, takes 2.97s. These figures are to be compared to 1.59s and 41.41s, respectively, for standard NEXTCLOSURE. The following table presents the performance scores of a selected set of algorithms that were designed within the study

Table 1. Performances of algorithms based on the merge framework: on closure computation (left) and on joint closure and implication computation (rigth)

Algotihm name	Description	CPU-time C_3
NEXTCLOSURE	Standard version	1.59s
AINTPROL	Intent computation	0.82s

Algotihm name	Description	CPU-time $C_3 + B_3$
NEXTCLOSURE	B_3-based variant	41.41s
AINTPROL	B_3, all tools	2.97s
AINTPROP	Only $B_{3/1,2}$ (no reduction), all tools	2.05s
AINTPROK	Only $B_{3/1,2}$, no light-weight tools	10.22s

reported here. Notice that the figures for all merge algorithms include the time taken by the subordinated NEXTCLOSURE calls.

We tend to read both performance results as an indication that reducing the candidates by using the subdirect product and local implications can be efficient. The fact that the efficiency gains with respect to NEXTCLOSURE are much larger for implications than for intents is hardly surprising, since the closure for implications is far more painful due to reiteration. Indeed, while NEXTCLOSURE scans only the vicinity of every intent in $\mathcal{L}_3$, our algorithm has to explore every node of the direct product $\mathcal{L}_{1,2}$, a structure whose size may easily grow up to a square function of the size of $\mathcal{L}_3$. However, AINTPROP spends only a tiny laps of time on most non-valid nodes, except on those forcing a change in the current local implication. In contrast, NEXTCLOSURE needs to compute closures in the context — or through the growing list of implications — for a number of candidates that is usually substantially bigger. Of course, both algorithms use local regularities in the listing orders for candidates to avoid redundant computation by skipping invalid nodes with minimal or no examination at all. In doing this, NEXTCLOSURE goes a step further and prunes candidates that have not even been explicitly listed. Nevertheless, in the light of the current test results, the introduction of the direct product and the underlying constraint to consider explicitly *every* candidate, seem still to pay back, essentially through the reduction of the effort to spend on each non-valid candidate.

7   Conclusion

The motivation of the current study was to push factor lattice merge to its efficiency limits. To that end we reshuffled the initial problem settings to limit the computation to strictly necessary elements and brought in the implications in order to take advantage of the decrease in the attribute dimension along subsequent splits. The latter innovation entails significant overhead in the algorithmic design as the initial merge scheme was *de facto* abandoned. The bid behind the new methods seems to be right, at least in the light shed by the initial performance studies. Indeed, in its current implementation, the optimized merge method clearly outperforms the standard version of NEXTCLOSURE. More extensive performance tests will be necessary in order to clarify this sensitive topic. Nevertheless, the available results are already very encouraging since the basic merge framework works equally well with iceberg lattices as with complete ones and therefore should easily adapt to data mining tasks.

Rather than the presenting the results of an accomplished study, the present paper reports on an intermediate step of a larger research project. In fact, we are currently looking not only at the practical performance aspects of the lattice merge problem, but also at a wide range of questions that are still to be answered. Recall that merge of factor results is a mere step in the global process that performs merges at every non-leaf node of the recursion tree. Effective divide-and-conquer methods based on the research reported here are yet to be designed, especially since these require more flexible decision criteria for halting the recursive splits and going into a batch mode. An orthogonal question that remains open is whether or not there is an inexpensive way of dividing the context into parts of unbound, i.e., not necessarily balanced, sizes so that the effort of merging the resulting factors is minimized. Pushing further on that idea, one may want to research other ways of dividing a context, possibly into sub-contexts of non-disjoint attribute and/or object dimensions, which allow for less expensive merge. Finally, key argument in favor of divide-and-conquer methods is the possibility to easily adapt them to parallel environments (see [7]).

References

[1] Agrawal, R., Srikant, R.: Fast algorithms for mining association rules. In: Proceedings of the 20th International Conference on Very Large Data Bases (VLDB 1994), September 1994, Santiago, Chile, pp. 487–499 (1994)

[2] Armstrong, W.: Dependency Structures of Data Base Relationships. In: Rosenfeld, J. (ed.) Proceedings of the 1974 IFIP Congress, Information Processing, Stockholm (SE), August 5-10, 1974, vol. 74, pp. 580–583 (1974)

[3] Birkhoff, G.: Lattice Theory. AMS, 3rd edn., vol. XXV. AMS Colloquium Publications (1967)

[4] Day, A.: The lattice theory of functional dependencies and normal decompositions. Intl. J. of Algebra and Computation 2(4), 409–431 (1992)

[5] Demetrovics, J., Libkin, L., Muchnik, I.B.: Functional dependencies in relational databases: A lattice point of view. Discrete Applied Mathematics 40(2), 155–185 (1992)

[6] Denecke, K., Erné, M., Wismath, S. (eds.): Galois connections and Applications. Kluwer Academic Publishers, Dordrecht (2004)

[7] Djoufak Kengue, J.F., Valtchev, P., Tayou Djamegni, C.: A parallel algorithm for lattice construction. In: Ganter, B., Godin, R. (eds.) ICFCA 2005. LNCS (LNAI), vol. 3403, pp. 248–263. Springer, Heidelberg (2005)

[8] Duquenne, V.: Contextual implications between attributes and some representation properties for finite lattices. In: Ganter, B., Wille, R., Wolf, K.-E. (eds.) Beiträge zur Begriffsanalyse, pp. 213–239, B.I.-Wiss. Verl., Mannheim (1987)

[9] Ganter, B.: Two basic algorithms in concept analysis. preprint 831, Technische Hochschule, Darmstadt (1984)

[10] Ganter, B., Wille, R.: Formal Concept Analysis, Mathematical Foundations. Springer, Heidelberg (1999)

[11] Guigues, J.L., Duquenne, V.: Familles minimales d'implications informatives résultant d'un tableau de données binaires. Mathématiques et Sciences Humaines 95, 5–18 (1986)

[12] Kautz, H.A., Kearns, M.J., Selman, B.: Horn approximations of empirical data. Artificial Intelligence 74(1), 129–145 (1995)

[13] Khardon, R.: Translating between horn representations and their characteristic models. Journal of Artificial Intelligence Research 3, 349–372 (1995)

[14] Maier, D.: Minimum covers in the relational database model. In: Eleventh Annual ACM Symposium on Theory of Computing (STOC 1979), pp. 330–337. ACM Press, New York (1979)

[15] Maier, D.: The theory of Relational Databases. Computer Science Press (1983)

[16] Pasquier, N., et al.: Efficient Mining of Association Rules Using Closed Itemset Lattices. Information Systems 24(1), 25–46 (1999)

[17] Pei, J., Han, J., Mao, R.: CLOSET: An Efficient Algorithm for Mining Frequent Closed Itemsets. In: Proc. SIGMOD Workshop DMKD 2000, Dallas (TX), pp. 21–30 (2000)

[18] Stumme, G.: Attribute exploration with background implications and exceptions. In: Bock, H.-H., Polasek, W. (eds.) Data Analysis and Information Systems. Statistical and Conceptual approaches. Studies in classification, data analysis and knowledge organization, vol. 7, pp. 457–469. Springer, Heidelberg (1996)

[19] Taouil, R., et al.: Mining bases for association rules using closed sets. In: Proc. of the ICDE-2000, San Diego (CA), p. 307. IEEE Computer Society Press, Los Alamitos (2000)

[20] Valtchev, P., Duquenne, V.: Towards scalable divide-and-conquer methods for computing concepts and implications. In: SanJuan, E., et al. (eds.) Proc. of the 4th Intl. Conf. Journées de l'Informatique Messine (JIM 2003): Knowledge Discovery and Discrete Mathematics, Metz (FR), INRIA, 3-6 September 2003, pp. 3–14 (2003)

[21] Valtchev, P., Missaoui, R., Lebrun, P.: A partition-based approach towards building Galois (concept) lattices. Discrete Mathematics 256(3), 801–829 (2002)

[22] Wild, M.: Computations with finite closure systems and implications. In: Li, M., Du, D.-Z. (eds.) COCOON 1995. LNCS, vol. 959, pp. 111–120. Springer, Heidelberg (1995)

[23] Wille, R.: Restructuring lattice theory: An approach based on hierarchies of concepts. In: Rival, I. (ed.) Ordered sets, pp. 445–470. Reidel, Dordrecht-Boston (1982)

[24] Zaki, M.J.: Generating Non-Redundant Association Rules. In: Proc. of the 6th Intl. Conf. on Knowledge Discovery and Data Mining (KDD 2000), Boston (MA), pp. 34–43 (2000)

[25] Zaki, M.J., Hsiao, C.-J.: ChARM: An Efficiently Algorithm for Closed Itemset Mining. In: Grossman, R., et al. (eds.) Proceedings of the 2nd SIAM International Conference on Data Mining (ICDM 2002) (2002)

Lattices of Rough Set Abstractions as P-Products

Bernhard Ganter

Institut für Algebra
Dresden University of Technology
D-01062 Dresden
Bernhard.Ganter@tu-dresden.de

Abstract. We show how certain lattices occurring in the theory of Rough Sets can be described in the language of Formal Concept Analysis. These lattices are obtained from generalised approximation operators forming a kernel-closure pair. We prove a general context representation theorem and derive first consequences. It becomes clear under which conditions the approximations can be interpreted as intervals in a lattice of "definable sets".

1 Introduction

The theory of Rough Sets [1] offers, among other things, an *interval arithmetic for sets*. The original setting, introduced by Z. Pawlak [4], assumed that sets are chosen from a *universe* U, but that elements of U can be specified only up to an *indiscernibility* equivalence relation $\sim$ on U. If a subset $A \subseteq U$ contains an element indiscernible from some element not in A, then A is *rough*. In other words, a subset of U is not rough iff it is a union of equivalence classes of $\sim$. Up to indiscernibility, a rough set A is described by two approximations:

- the **upper approximation** $\overline{R}(A) := \{u \in U \mid \exists_{y \sim u} \; y \in A\}$, and
- the **lower approximation** $\underline{R}(A) := \{u \in U \mid \forall_{y \sim u} \; y \in A\}$.

Clearly $\underline{R}(A) \subseteq A \subseteq \overline{R}(A)$. The pair $(\underline{R}(A), \overline{R}(A))$ sometimes is considered a **rough set abstraction**, no matter if A is rough or not. These abstractions, being pairs of sets, are naturally ordered by applying the $\subseteq$–order componentwise. They thereby form a complete lattice, which is a **Stone algebra**, a mild generalisation of a Boolean algebra.

Many authors, including Pawlak himself [5], saw a need for generalised approximation operators, not necessarily built from an equivalence relation. For such a generalised setting it is not automatically the case that the abstractions form a lattice. And even if they do, its lattice structure requires investigation.

We attack these tasks here, studying generalisations for which the lower approximation operator $\underline{R}(\cdot)$ is an *interior operator* and the upper approximation operator $\overline{R}(\cdot)$ is a *closure operator*. For either operator type, the set of images forms a complete lattice, since they constitute a *kernel system* and a *closure system*, respectively.

R. Medina and S. Obiedkov (Eds.): ICFCA 2008, LNAI 4933, pp. 199–216, 2008.

The abstractions, i.e., the pairs $(\underline{R}(A), \overline{R}(A))$, generate (but do not necessarily form) a sublattice of the direct product of the two lattices, of interior sets and of closures.

Proposition 1. Let U be a set, let $X \mapsto \underline{R}(X)$ be an interior operator and $X \mapsto \overline{R}(X)$ be a closure operator on U. Then the pairs

$$(\underline{R}(X), \overline{R}(Y)), \quad X, Y \subseteq U,$$

form a complete lattice with the operations

$$\bigvee_{t \in T} (X_t, Y_t) := (\bigcup_{t \in T} X_t, \overline{R}(\bigcup_{t \in T} Y_t))$$

$$\bigwedge_{t \in T} (X_t, Y_t) := (\underline{R}(\bigcap_{t \in T} X_t), \bigcap_{t \in T} Y_t).$$

This is obvious. The complete sublattice generated by the pairs $(\underline{R}(X), \overline{R}(X))$ is called the **lattice of rough set abstractions**. The reader must be warned that this notion may be misleading:

The pairs $(\underline{R}(X), \overline{R}(X)), X \subseteq U$ can naturally be ordered by component-wise set inclusion. It can easily happen that this ordered set is a lattice (for example, when the mapping $X \mapsto (\underline{R}(X), \overline{R}(X))$ is an order embedding), but is not a sublattice of the direct product. That lattice structure is not what we study here. We use the name "lattice of rough set abstractions" for the sublattice of the direct product that is generated by these pairs. In most cases, this lattice contains pairs that are *not* of the form $(\underline{R}(X), \overline{R}(X))$ for some $X \subseteq U$. It is the aim of the present article to describe this lattice in terms of a formal context construction.

Closure- and kernel systems can conveniently be described in the language of Formal Concept Analysis [3]. In fact, we can find **representing contexts** (U, M, I) and (U, N, J) such that the closures $\overline{R}(A)$ are precisely the extents of (U, N, J) and the interior sets $\underline{R}(A)$ are precisely the extent-complements of (U, M, I). The lattice of closed sets then is isomorphic to $\underline{\mathfrak{B}}(U, N, J)$, while the lattice of interior sets is isomorphic to $\underline{\mathfrak{B}}(M, U, I^d)$, and the lattice of rough set abstractions is isomorphic to a complete sublattice of the direct product

$$\underline{\mathfrak{B}}(M, U, I^d) \times \underline{\mathfrak{B}}(U, N, J).$$

As we shall see, this sublattice can be characterised as a P-product of the factor lattices.

2   P-Products

One of the standard constructions in Formal Concept Analysis is that of the *context sum*, corresponding to the direct product of the two concept lattices. Let

(G, M, I) and (H, N, J) be formal contexts with $G \cap H = \emptyset = M \cap N$. The **sum** of (G, M, I) and (H, N, J) is the formal context

$$(G \cup H, M \cup N, I \cup J \cup (G \times N) \cup (H \times M)),$$

see Figure 1.

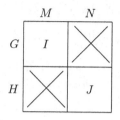

Fig. 1. The context sum

Proposition 2 (see [3]). *The concept lattice of the sum is isomorphic to the direct product of the respective concept lattices. (A, B) is a formal concept of the sum if and only if $(A \cap G, B \cap M)$ is a formal concept of (G, M, I) and $(A \cap H, B \cap N)$ is one of (H, N, J).*

Remark 1. *The definition of the context sum can easily be modified so that the disjointness assumptions can be abandoned, and Proposition 2 still holds. To achieve this, (G, M, I) and (H, N, J) are simply replaced by disjoint isomorphic copies. In the sequel, our notation will ignore this technical difficulty. We shall use the construction as if the contexts were disjoint, even if they are equal. Figure 3 e.g. has to be interpreted in this sense.*

Let P be some fixed set. A P-**lattice** (L, α) consists of a complete lattice L together with a mapping $\alpha : P \to L$, the image of which generates L. If (L_1, α_1) and (L_2, α_2) are P-lattices, then their P-**product** is the complete sublattice of $L_1 \times L_2$ that is generated by the pairs

$$\{(\alpha_1(p), \alpha_2(p)) \mid p \in P\}.$$

A P-product automatically is a subdirect product, because each component contains a generating set of the respective factor.

   P-products of concept lattices correspond to P-**fusions** of their contexts. We briefly sketch the construction here; details can be found in [3]. P-fusions are built from *bonds*. Therefore we introduce these first.

   Let (G, M, I) and (H, N, J) be formal contexts. A **bond** from (G, M, I) to (H, N, J) is a relation $R \subseteq G \times N$ with the property that

– g^R is an intent of (H, N, J) for every $g \in G$, and
– n^R is an extent of (G, M, I) for every $n \in N$.

The set of all bonds from (G, M, I) to (H, N, J) is closed under intersection, because the set of all intents and the set of all extents are. Thus for every subset $S \subseteq G \times N$ there is a smallest bond containing S.

Now let P be some set, let (G, M, I) and (H, N, J) be formal contexts and let $\alpha_1 : P \to \underline{\mathfrak{B}}(G, M, I)$ and $\alpha_2 : P \to \underline{\mathfrak{B}}(H, N, J)$ be mappings onto generating sets of $\underline{\mathfrak{B}}(G, M, I)$ and $\underline{\mathfrak{B}}(H, N, J)$, respectively. Let

$$\alpha_1(p) =: (A_1^p, B_1^p), \qquad \alpha_2(p) =: (A_2^p, B_2^p)$$

be the concepts to which $p \in P$ is mapped under α_1 and α_2, respectively.

The *P*-fusion of the two *P*-contexts $((G, M, I), \alpha_1)$ and $((H, N, J), \alpha_2)$ is defined as the formal context

	M	N
G	I	$R_{1,2}^\beta$
H	$R_{2,1}^\beta$	J

,

where for $\{i, j\} = \{1, 2\}$ the relation $R_{i,j}^\beta$ is the smallest bond containing

$$R_{i,j} := \bigcup_{p \in P} A_i^p \times B_j^p.$$

Theorem 1 (see [3]). Let $((G, M, I), \alpha_1)$ and $((H, N, J), \alpha_2)$ be *P*-contexts. The concept lattice of their *P*-fusion is isomorphic to the *P*-product of their concept lattices.

The formal concepts of the *P*-fusion are precisely the pairs $(A_1 \cup A_2, B_1 \cup B_2)$, where (A_1, B_1) and (A_2, B_2) are formal concepts of (G, M, I) and of (H, N, J), respectively, such that the pair $((A_1, B_1), (A_2, B_2))$ is an element of the *P*-product.

3   The Lattice of Abstractions

The notion of a *P*-product can naturally be applied to the rough set approximations described in the first section. For the set P we will use the power set of the universe U. The mappings $X \mapsto \underline{R}(X)$ and $X \mapsto \overline{R}(X)$ are, of course, onto their images and thus clearly make the lattice of lower and the lattice of upper approximations *P*-lattices. Since the lattice of rough set abstractions is generated by the pairs $(\underline{R}(X), \overline{R}(X))$, it is their *P*-product and isomorphic to the concept lattice of the *P*-fusion

	U	N
M	I^d	$R^\beta_{1,2}$
U	$R^\beta_{2,1}$	J

However, in the setting considered here, the relations $R^\beta_{1,2}$ and $R^\beta_{2,1}$ can be determined explicitly. This is stated in the following theorem:

Theorem 2. Let (U, M, I) and (U, N, J) be formal contexts with $U \cap (M \cup N) = \emptyset$ and let for every $X \subseteq U$

$$\overline{R}(X) := \text{ smallest extent of } (U, N, J) \text{ containing } X, \text{ and}$$
$$\underline{R}(X) := \text{ largest extent-complement of } (U, M, I) \text{ contained in } X.$$

Then the lattice of rough set abstractions is isomorphic to the concept lattice of the formal context shown in Figure 2 and explained in Propositions 3, 4, and 5.

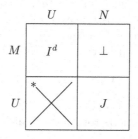

Fig. 2. The P-fusion corresponding to the lattice of rough set abstractions. The relation in the upper right quadrant is defined by $m \perp n$: $\iff$ $m^I \cup n^J = U$, the relation in the lower left quadrant is $U \times U \setminus \{(g,g) \mid g$ is an extremal point of (U, M, I) and of $(U, N, J)\}$.

Proof. Three propositions will be given which together prove Theorem 2.

In order to compute $R^\beta_{i,j}$, recall that for any given set $X \subseteq U$ the lower approximation $\underline{R}(X)$ is a complement of an intent of (M, U, I^d). More precisely, it is the largest intent-complement of (M, U, I^d) that is contained in X, and therefore the complement of the smallest intent containing X^C, the complement of X. The upper approximation of X is the smallest extent of (U, N, J) containing X, thus equal to X^{JJ}. So we get that

$$\underline{R}(X) = ((X^C)^{II})^C, \qquad \overline{R}(X) = X^{JJ}.$$

The corresponding formal concepts are

$$((X^C)^I, (X^C)^{II}) =: (A_1^X, B_1^X) \quad \text{and} \quad (X^{JJ}, X^J) =: (A_2^X, B_2^X).$$

So we obtain

$$R_{1,2} = \bigcup_{p \in P} A_1^p \times B_2^p = \bigcup_{X \subseteq U} (X^C)^I \times X^J \subseteq M \times N$$

and

$$R_{2,1} = \bigcup_{p \in P} A_2^p \times B_1^p = \bigcup_{X \subseteq U} X^{JJ} \times (X^C)^{II} \subseteq U \times U.$$

When is $(m,n) \in R_{1,2}$? This is the case iff there is some $X \subseteq U$ satisfying $m \in (X^C)^I$ and $n \in X^J$, which is equivalent to

$$X^C \subseteq m^I \quad \text{and} \quad X \subseteq n^J.$$

Obviously, this is the case if and only if $m^I \cup n^J = U$. Abbreviating this by $\perp$ we get

Proposition 3

$$R_{1,2} = \perp := \{(m,n) \mid m^I \cup n^J = U\}.$$

Similarly we can determine $R_{2,1}$. We have that $(g,h) \in R_{2,1}$ iff there is some $X \subseteq U$ with $g \in X^{JJ}$ and $h \in (X^C)^{II}$. Choosing $X := \{g\}$ we get that $X^C = U \setminus \{g\}$. Then $(X^C)^{II}$ is either U or $U \setminus \{g\}$. As a result, we find that (g,h) is always in $R_{2,1}$, with the only possible exception of $h = g$. And $(g,g) \notin R_{2,1}$ can only be true if g is an **extremal point** of (U, M, I), i.e., if $U \setminus \{g\}$ is an extent. But this is not sufficient: Consider the other extreme case, that $X := U \setminus \{g\}$. Then

$$X^{JJ} \times (X^C)^{II} = (U \setminus \{g\})^{JJ} \times \{g\}^{II},$$

which contains (g,g) except if $g \notin (U \setminus \{g\})^{JJ}$ i.e., if g is an extremal point of (U, N, J). The two conditions together are in fact sufficient, so that we can conclude:

Proposition 4

$$R_{2,1} = U \times U \setminus \{(g,g) \mid g \text{ is an extremal point of } (U, M, I) \text{ and of } (U, N, J)\}.$$

So $R_{2,1}$ essentially is the total relation $U \times U$, consisting of all possible pairs. Only in exceptional cases, there may some "diagonal" pairs (g,g) be missing. We denote this relation by the symbol

$$\overset{*}{\times}.$$

Note that the missing pairs (g,g) are exactly those for which there are $m \in M$ and $n \in N$ such that

$$m^I = U \setminus \{g\} = n^J.$$

Such g satisfy the condition

$$g \in \overline{R}(X) \Rightarrow g \in \underline{R}(X).$$

For simplicity, we shall call such elements *isolated points* of the universe.

Proposition 5. $R_{1,2}$ and $R_{2,1}$ are bonds.

Proof. It is easy to see that $R_{2,1}$ is a bond, since each row and each column of $R_{2,1}$ represents a subset of U which is either equal to U or to $U \setminus \{g\}$, where the latter only happens if g is an isolated point. Clearly these sets are all closed.

To see that $R_{1,2}$ is a bond, consider some $m \in M$. Then

$$\{n \in N \mid (m, n) \in R_{1,2}\} = \{n \in N \mid m \perp n\} = ((m^I)^C)^J$$

is an intent of (U, N, J), and dually. $\qquad\square$

The three propositions show that the parts of the P-fusion are as claimed in Theorem 2. $\qquad\square$

4   The Case of an Equivalence Relation

In Pawlak's original definition, as described in the introduction, the approximation operators are defined with reference to some equivalence relation $\sim$ on U. We first verify that Theorem 2 gives the (well known) characterisation of the approximation lattices in this basic case.

The formal contexts for the lower and the upper approximation are easily seen to be the same, both identical to

$$(U, U, \nsim),$$

where $\nsim$ is the complement of the equivalence relation $\sim$. Such contexts are well understood, their extents (and intents) are precisely those subsets of U which are unions of equivalence classes of $\sim$ (the "definable sets"), and their concept lattices are isomorphic to the power set lattices of the partitions $U/\sim$.

Since $g^I = \{h \in U \mid g \nsim h\}$, we find that $g \perp h \iff g \nsim h$. The (doubly) extremal points are also easy to determine: They are those elements $g \in U$ for which $\{g\}$ is a one-element equivalence class, because exactly then the complement is a union of equivalence classes. From Theorem 2 we infer

Corollary 1. If $\sim$ is some equivalence relation on U, and if the approximation operators are defined for all $X \subseteq U$ by

$$\overline{R}(X) := \{u \in U \mid \exists_{y \sim u}\ y \in X\}$$
$$\underline{R}(X) := \{u \in U \mid \forall_{y \sim u}\ y \in X\},$$

then the lattice of rough set approximations is isomorphic to the concept lattice of the formal context in Figure 3, where

$$\overset{*}{\times} = U \times U \setminus \{(u, u) \mid \{u\}\text{ is an equivalence class of }\sim\}.$$

The formal context in Figure 3 has structural properties which are worth mentioning:

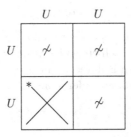

Fig. 3. The P-fusion of the "classical" approximations. See Remark 1 for the notation.

Corollary 2. The P-fusion in the case of an equivalence indiscernibility is symmetric. The lattice of rough set approximations therefore has a natural anti-automorphism $(A, B) \mapsto (B, A)$ corresponding to rough set *complementation*.

A second property can be derived using results proved in [2].

Corollary 3. If there are no one-element equivalence classes, then the P-fusion is of the form shown in Figure 5. The concept lattice of such a context is isomorphic to the *order relation* of $\underline{\mathfrak{B}}(G, M, I)$, considered as a complete sublattice of the square of $\underline{\mathfrak{B}}(G, M, I)$. The lattice of rough set approximations then is just the lattice of all *intervals* of $\underline{\mathfrak{B}}(G, M, I)$, ordered by their lower and their upper bounds simultaneously.

Indeed, when indiscernibility is an equivalence $\sim$, then the rough set approximations correspond to the intervals of the power set lattice of $U/\sim$. From every

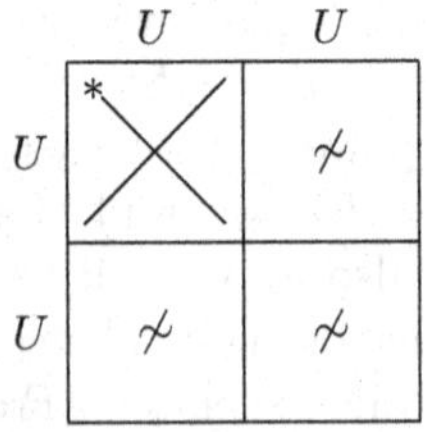

Fig. 4. The same context as in Figure 3, but with upper and lower part interchanged to demonstrate symmetry

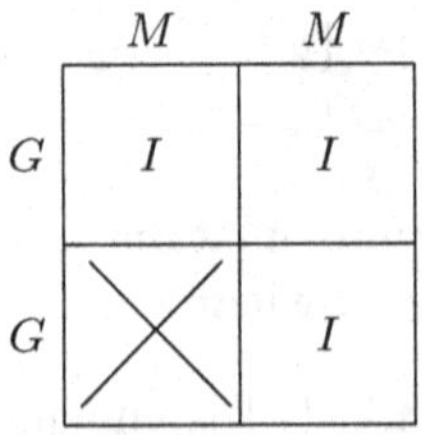

Fig. 5. The concept lattice of this formal context is isomorphic to the *order relation* of $\underline{\mathfrak{B}}(G, M, I)$, considered as a complete sublattice of the square of $\underline{\mathfrak{B}}(G, M, I)$

rough set $X \subseteq U$, we obtain the interval having $\underline{R}(X)/\sim$ as its smallest and $\overline{R}(X)/\sim$ as its largest element. Conversely, given sets $S \subseteq T \subseteq U/\sim$, the way construct a set X with $\underline{R}(X) = S$ and $\overline{R}(X) = T$ is to include in X all elements from sets in S and one element (but not all) from each set in $T \setminus S$. This leads to the desired result except when $T \setminus S$ contains singleton equivalence classes.

When studying generalisations, we shall investigate if these properties can be preserved.

The special form of this fusion can also be used to unravel the lattice structure: When indiscernibility is an equivalence, then the lattice of all rough set approximations is a direct product of chains with three or two elements. The two element chains correspond to the exceptional singleton equivalence classes.

5 Granules

A popular interpretation is to understand the indiscernibility classes as "information granules". A possible generalisation of Pawlak's original approach allows that granules are not necessarily disjoint.

So let us start with an arbitrary family $\mathcal{F}$ of subsets of the universe U, the elements of which we call **granules**. As the lower approximation of each subset $X \subseteq U$ we take the union of the granules contained in X. This will make $\underline{R}(X)$ an interior operator. The upper approximation will consist of all elements that are not contained in a granule disjoint from X. In this way $\overline{R}(X)$ will become a closure operator. Formally,

$$\overline{R}(X) := U \setminus \bigcup \{F \in \mathcal{F} \mid F \cap X = \emptyset\},$$
$$\underline{R}(X) := \bigcup \{F \in \mathcal{F} \mid F \subseteq X\}.$$

The context representing the two approximation operators can easily be determined.

Theorem 3. *If lower and upper approximations are based on a family $\mathcal{F}$ of granules, as defined above, then the lattice of rough set approximations is isomorphic to the concept lattice of the formal context displayed in Figure 6.*

Proof. We first determine the representing formal contexts for the approximation operators. The lower approximations are exactly the unions of granules. Their complements therefore are exactly the intersections of granule-complements. A formal context having these as extents is $(U, \mathcal{F}, \not\ni)$. So we may take this as the representing context for the lower approximation. Upper approximations are precisely the complements of lower approximations. So again, they are exactly the extents of $(U, \mathcal{F}, \not\ni)$. The two representing contexts are equal.

Given a granule $f \in \mathcal{F}$, then $f' = \{u \in U \mid f \not\ni u\}$ is just the complement of f in U. So two granules f, g are in relation $\perp$ iff the union of their complements covers U, that is, iff they are disjoint. A point $u \in U$ is extremal iff its complement is an extent of $(U, \mathcal{F}, \not\ni)$, which means that $\{u\}$ must be a union of granules. So $\{u\}$ must be a granule. $\qquad\qquad\square$

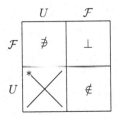

Fig. 6. The P-fusion when the approximation operators are based on a family $\mathcal{F}$ of *granules*. Two granules are in relation $\perp$ iff they are disjoint. Extremal points correspond to singleton granules.

The representing context $(U, \mathcal{F} \not\ni)$ for a set $\mathcal{F}$ of granules is by no means special. In fact, whenever (U, M, I) is some attribute-reduced formal context with object set U, then letting

$$\mathcal{F} := \{U \setminus m' \mid m \in M\}$$

yields a formal context $(U, \mathcal{F}, \not\ni)$ which is isomorphic to (G, M, I). Therefore the case of granules is, up to the trivial possibility of "repeated attributes" (with identical attribute extents) the same as the general case of two identical representing contexts. We have indicated that situation in Figure 7, where, as in Figure 4, we have interchanged the lower and the upper half to exhibit symmetry.

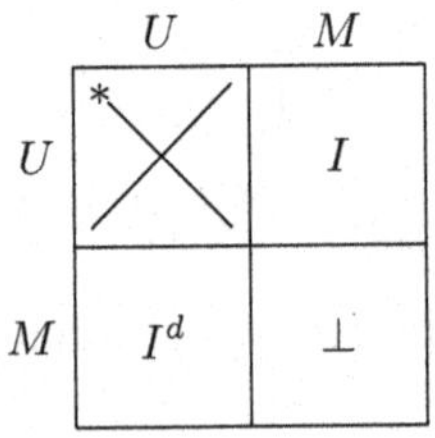

Fig. 7. Whenever the two representing contexts are equal, the P-fusion becomes symmetric

Proposition 6. When the representing contexts for the lower and the upper approximation are the same, then the lattice of rough set approximations has an involutory anti-automorphism. If there are no isolated points and $U^I = \emptyset$, then $I = \perp = I^d$ implies that $I = \not\sim$ for some equivalence relation $\sim$ in U.

Proof. The first claim can immediately be read off from Figure 7: The P-fusion is symmetric and thus $(A, B) \mapsto (B, A)$ is an involutory concept lattice anti-automorphism. If $I = I^d$, then I is symmetric. The condition that $I = \perp$ requires that two elements are in relation I iff their neighbourhoods cover U. This is equivalent to I being the complement of an equivalence relation. □

6 Inner Granules, Outer Granules

A further generalisation step is to allow different families of granules for the two approximation operators. Let $\mathcal{E}$ and $\mathcal{F}$ be families of subsets of U ("inner granules" and "outer granules") and let

$$\overline{R}(X) := U \setminus \bigcup\{F \in \mathcal{F} \mid F \cap X = \emptyset\}$$

$$\underline{R}(X) := \bigcup\{E \in \mathcal{E} \mid E \subseteq X\}.$$

It is immediate from the results above how to construct the P-fusion for the lattice of rough set approximations, the result is depicted in Figure 8. Note that

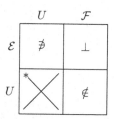

Fig. 8. The P-fusion for two granule families

this is essentially the most general case, because up to "repeated attributes" every formal context with object set U can be written as $(U, \mathcal{E}, \not\subseteq)$ or as $(U, \mathcal{F}, \not\subseteq)$. So except for a technical detail, the contexts shown in Figures 2 and 8 are the same.

It may be asked under which conditions the properties of Pawlak's original construction that were mentioned in Corollaries 2 and 3 hold for this generalised construction. While the quest for *symmetry*, at least in its straightforward form, leads back to Section 5, the property described in Corollary 3 seems more promising. Indeed, we obtain a classification by means of *idempotent relations*, i.e., relations C on U satisfying

$$C \circ C = C.$$

Let us call $g \in U$ an **isolated point** for an idempotent relation C if $(g, g) \in C$, but $(v, g) \notin C$ and $(g, v) \notin C$ for all $v \neq g$.

Proposition 7. The P-fusion for the lattice of rough set approximations, as shown in Figure 2, is of the form shown in Figure 5 iff the representing contexts are dual to each other and isomorphic to (U, U, J^d) and (U, U, J), respectively, where J is the complement of an idempotent relation without isolated points.

Proof. The proposition characterises when $I^d = \bot = J$ in Figure 2. This clearly requires $M = U = N$ and $I = J^d$, and under these conditions is equivalent to $J = \bot$. Expanding the definition of $\bot$, the latter becomes

$$u \, J \, v \iff u \perp v \iff u^I \cup v^J = U.$$

We get a chain of equivalent conditions:

$$u\,J\,v \iff u^I \cup v^J = U$$
$$u\,J\,v \iff u^{J^d} \cup v^J = U$$
$$u\,J\,v \iff \forall_{x \in U}\ u\,J\,x \text{ or } x\,J\,v.$$

Let C be the complementary relation to J. Then this chain continues to

$$u\,C\,v \iff \exists_{x \in U}\ u\,C\,x \text{ and } x\,C\,v$$
$$C = C \circ C.$$

It remains to check for extremal points. $g \in U$ is extremal for (U, U, J) iff there is some element $u_1 \in U$ such that $u\,J\,u_1 \iff u \ne g$, and extremal for (U, U, J^d) iff dually there is some element $u_2 \in U$ such that $u_2\,J\,u \iff u \ne g$. For the complement C of J this amounts to

$$u\,C\,u_1 \iff u = g \iff u_2\,C\,u.$$

Thus $g\,C\,u_1$. And since C is idempotent, there must be some $u \in U$ with $g\,C\,u$ and $u\,C\,u_1$. But $u\,C\,u_1$ implies $u = g$ and thus $g\,C\,g$. Now assume $v\,C\,g$. Since idempotent relations are transitive, we conclude that $v\,C\,u_1$. Again, this implies $v = g$. Dually $g\,C\,v$ implies $v = g$. Thus g must be an isolated point, and this is also sufficient. $\qquad\square$

Idempotent relations play a rôle in Domain Theory, where they have been studied under the name **infosys** (Vickers [6]). The use of infosys in general for rough sets remains to be investigated. The special case of a **quasi-order**, i.e., a reflexive and transitive relation, has a natural interpretation in terms of inner and outer granules, as Theorem 4 will show.

Corollary 4. Let $\le$ be an idempotent relation on U (not necessarily a quasi-order) without isolated points, let $(U, U, \not\le)$ and $(U, U, \not\ge)$ be the representing contexts for the approximation operators $\underline{R}(\cdot)$ and $\overline{R}(\cdot)$. Then the lattice of rough set abstractions is equal to the order relation of the lattice of extents of $(U, U, \not\ge)$.

A quasi-order $\le$ on the universe U may be interpreted as a *specialisation order*, so that $v \le u$ may be read as "v is a specialisation of u". This may also be understood as a non-symmetric indiscernibility, where an element u is indiscernible from all its specialisations, but not necessarily vice versa. For a universe U with such an *indiscernibility quasi-order*, there is a natural definition of the approximation operators: The upper approximation of a set X must contain X and all specialisations of elements of X, whereas an element u belongs to the lower approximation of X if u and all specialisations of u are in X. Let us denote for $e \in U$ by

$$\downarrow e := \{u \in U \mid e \ge u\}$$

the quasi-order ideal generated by e, formed by the set of all specialisations of e. Dually define for $f \in U$

$$\uparrow f := \{u \in U \mid u \geq f\}$$

the quasi-order filter generated by f, consisting of all elements which f is a specialisation of. Let

$$\mathcal{E} := \{\downarrow e \mid e \in U\}, \qquad \mathcal{F} := \{\uparrow f \mid f \in U\}.$$

Using $\mathcal{E}$ and $\mathcal{F}$ as families of inner and outer granules in Figure 8, and observing that

$$- \downarrow e \not\ni u \iff e \not\geq u,$$
$$- u \notin \uparrow f \iff u \not\geq f,$$
$$- e \perp f \iff (U \backslash \downarrow e) \cup (U \backslash \uparrow f) = U \iff e \not\geq f,$$

we get that the fusion context in Figure 8 becomes as shown in Figure 9.

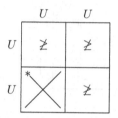

Fig. 9. The P-fusion for an indiscernibility quasi-order

Theorem 4. Let $\leq$ be a quasi-order on the universe U, having no isolated points. For $X \subseteq U$ define

$$- \overline{R}(X) := \{u \in U \mid \exists_{x \geq u} \ x \in X\}$$
$$- \underline{R}(X) := \{u \in U \mid \forall_{x \leq u} \ x \in X\}.$$

Then the lattice of rough set approximations equals the order relation of the lattice of order ideals of the quasi-ordered set $(U, \leq)$.

This is indeed a natural and meaningful generalisation of the "classical" case when indiscernibility is an equivalence relation, and it shares many properties with that case. The sets of images of the two operators are equal, because for given $X \subseteq U$

- the upper approximation $\overline{R}(X)$ is the smallest order ideal containing X, and
- the lower approximation $\underline{R}(X)$ is the largest order ideal contained in X.

Therefore both lattices are equal to the distributive lattice of all order ideals of $(U, \leq)$. The rough set approximations form a sublattice of its square, and this lattice is distributive as well.

7 A Toy Example

For data of realistic size the lattice of rough set approximations will be too large to be displayed in a single readable diagram. This lattice mainly serves as a background structure for computations. The small example which we discuss now is not meant to be realistic, but to make our findings of the previous sections more transparent. We use a four-element universe

$$U := \{\text{father, mother, parent, woman}\},$$

abbreviated $U = \{f, m, p, w\}$, which is endowed with the order relation given in Figure 10. We obtain an operator pair $(\underline{R}(\cdot), \overline{R}(\cdot))$ by using the order ideals as inner granules and the order filters as outer granules, as it is described in Section 6.

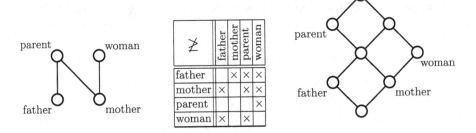

$\not\geq$	father	mother	parent	woman
father		×	×	×
mother	×		×	×
parent				×
woman	×		×	

Fig. 10. An ordered set, the contraordinal scale, and its concept lattice (with object labels only)

The inner granules are the order ideals and therefore are the extents of the corresponding contraordinal scale, also displayed in Figure 10 together with its concept lattice. Since the union of order ideals is again an order ideal, this set is already the set of all possible lower approximations.

The upper approximations are intersections of complements of the order filters. However, complements of order filters are order ideals, and the set of order ideals is closed under intersections. Therefore the set of all upper approximations is the same as that of all lower ones. Note that in both cases the lattice operations are simply set-theoretic union and intersection.

Sixteen pairs $(\underline{R}(X), \overline{R}(X))$ can be built from the sixteen choices of

$$X \subseteq \{f, m, p, w\},$$

and they turn out to be all different. These pairs are pairs of extents of the contraordinal scale, i.e., both $\underline{R}(X)$ and $\overline{R}(X)$ are extents of the lattice in Figure 10. Therefore these pairs form a subset of the cartesian square of that lattice, see Figure 11. The mapping

$$X \mapsto (\underline{R}(X), \overline{R}(X))$$

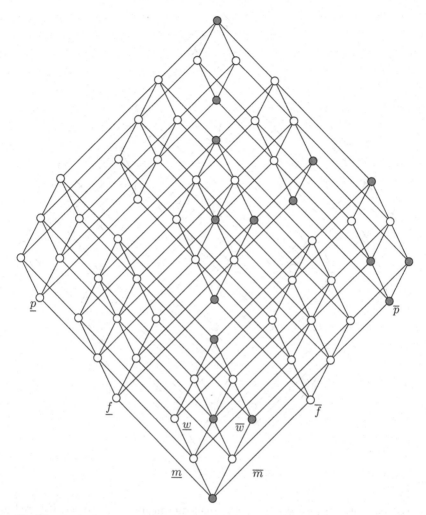

Fig. 11. The cartesian square of the concept lattice in Figure 10. The abbreviated object labels are underlined for the first factor and overlined for the second. The shaded elements correspond to the sixteen pairs $(\underline{R}(X), \overline{R}(X))$, $X \subseteq \{f, m, p, w\}$. As an ordered set, they are isomorphic to the power set of $\{f, m, p, w\}$, but they do not form a sublattice of the depicted lattice. The eight elements on the spine do form a sublattice. It is isomorphic to the concept lattice in Figure 10.

in this case is indeed an order embedding. Thus the set of images is order isomorphic to the power set lattice of $\{f, m, p, w\}$. But it is not a sublattice of the lattice of all pairs, as can easily be seen. In fact, the sublattice generated by these elements is almost twice as large. It is shown in Figures 12 and 14.

There are three ways to achieve this lattice. One is to generate the sublattice from the shaded elements in Figure 11. This leads to Figure 12. The second is to apply Theorem 4 to show that it is the concept lattice of a formal context as

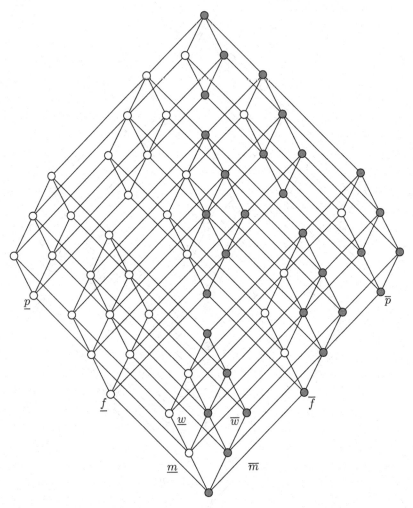

Fig. 12. The sublattice generated by the shaded elements in Figure 11. It is redrawn in Figure 14.

displayed in Figure 9. Since the order under investigation has no isolated points, we get the formal context shown in Figure 13. Its concept lattice is the lattice in Figure 14. Finally, we know (also from Theorem 4) that the lattice must be the order relation of the lattice of order ideals shown in Figure 10. As a quick check for agreement, we count the size of the order by adding up the sizes of the principal filters $\uparrow a$. This gives for each element a of the lattice in Figure 10 the number of pairs (a, x) in the relation. Reading the diagram from top to bottom and from left to right, we get

$$1 + 2 + 2 + 4 + 3 + 5 + 6 + 8 = 31,$$

and this is indeed the size of the lattice in Figure 14.

	$\underline{f}$	$\underline{m}$	$\underline{p}$	$\underline{w}$	$\overline{f}$	$\overline{m}$	$\overline{p}$	$\overline{w}$
$\underline{f}$		×	×	×		×	×	×
$\underline{m}$	×		×	×	×		×	×
$\underline{p}$				×				×
$\underline{w}$	×		×		×		×	
$\overline{f}$	×	×	×	×		×	×	×
$\overline{m}$	×	×	×	×	×		×	×
$\overline{p}$	×	×	×	×				×
$\overline{w}$	×	×	×	×	×		×	

Fig. 13. The formal context for the lattice of rough set approximations as provided by Theorem 4. The concept lattice is shown in Figure 14.

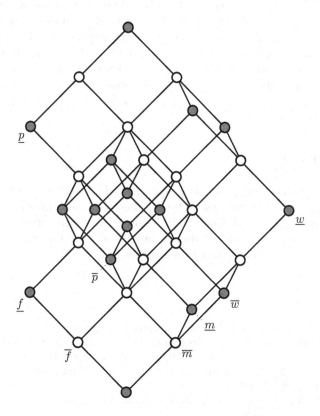

Fig. 14. The lattice of rough set approximations. The shaded elements are the same as in Figure 11. They correspond to the pairs $(\underline{R}(X), \overline{R}(X))$, $X \subseteq \{f, m, p, w\}$.

One might wonder if the non-shaded elements in Figure 14 are somehow meaningful. There is in fact a simple explanation for them. Suppose that we modify the ordered set in Figure 10, making it a quasi-order by simply doubling each element and making the elements of each pair equivalent (that is, comparable in both directions). The formal contexts in Figures 10 and 13 will double in size, simply because each row and each column is repeated. Up to clarification, the contexts will not change at all. The lattice of Figure 14 remains unchanged, except that now all elements are shaded.

Finally we come back to Proposition 6, which gives the existence of an involutory anti-automorphism in case that the two representing contexts are the same. In our case, they are dual to each other (the representing contexts are $(U, U, \not\leq)$ and $(U, U, \not\geq)$), but since the ordered set under consideration is self-dual, the two contexts are also equal. Therefore the lattice of rough set approximations has an involutory anti-automorphism, expressed in the diagram in Figure 14 as a mirror symmetry at the horizontal middle axis.

Another way to understand this anti-automorphism is to construct the mapping directly. The ordered set $(U, \leq)$ is self-dual, and so to each order ideal there corresponds its dual order filter. This induced an anti-automorphism, say α, of the lattice of order ideals (the lattice in Figure 10). α maps each order ideal to the complement of its dual order filter. The anti-automorphism α then induces an anti-automorphism of the order relation: Every pair (x, y) with $x \leq y$ is mapped to the pair $(\alpha(y), \alpha(x))$. Suppose that $(x_1, y_1) \leq (x_2, y_2)$. Then $x_1 \leq x_2$ and $y_1 \leq y_2$. Because of $\alpha(x_2) \leq \alpha(x_1)$ and $\alpha(y_2) \leq \alpha(y_1)$ we obtain $(\alpha(x_2), \alpha(y_2)) \leq (\alpha(x_1), \alpha(y_1))$.

8   Conclusion

Generalised rough set approximation operators that form a kernel-closure pair lead to lattices of rough set abstractions. These can conveniently be described using the P-fusion of formal contexts. If the generalised indiscernibility relation is a quasi-order, then the approximations can naturally be interpreted as intervals of "definable sets", the latter being quasi-order ideals.

References

1. http://roughsets.home.pl/ for an extensive bibliography
2. Ganter, B.: Relational Galois Connections. In: Kuznetsov, S.O., Schmidt, S. (eds.) ICFCA 2007. LNCS (LNAI), vol. 4390, pp. 1–17. Springer, Heidelberg (2007)
3. Ganter, B., Wille, R.: Formal Concept Analysis – Mathematical Foundations. Springer, Heidelberg (1999)
4. Pawlak, Z.: Rough Sets. International Journal of Computer and Information Sciences 11, 341–356 (1982)
5. Pawlak, Z.: Rough Sets: Theoretical Aspects of Reasoning About Data. Kluwer Academic Publishing, Dordrecht (1991)
6. Vickers, S.: Information Systems for continuous posets. Theoretical Computer Science 114, 204–221 (1993)

Scale Coarsening as Feature Selection

Bernhard Ganter and Sergei O. Kuznetsov

Institut für Algebra
Dresden University of Technology
D-01062 Dresden, Germany
Bernhard.Ganter@tu-dresden.de
Department of Applied Mathematics
Higher School of Economics
Kirpichnaya ul. 33/5
105679 Moscow, Russia
skuznetsov@hse.ru

Abstract. We propose a unifying FCA-based framework for some questions in data analysis and data mining, combining ideas from Rough Set Theory, JSM-reasoning, and feature selection in machine learning. Unlike the standard rough set model the indiscernibility relation in our paper is based on a quasi-order, not necessarily an equivalence relation. Feature selection, though algorithmically difficult in general, appears to be easier in many cases of scaled many-valued contexts, because the difficulties can at least partially be projected to the scale contexts. We propose a heuristic algorithm for this.

1 Introduction

A paper recycling company gets vast amounts of material delivered for recycling every day. The first step in their process is to separate the waste from the recyclable part. This is done automatically: A machine performs certain optical measurements on every single piece and then decides which fraction it goes to. We are interested in the rules by which these decisions are made.

The situation is typical for applications of *Machine Learning* [14], and most likely the decision rules were obtained from a *training data* set, using a method of *supervised learning*. Machine Learning offers powerful algorithms, in particular when the data is numerical in nature. Here we concentrate on the more general case of *qualitative data*, and formalise the learning scenario as follows: We are given a formal context (G, M, I) [8], describing the "observations" or "measurements", together with a set $G_+ \subseteq G$, comprising the objects of interest, also called the *positive examples*. Objects from the complement $G_- = G \setminus G_+$ are called negative examples. The task then is to give a characterisation of G_+ in terms of (G, M, I) (a similar problem may be stated for G_-). The nicest case, of course, is that membership in G_+ is equivalent to some attribute combination, i.e., that G_+ is a concept extent of (G, M, I). But even if that is not the case, often a classification is desired. The second best choice then is to find attribute

R. Medina and S. Obiedkov (Eds.): ICFCA 2008, LNAI 4933, pp. 217–228, 2008.

combinations ("*classifiers*") that are *sufficient* for membership in G_+. And ideally there should be enough such classifiers to cover all elements of G_+. This motivates our first definition:

Definition 1. *Let (G, M, I) be a formal context. A set*

$$G_+ \subseteq G$$

*is called **grounded** or, equivalently, **definable**, iff*

$$G_+ = \bigcup \{P' \mid P \subseteq M, P' \subseteq G_+\}.$$

The word "grounded" is used in JSM-theory of inductive reasoning [2,3,4], and it is defined there in a slightly different manner: The sets P in the above definition are required to be **(positive) hypotheses** for G_+, that is, *concept intents P with $P' \subseteq G_+$* (see FCA formalization [5] of [4]). But it is easy to see that this causes no additional difficulty, simply replacing each P by its closure P''.

The word "definable" comes from Rough Set Theory [15,16,18,19], where it is defined in terms of an **indiscernibility relation**, usually an equivalence relation. Our approach generalises this. The role of the indiscernibility will be taken by the *object quasi-order* of the formal context and will not necessarily be symmetric. This is unfolded in Theorem 1 below. The relation between FCA, JSM-reasoning and Rough Set Theory was first studied in [17], but for specific hypotheses from [10].

2   Definability and the Object Quasi-order

The **object quasi-order** of a formal context (G, M, I) is defined by

$$g \leq h :\iff g' \supseteq h' \quad (g, h \in G).$$

It is indeed reflexive and transitive, but not necessarily anti-symmetric. This makes it a quasi-order (called a *preorder* by some authors). The notion of an **order ideal** is the same as for ordered sets: a subset $S \subseteq G$ such that $h \in S$ and $g \leq h$ always implies $g \in S$. The quasi-order ideals are precisely the extents of the formal context $(G, G, \not\geq)$, as in the case of an ordered set.

Theorem 1. *The definable object sets of (G, M, I) are precisely the quasi-order ideals of the object quasi-order. Each subset $G_+ \subseteq G$ contains a largest definable set $\underline{R}(G_+)$, and has a smallest definable set containing it, denoted $\overline{R}(G_+)$.*

Proof. Let $G_+ \subseteq G$ be some subset and let $g \in G$. There exists some $P \subseteq M$ with

$$g \in P' \subseteq G_+$$

iff $g'' \subseteq G_+$. But since

$$g'' = \{h \mid h' \supseteq g'\} = \{h \mid h \leq g\},$$

this is equivalent to G_+ being a quasi-order ideal. Since the family of quasi-order ideals is closed under set union and intersection, the rest of the proposition is immediate.

Several algorithmic questions about the object quasi-order will arise in the sequel. We mention here three of them:

Feature Sets: Given a definable set $G_+ \subseteq G$, which subsets $F \subseteq M$ suffice to make G_+ definable? In other words, what are the subsets $F \subseteq M$ for which $G_+ = \bigcup \{P' \mid P \subseteq F, P' \subseteq G_+\}$?

Global Reducts: Which subsets of the attribute set of a formal context (G, M, I) induce the same definability? In other words, what are the subsets $F \subseteq M$ for which it is true that

$$g \leq h \iff g' \cap F \supseteq h' \cap F \qquad \text{for all } g, h \in G?$$

Separability: Given subsets $L, U \subseteq G$ such that $u \leq l$ holds for no $u \in U$ and no $l \in L$. Which sets also separate L from U, i.e., for which $E \subseteq M$ is it true that

$$u' \cap E \supseteq l' \cap E$$

holds for no $u \in U$ and no $l \in L$?

In what follows we will show that these problems are algorithmically difficult if we require the respective subsets of attributes to be minimal.

The **lower** and **upper approximation operators**, as the operators $\underline{R}(\cdot)$ and $\overline{R}(\cdot)$ occurring in the theorem are called in Rough Set Theory, are given as follows:

$$\underline{R}(G_+) = \bigcup \{P' \mid P \subseteq M, P' \subseteq G_+\}$$
$$\overline{R}(G_+) = \bigcup \{g'' \mid g \in G_+\}.$$

The operators can also be given in terms of the JSM-method. In [6,7] we have introduced the notion of a **hopeless example**, by which we meant a positive example $g \in G_+$ which cannot be classified because there is some object $h \notin G_+$ having all attributes of g. That is $g \in G_+$ is hopeless iff there is some $h \notin G_+$ such that $h \leq g$. In that language then

$$\underline{R}(G_+) = \{g \in G_+ \mid g \text{ is not hopeless}\}$$
$$\overline{R}(G_+) = \underline{R}(G_+) \cup \{h \mid h \leq g \text{ for some hopeless } g \in G_+\}.$$

The next proposition is now immediate.

Proposition 1. *The following conditions are equivalent:*

1. G_+ *is definable,*
2. $\overline{R}(G_+) = G_+,$
3. $\underline{R}(G_+) = G_+.$

Example. We illustrate our definitions by means of an artificial example. Consider the following context (G, M, I) where positive examples are fruits. This information is given by the target attribute "fruit", which does not belong to the set of attributes M.

	color	firm	smooth	form	fruit
apple	yellow	no	yes	round	+
grapefruit	yellow	no	no	round	+
kiwi	green	no	no	oval	+
plum	blue	no	yes	oval	+
toy cube	green	yes	yes	cubic	−
egg	white	yes	yes	oval	−
tennis ball	white	no	no	round	−

Consider a natural scaling of the context

	white	yellow	green	blue	firm	nonfirm	smooth	nonsmooth	round	nonround	fruit
apple		×				×	×		×		+
grapefruit		×				×		×	×		+
kiwi			×			×		×		×	+
plum				×		×	×			×	+
toy cube			×		×		×			×	-
egg	×				×		×			×	-
tennis ball	×					×		×	×		-

- A minimal feature set for G_+ is the set {yellow, nonfirm, nonround}.
- A minimal feature set for G_- is {white, firm}.
- A minimal global reduct is, e.g., the set
    {white, yellow, green, smooth, nonsmooth, round, nonround}.
- The set {white, firm} is a minimal set separating G_- from G_+.
- The set of positive examples is definable, since
    $\overline{R}(G_+) = \underline{R}(G_+) = G_+ = \{apple,\ grapefruit,\ kiwi,\ plum\}$.
- Consider another positive example *orange*, which is orange, nonfirm, nonsmooth and round. Under the scaling chosen,
    $orange' = \{nonfirm, nonsmooth, round\}$.
  Thus this example is hopeless for the scaling, since $orange' \subseteq tennis\ ball'$.
- For the extended data set we have
    $\overline{R}(G_+) = \{apple,\ grapefruit,\ kiwi,\ plum,\ tennis\ ball,\ orange\}$.
    $\underline{R}(G_+) = \{apple,\ grapefruit,\ kiwi,\ plum\}$ and
  Thus including *orange* in the set G_+ of positive examples makes G_+ undefinable (for the given scaling).

3   Feature Sets

Not all attributes in the attribute set M may be necessary for the classification, often a subset may suffice. Such subsets are called *feature sets*. In Rough Set

Theory, minimal feature sets are called *reducts*. The process of thinning out the attributes to obtain a feature set is called *feature selection* [13]. In relation to FCA-based hypotheses this was studied in [6,1]. To relate these issues to the Rough Set Theory, we introduce for arbitrary subsets $N \subseteq M$ the **relative approximation operators**:

$$\underline{R}_N(G_+) = \bigcup\{P' \mid P \subseteq N, P' \subseteq G_+\}$$
$$\overline{R}_N(G_+) = \bigcup\{(g' \cap N)' \mid g \in G_+\}.$$

So the relative approximation operators are simply the approximations operators for the shortened formal context $(G, N, I \cap G \times N)$. They therefore share the properties of approximation operators. We say that G_+ is *definable in terms of N*, shortly *N*-**definable** or *N*-**grounded** iff

$$\underline{R}_N(G_+) = G_+ = \overline{R}_N(G_+),$$

where again each of the two equalities implies the other.

If G_+ has only one element, we omit the set brackets and write $\overline{R}_N(g)$ instead of $\overline{R}_N(\{g\})$.

Proposition 2. $\overline{R}_A(g) \cap \overline{R}_B(g) = \overline{R}_{A \cup B}(g)$.

Proof. $\overline{R}_A(g) \cap \overline{R}_B(g) = (g' \cap A)' \cap (g' \cap B)' = ((g' \cap A) \cup (g' \cap B))' = (g' \cap (A \cup B))'$.

There are two different ways to formally define the notion of a feature set. In the *global* view, we look for sets inducing the same definable sets as M does. We call $F \subseteq M$ a **global feature set** if for *all* subsets $S \subseteq G$ it holds that

$$\underline{R}_F(S) = \underline{R}(S) \quad \text{and} \quad \overline{R}_F(S) = \overline{R}(S),$$

which is equivalent to the condition that

$$S \text{ is } F\text{-definable iff } S \text{ is definable.}$$

Finding global feature sets is equivalent to the global reduct problem mentioned above. Its complexity will be treated in the next section.

Our focus here is more on finding feature sets for a given target set G_+ of positive examples. So we are interested in finding, for a fixed given definable set $G_+ \subseteq G$ sets $F \subseteq M$ such that

$$\underline{R}_F(G_+) = G_+ = \overline{R}_F(G_+).$$

Such a set will be called a **feature set** for G_+. Note that we do not restrict ourselves to *minimal* such sets. But finding small ones is indeed intractable, as it is for reducts in the case of Rough Sets:

Proposition 3. *The problem of finding small feature sets, given by*

INSTANCE: *A formal context (G, M, I), a definable set $G_+ \subseteq G$, and a natural number k.*

QUESTION: *Is there a feature set for G_+ of size $\leq k$, i.e., a subset $F \subseteq M$ such that $\underline{R}_F(G_+) = G_+$ and $|F| \leq k$?*

is $\mathcal{NP}$–complete.

Proof. The problem is in $\mathcal{NP}$, because for testing if a given $F \subseteq M$ is a feature set we only need to check if $\bigcup\{(g' \cap F)' \mid g \in G_+\} = G_+$. This can clearly be done in polynomial time.

To show that the problem is $\mathcal{NP}$-hard, we reduce it to a problem well known to be $\mathcal{NP}$-complete: Finding transversals of a family of sets:

INSTANCE: A set M, a family $S_t, t \in T$ of nonempty proper subsets of M (here T is some index set), and an integer k.
QUESTION: Is there a subset $F \subseteq M$, $|F| \leq k$, such that $F \cap S_t \neq \emptyset$ for all $t \in T$?

Given an instance of the transversal problem, we can construct a formal context (G, M, I) by letting $G := T \cup \{g_0\}$, $t' := S_t$ for $t \neq g_0$ and $g_0' := \emptyset$. Moreover, we set $G_+ := T$. It is easy to check that $F \subseteq M$ is a feature set for G_+ iff it is a transversal for $\{S_t \mid t \in T\}$.

Our approach to finding feature sets for G_+ is an indirect one. Rather than building such sets bottom-up, we assume that we are already given one, say F, where $F = M$ is a possible choice. Then we try thinning F, using the following strategy: We consider some subset of F which is not a feature set for G_+ and investigate which elements of F must be added to extend that subset to a feature set for G_+. There will be no unique answer to this question. Our goal is to describe all possible solutions.

More formally, let F be a feature set for G_+, so that

$$\underline{R}_F(G_+) = G_+ = \overline{R}_F(G_+).$$

Fix some subset $N \subseteq F$ which is not a feature set, so that

$$\underline{R}_N(G_+) \subsetneq G_+ \subsetneq \overline{R}_N(G_+).$$

Then both the **lower boundary**

$$L := G_+ \setminus \underline{R}_N(G_+)$$

and the **upper boundary**

$$U := \overline{R}_N(G_+) \setminus G_+$$

are nonempty sets. The lower boundary consists of those elements $g \in G_+$ which are not in the extent of any hypothesis $H \subseteq N$ with $H' \subseteq G_+$.

Theorem 2. *Let $N, E \subseteq M$ and let L, U denote the lower and upper boundary with respect to N. Then $N \cup E$ is a feature set for G_+ iff for all $g \in L$ it holds that*

$$\overline{R}_N(g) \cap \overline{R}_E(g) \subseteq G_+.$$

A sufficient condition is

$$\overline{R}_E(L) \cap U = \emptyset.$$

Proof. $N \cup E$ is a feature set for G_+ iff each object $g \in G_+$ is implied by some $P \subseteq N \cup E$ with $P' \subseteq G_+$. For objects in $\underline{R}_N(G_+)$ this is clear anyway, so it suffices to consider objects from the lower boundary $L = G_+ \setminus \underline{R}_N(G_+)$. For every such object $g \in L$ we must have that

$$\overline{R}_{N \cup E}(g) \subseteq G_+.$$

By Proposition 2, this is equivalent to

$$\overline{R}_N(g) \cap \overline{R}_E(g) \subseteq G_+ \quad \text{for all } g \in L.$$

Since $\overline{R}_N(g) \subseteq \overline{R}_N(G_+)$ holds anyway, it suffices that

$$\overline{R}_E(g) \cap U = \emptyset$$

holds for all $g \in L$. But because of $\overline{R}_E(L) = \bigcup_{g \in L} \overline{R}_E(g)$ this can be summarised to

$$\overline{R}_E(L) \cap U = \emptyset.$$

4   Global Reducts and Separators

Finding minimal global reducts may be hard, which is expressed by the following

Proposition 4. *The following problem is $\mathcal{NP}$-complete[1]:*

INSTANCE: *A formal context (G, M, I) and a natural number k.*
QUESTION: *Is there a subset $F \subseteq M$, $|F| \leq k$, such that*

$$g \leq h \iff g' \cap F \supseteq h' \cap F \qquad \text{for all } g, h \in G?$$

Proof. We reduce "3-dimensional matching", a well-known $\mathcal{NP}$-complete problem [9], to our problem. It requires to decide, for given disjoint sets X, Y, and Z of equal cardinality k and a set $T \subseteq X \times Y \times Z$, if T contains a *matching*, that is, a subset $T' \subseteq T$ such that $|T'| = k$ and no two elements of T' agree in any coordinate. Such a matching can of course only exist if the coordinates of T cover the sets X, Y, and Z, respectively, so this can be assumed as additional precondition.

Given such an instance T for some $k > 1$, we can construct a formal context having a global reduct of size $\leq k$ if and only if the instance contains a matching. The construction is as follows. Let

$$G_0 := \{(w, 0) \mid w \in X \cup Y \cup Z\}, \text{ and}$$
$$G_1 := \{(w, 1) \mid w \in X \cup Y \cup Z\}.$$

We investigate the formal context $(G, T \,\dot\cup\, \{m_X, m_Y, m_Z\}, I)$ with $G := G_0 \cup G_1$, where the incidence is defined as follows:

---

[1] See the acknowledgements in Section 7 below.

$$m'_X := X \times \{0,1\}, \; m'_Y := Y \times \{0,1\}, \; m'_Z := Z \times \{0,1\},$$

and, for each $t =: (x, y, z) \in T$,

$$t' := G_0 \setminus \{(x,0), (y,0), (z,0)\} \cup \{(x,1), (y,1), (z,1)\}.$$

When is $g \le h$ in this formal context? Recall that objects are pairs (w, i), where $w \in X \cup Y \cup Z$ and $i \in \{0,1\}$. An analysis of the possible cases shows that $(w_1, i_1) \le (w_2, i_2)$ holds if and only if w_1 and w_2 are from the same set (that is, $\{w_1, w_2\}$ is a subset of either X or Y or Z), $w_1 \ne w_2$, $i_1 = 0$ and $i_2 = 1$. Actually, this order is obtained exactly from those subsets of T containing triples such that each element of $X \cup Y \cup Z$ occurs at least once as a component. Such a subset has cardinality $\le k$ if and only if it is a matching. Therefore the existence of a 3-dimensional matching is reduced to the problem of finding a global reduct with $\le k$ attributes.

A similar result holds for the problem of finding a minimal separator, i.e., a minimal set of attributes separating a set of objects from another one, as stated by the following

Proposition 5. *The following minimal separator problem is $\mathcal{NP}$-complete:*

INSTANCE: *A formal context (G, M, I), two sets of objects $L, U \subseteq G$ such that $u \le l$ holds for no $l \in L$, $u \in U$, and a natural number k.*

QUESTION: *Is there a subset $F \subseteq M$, $|F| \le k$ such that*

$$u' \cap F \supseteq l' \cap F \qquad \text{holds for no } u \in U, l \in L?$$

Proof. We reduce the minimal transversal problem

INSTANCE: A set M, a family $S_t, t \in T$ of nonempty proper subsets of M (here T is some index set), and an integer k.

QUESTION: Is there a subset $F \subseteq M$, $|F| \le k$, such that $F \cap S_t \ne \emptyset$ for all $t \in T$?

Given an instance of the transversal problem, we can construct a formal context (G, M, I) by letting $G := T \cup \{g_0\}$, $t' := M \setminus S_t$ for $t \ne g_0$ and $g'_0 := M$. Let $L = \{g_0\}$, $U = T$. It is easy to check that $F \subseteq M$ separates $L = \{g_0\}$ from $U = T$ iff F is a transversal for $\{S_t \mid t \in T\}$. The reduction is completed, its polynomiality, as well as the membership of the minimal separator problem in $\mathcal{NP}$ are obvious.

5   Scale Coarsening

Theorem 2 was tailored for applications to *scaled many-valued contexts*. For understanding this article it is not required to recall the precise definitions (which can be found in [8]). It suffices to understand that these are formal contexts (G, M, I) for which the attribute set M can be subdivided into subsets $M_s, s \in S$, such that each such M_s comes from a standardised formal context

$\mathbb{S}_s := (G_s, M_s, I_s)$, a "scale". Some scales are used frequently because of their interpretation and their particularly simple structure, like "nominal", "ordinal" or "interordinal" scales. For these, the algorithmic problems mentioned above are easy to solve.

The heuristic procedure that we suggest for feature selection in scaled many valued contexts builds on this. Feature selection will result in coarser scales, because some scale attributes will not be used. We propose the following strategy:

- Start with some feature set F, for example $F := M$.
- Then pick a scale, one after another, and
  1. remove the set M_s of scale attributes from the feature set.
  2. The result $N := F \setminus M_s$ may fail to be a feature set. In that case, use Theorem 2 to find an appropriate set $E \subseteq F \setminus N$ such that $N \cup E$ is a feature set.
  3. Replace F by $N \cup E$, and continue.

Note that choosing E can be done in two ways, according to Theorem 2. Either we use the equivalence stated in the first part of the theorem, which gives the precise results. Or we use the sufficient condition given in the second part. Note that this amounts to solving the separation problem stated above, but only of the formal context $(G, E, I \cap (G \times E))$.

This is, as already said, a heuristic procedure. Its result depends on the sequence in which the scales are handled, and even if the set E is chosen minimal in each step, we do not claim that the result is a minimal reduct. This heuristic can be useful for data with very large sets of attributes like those described in [12], where standard context reduction [8] is difficult because it is hard even to keep the context in the memory.

However, we expect that the method leads to reasonably small feature sets in a reasonable computing time, since the application of Theorem 2 avoids exhaustive search in testing whether a subset of attributes is good (but not in finding the minimal reduct itself) by projecting the problem to standardised scales.

But more importantly, the method is flexible enough to include other criteria into the search for good feature sets. Small size is not always the most desirable property, and other aspects may be more important. The next section gives an example of this.

6 To Avoid Overfitting

Recall the example that was mentioned in the introduction, where paper samples were to be classified based on the spectra of the light spectra they emit. We are actually working on such a data set (it is too large to be discussed here in detail). There the spectra are given with such a precision that virtually *every* subset of the training data set is grounded, simply because no two of the spectra coincide precisely in every decimal digit. Thus the condition of definability,

$$G_+ = \bigcup \{P' \mid P \subseteq M, P' \subseteq G_+\},$$

is satisfied because for each $g \in G_+$ we get as a classifying attribute set $P := g'$, with $P' = \{g\}$.

However, such a classification will probably be useless when the classification rules learnt from the training set are to be applied to other data. Then, since the positive examples in the training set have been described so precisely, their descriptions will most likely not fit new examples outside the training set. This effect is called **overfitting**. There are many suggestions how this can be avoided.

In the original version of the JSM-method, for example, it is required that only rules are used for classification that apply to at least two positive examples. A set $G_+ \subseteq G$ is called **sufficiently grounded** if for each g in G_+ there exists some $h \in G_+$ such that

$$\{g, h\}'' \subseteq G_+.$$

This is the case if and only if

$$G_+ = \bigcup \{P' \mid P \subseteq M, P = P'', P' \subseteq G_+, |P'| \geq 2\}.$$

Note that the requirement $P = P''$ can be omitted here.

If the set of positive training examples is sufficiently grounded, it is possible to allowing only attribute sets $P \subseteq M$ as classifiers whose **support** $|P'|$ is at least 2. It is reasonable that this restriction lowers the effect of overfitting, because an attribute combination that applies to at least two different objects is more likely to apply to other objects as well. This approach can, of course, be varied by replacing 2 by other thresholds and so on. We are not going into such details here. Instead, we shall study the following problem: Call $F \subseteq M$ a **strong feature set** for G_+ if

$$G_+ = \bigcup \{P' \mid P \subseteq F, P' \subseteq G_+, |P'| \geq 2\}.$$

Clearly G_+ is sufficiently grounded if and only if there is a strong feature set for G_+. However, even if G_+ is sufficiently grounded, not every feature set for G_+ must be strong. The question to investigate therefore is: How can the feature selection procedure described in Section 5 be modified to obtain strong feature sets? Unfortunately, the necessary modification of Theorem 2 is not very elegant:

Proposition 6. *Suppose that $F \subseteq M$ is a strong feature set for G_+ and that $N, E \subseteq F$. Then $N \cup E$ is a strong feature set for G_+ iff for each $g \in G_+$ there is some h such that*

$$(\{g, h\}' \cap N)' \cap (\{g, h\}' \cap E)' \subseteq G_+.$$

This is rather obvious. Not so obvious, but not a hopeless task, is how this can be made efficient in an algorithm. We pose this as a problem.

7   Conclusion

We considered a framework for selecting important subsets of attributes (or attribute values) in FCA-based knowledge discovery. This framework uses the ideas of reducts, upper and lower approximations of the Rough Set Theory, at the same time generalizing the latter by allowing for a quasi-order (not necessarily equivalence) indiscernibility relation. We showed that choosing smallest representations (global reducts, feature sets) is intractable (NP-complete) in general settings, and propose a heuristic based on coarsening the set of attribute values.

Acknowledgements

We thank Graham Brightwell (LSE, London) and William T. Trotter (Georgia Tech, Atlanta) for giving us most useful hints for proving NP-completeness of the minimal global reducts problem.

References

1. Ferré, S., Ridoux, O.: The Use of Associative Concepts in the Incremental Building of a Logical Context. In: Priss, U., Corbett, D.R., Angelova, G. (eds.) ICCS 2002. LNCS (LNAI), vol. 2393, pp. 299–313. Springer, Heidelberg (2002)
2. Finn, V.K.: On Machine-Oriented Formalization of Plausible Reasoning in the Style of F. Backon–J. S. Mill (in Russian). Semiotika Informatika 20, 35–101 (1983)
3. Finn, V.K.: Plausible Reasoning in Systems of JSM Type (in Russian). Itogi Nauki i Tekhniki, Seriya Informatika 15, 54–101 (1991)
4. Finn, V.K.: Synthesis of cognitive procedures and the problem of induction (in Russian). Nauchno-Tekhnicheskaya Informatsiya 2(1-2), 8–44 (1999)
5. Ganter, B., Kuznetsov, S.O.: Formalizing Hypotheses with Concepts. In: Ganter, B., Mineau, G.W. (eds.) ICCS 2000. LNCS, vol. 1867, pp. 342–356. Springer, Heidelberg (2000)
6. Ganter, B., Kuznetsov, S.O.: Pattern Structures and Their Projections. In: Delugach, H.S., Stumme, G. (eds.) ICCS 2001. LNCS (LNAI), vol. 2120, pp. 129–142. Springer, Heidelberg (2001)
7. Ganter, B., Kuznetsov, S.O.: Hypotheses and Version Spaces. In: Ganter, B., de Moor, A., Lex, W. (eds.) ICCS 2003. LNCS, vol. 2746, pp. 83–95. Springer, Heidelberg (2003)
8. Ganter, B., Wille, R.: Formal Concept Analysis: Mathematical Foundations. Springer, Heidelberg (1999)
9. Garey, M.R., Johnson, D.S.: Computers and Intractability. A Guide to the Theory of NP-Completeness. W.H. Freeman, New York (1979)
10. Grigoriev, P.A., Kuznetsov, S.O., Obiedkov, S.A., Yevtushenko, S.A.: On a Version of Mill's Method of Difference. In: Proc. ECAI 2002 Workshop on Concept Lattices in Data Mining, Lyon, pp. 26–31 (2002)
11. Kuznetsov, S.O.: JSM-method as a Machine Learning System. Itogi Nauki i Tekhniki, ser. Informatika 15, 17–54 (1991)
12. Kuznetsov, S.O., Samokhin, M.V.: Learning Closed Sets of Labeled Graphs for Chemical Applications. In: Kramer, S., Pfahringer, B. (eds.) ILP 2005. LNCS (LNAI), vol. 3625, pp. 190–208. Springer, Heidelberg (2005)
13. Liu, H., Motoda, H.: Feature Selection for Knowledge Discovery and Data Mining. In: The Springer International Series in Engineering and Computer Science, vol. 454, Springer, New York (1998)

14. Mitchell, T.: Machine Learning, The McGraw-Hill Companies (1997)
15. Pawlak, Z., Wong, S.K.M., Ziarko, W.: Rough sets: Probabilistic versus deterministic approach. International Journal of Man-Machine Studies 29, 81–95 (1988)
16. Pawlak, Z.: Rough Sets: Theoretical Aspects of Reasoning about Data. Kluwer Academic Publishing, Dordrecht (1991)
17. Wolski, M.: Galois connections and data analysis. Fundamenta Informatica 60, 401–415 (2004)
18. Ziarko, W.: Rough sets as a methodology for data mining. In: Rough Sets in Knowledge Discovery 1: Methodology and Applications, pp. 554–576. Physica-Verlag, Heidelberg (1998)
19. Ziarko, W., Shan, N.: Discovering attribute relationships, dependencies and rules by using rough sets. In: Proc. 28th Annual Hawaii International Conference on System Sciences (HICSS 1995), pp. 293–299 (1995)

Formal Concept Analysis for the Identification of Combinatorial Biomarkers in Breast Cancer

Susanne Motameny[1], Beatrix Versmold[2], and Rita Schmutzler[2]

[1] ZAIK, Center for Applied Computer Science, University of Cologne, Germany
[2] Division of Molecular Gyneco-Oncology, Department for Obstetrics and Gynecology, University of Cologne, Germany
motameny@zpr.uni-koeln.de

Abstract. When cancer breaks out, central processes in the cell are disturbed. These disturbances are often due to abnormalities in gene expression. The microarray technology allows to monitor the expression of thousands of genes in human cells simultaneously. It is common knowledge that tumor cells show different gene expression profiles compared to normal tissue but also to tissue obtained from metastases. However, the identification of biomarkers, that is sets of genes whose expression change is highly correlated with the disease, poses a great challenge. Increasingly important is the extraction of combinatorial biomarkers. Here, the correlation to the disease is a result of the joint expression of several genes, whereas the single genes do not necessarily distinguish well between healthy and diseased tissue types. In this paper we describe how formal concept analysis can be used to identify gene combinations that are able to distinguish between tumor- and metastasis tissue in breast cancer based on microarray gene expression data.

Keywords: Gene expression, formal concept analysis, breast cancer, classification.

1 Introduction

Complex diseases like cancer have become the major challenge of medical research of our time. Many of these diseases are known to have a genetic component. That is why DNA sequence analysis and gene expression measurements play an increasing role both in fundamental research and clinical treatment of such diseases. In this paper, the focus is on combinatorial biomarkers that can be used for diagnosis or prognosis of a disease. A combinatorial biomarker is a set of genes that is able to distinguish reliably between two classes (for example healthy and cancerous tissue or metastasis and primary tumor) while no single member gene is required to do so. This means that a combinatorial biomarker can consist of genes whose individual behavior is rather uncorrelated to the classes of samples – it is the combination of the genes that provides the discriminative power. Up to now, many single genes have been identified the mutation or abnormal activity of which promotes the outbreak of cancer. In the case of

R. Medina and S. Obiedkov (Eds.): ICFCA 2008, LNAI 4933, pp. 229–240, 2008.

breast cancer for example, several studies have found that a mutation in one of the genes *BRCA1* or *BRCA2* (BRCA stands for BReast CAncer) implies an increased risk to develop the disease (see for example [4,10,12]). However, there are still many cases that cannot be explained by abnormalities of single genes. Combinatorial biomarkers are an obvious extension of the single gene approach and a possibility to gain a better understanding of complex diseases.

There exists a method called logical analysis of data (LAD) that has shown its ability to find combinatorial biomarkers of satisfying quality in several applications (see [1,2,3]). LAD takes a formal context as input and searches for prime patterns, which are minimal sets of attributes describing only objects from a single class. These patterns are then combined to obtain a combinatorial biomarker that applies to all objects of the class of interest. The patterns in LAD are essentially minimal generators in the sense of [5] of special concept intents of the input context. Therefore it is obvious that formal concept analysis (FCA) should be able to identify high-quality combinatorial biomarkers as well. The attribute combinations in formal concept intents are each associated to a unique set of objects. This principle is taken advantage of to identify combinatorial biomarkers that are able to distinguish primary tumor samples from metastasis samples based on gene expression data. This is done via a standard classification approach followed by a heuristic biomarker selection. The paper is organized as follows: In Section 2 we will shortly review the biological background and explain what gene expression data are. Section 3 contains the description of a straightforward FCA classification method that produces a candidate pool of genes from which combinatorial biomarkers are then extracted by a simple heuristic. We apply the method to generate combinatorial biomarkers that are able to distinguish primary tumors from metastases in a breast cancer data set in Section 4. The paper closes with a discussion of the results and an outlook in Section 5.

2   Preliminaries

In this section we explain the biological background and recall the notions from FCA that will be used in the paper.

2.1   Gene Expression

The processes in a living cell are based on chemical reactions between molecules. Proteins are molecules that are produced by each cell from blueprints encoded on the DNA. A piece of DNA that encodes the blueprint of a protein is called a **gene**. By changing the amount or composition of the proteins present inside the cell, the chemical reactions that can take place can be controlled. The mechanism that produces a protein from its gene is called **gene expression**. It consists of two steps: transcription and translation. Figure 1 illustrates the gene expression process. In the first step, the transcription, a copy of the gene on the DNA is produced. The "hardware" of this copy is an RNA molecule, called **messenger RNA** (or mRNA for short). From the mRNA copy of the gene, the protein is

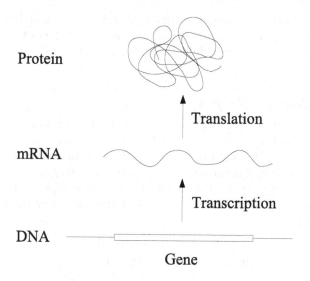

Fig. 1. Illustration of the gene expression process

produced in the translation step. The more mRNA copies of a gene are present in the cell, the more proteins can be produced from them. Therefore, the state of a cell can be inferred from measurements of mRNA abundance. On the human DNA there are approximately 25.000 genes (see [7]). Modern microarray technology allows to measure the abundance of mRNA copies for all genes simultaneously. Due to the detection limit of microarrays, mRNA concentrations of single cells are usually not measured. Instead, one takes homogeneous samples of cell populations (for example a piece of homogeneous tissue from a biopsy), extracts the mRNA molecules contained therein and measures their abundances. The result of such a microarray measurement is an n-dimensional vector e where e_g, $g \in \{1, \ldots n\}$ reflects the mRNA abundance of gene g in the sample: The higher e_g is, the more mRNA transcripts are present. The value e_g is called **expression value** of gene g. A **gene expression data set** is a collection of such sample measurements summarized in an $n \times m$ matrix E where each row corresponds to a gene and each column corresponds to a sample, so e_{gs} is the expression value of gene g in sample s, $s \in \{1, \ldots, m\}$. Our goal in this paper is to find gene combinations that are able to discriminate between two classes of samples. For this purpose it is often more convenient to consider the expression changes of genes between different samples rather than their absolute expression values. The common measure for the expression change of a gene g between two samples s and t is the **log ratio**. It is defined by

$$l_g = \log_2 \left(\frac{e_{gs}}{e_{gt}} \right). \tag{1}$$

If a gene has a log ratio of 1, then its expression value is two times higher in sample s compared to t. A log ratio of -1 means that the expression value of the gene is two times lower in sample s than in sample t. It is a usual and unfortunate situation that gene expression matrices contain many more genes than samples. This imbalance introduces great difficulties for the statistical analysis of gene expression data.

2.2 Formal Concept Analysis

Our notation follows essentially that of [8]. Let O be a set of objects, P be a set of properties (also called attributes), and $I \subseteq O \times P$ a binary relation. We use the letters O and P for the set of objects and attributes to avoid confusion with the set of genes G that will occur later on in the paper. If $(o, p) \in I$ we also write for short oIp and read "object o has the property p". The triple $K = (O, P, I)$ is called a formal context. We consider the usual derivation operators: Let $A \subseteq O$ and $B \subseteq P$. Then

$$A' = \{p \in P \ : \ oIp \ \forall o \in A\},$$
$$B' = \{o \in O \ : \ oIp \ \forall p \in B\}. \tag{2}$$

For sets containing only a single object o we will use the shorter notation o' instead of $\{o\}'$ and similarly for attributes. The pair (A, B) is a **concept** of $K = (O, P, I)$ if $A' = B$ and $B' = A$. The set A is called **extent** and B is called **intent** of (A, B). A formal context as defined above is one-valued, that is an object either does have the (one value of the) attribute or does not have the (one value of the) attribute. For our discussion we need many-valued contexts because our attributes will have several values that can apply to an object. Let W be the set of values attributes can take. Then the quadruple $K = (O, P, W, I)$ is called a many-valued context where I is now a ternary relation ($I \subseteq O \times P \times W$). If $(o, p, w) \in I$ we also write shortly $p(o) = w$ and read "the value of property p for object o is w". Furthermore, if o is an object and p a property, then $p(o) = w$ and $p(o) = v$ must imply that $w = v$. Before one can find concepts in a many-valued context, one has to turn it into a one-valued one by discretizing the many-valued attributes. This procedure is called **conceptual scaling** in FCA.

3   FCA Method for Combinatorial Biomarkers

The method we propose to identify combinatorial biomarkers can be applied both to absolute gene expression data or log ratios as input data. In both cases the input data are considered as a many-valued formal context $K = (S, G, W, I)$ with S being the set of samples, G being the set of genes, and $(s, g, w) \in I$ (or $g(s) = w$) if in sample s gene g has expression value or log ratio w, respectively. Furthermore the samples belong to two disjoint classes, a target class C_t (say diseased tissue) and a background class C_b (healthy tissue). The task is to find a set of genes that distinguishes between these two classes, while no single member gene is required to do so. We assume that the class memberships of the samples in the gene expression data set are known. The task of finding a combinatorial biomarker thus is a classical classification problem with some additional constraints and is tackled as such.

3.1   Training- and Validation Context

First, the many-valued context $K = (S, G, W, I)$ is divided into two subcontexts $K_T = (S_T, G, W, I)$ and $K_V = (S_V, G, W, I)$ with S_T and S_V being disjoint and $S_T \cup S_V = S$. K_T is called **training context** and K_V **validation context**. The training context will be used to identify combinatorial biomarkers by means of an FCA classification method while K_V serves for validation purposes. It is necessary that S_T contains samples from both classes and desirable that S_V does so, too.

3.2   Scaling

To identify combinatorial biomarkers for the target class, we first have to scale the training context. Note that the set of genes G here plays the role of the attribute set of K_T. Depending on the nature of the input data (absolute gene expression values or log ratios) we suggest two different scaling procedures.

Absolute expression values. We suggest that the scaling procedure be guided by the peculiarities exhibited by gene expression data sets in general. These are

1. the incomparability of expression values between genes and
2. a rather high noise level.

We want to use a dichotomic scale of the form

$$\mathbb{S}_g := \begin{array}{c|c|c} & \leqslant t_g & > t_g \\ \hline \leqslant t_g & X & \\ \hline > t_g & & X \end{array} \tag{3}$$

for each gene g. This means that we choose a threshold t_g for gene g and replace the many-valued attribute g by the two one-valued attributes "expression value of $g \leqslant t_g$" and "expression value of $g > t_g$". The threshold value t_g must be chosen specifically for each gene g due to the incomparability of expression values between genes. Furthermore, the threshold should be robust against the noise that is present in gene expression data. More precisely, if the noise level in the data is assumed to be ℓ, then we require that

$$\max\{g(s) \ : \ g(s) \leqslant t_g, s \in S_T\} \leqslant t_g - \ell \quad \text{and}$$
$$\min\{g(s) \ : \ g(s) > t_g, s \in S_T\} > t_g + \ell. \tag{4}$$

This condition assures that even if the training context is perturbed by noise as high as ℓ, the use of the thresholds t_g results in the same one-valued context as obtained for the original training context and the subsequent combinatorial biomarker identification remains the same. We implemented the scaling procedure in such a way that we sort the expression values of gene g in the validation context so that $e_{g1} \leqslant e_{g2} \leqslant \cdots \leqslant e_{g|S_T|}$ and then look for the largest interval $[e_{gi}, e_{gi+1}]$, $i \in \{1, \ldots, |S_T| - 1\}$. Let this interval be $[e_{gk}, e_{gk+1}]$ (in the case that there are several largest intervals, then one of them is selected randomly).

If $e_{gk+1} - e_{gk} \geqslant 2\ell$, then condition (4) is satisfied and the threshold t_g is chosen to be

$$t_g = \frac{e_{gk+1} + e_{gk}}{2}.$$ (5)

All genes which do not possess a threshold t_g satisfying condition (4) are dropped from the context with the intention to enhance the robustness against noise of the combinatorial biomarker. Thus, the scaling procedure implicitly includes a feature selection method, the strictness of which can be controlled via the parameter ℓ.

Log ratios. In the case of log ratios we can use a single scale for all attributes. It is given by

$$\mathbb{S} := \begin{array}{|c||c|c|} \hline & \leqslant -t & \geqslant t \\ \hline\hline \leqslant -t & X & \\ \hline \geqslant t & & X \\ \hline \end{array}$$ (6)

Here, t is a threshold value from $\mathbb{R}^+$. This scaling replaces each many-valued attribute g by the two one-valued attributes "log ratio of $g \leqslant -t$" and "log ratio of $g \geqslant t$". A typical threshold that is used in gene expression data analysis is 1, that is the gene must show a fold change of at least $2^1 = 2$ in its expression. Depending on the noise in the data set and the desired selectivity, t can be chosen smaller or greater. The greater t is chosen, the fewer samples will satisfy the scaled attributes and the subsequent search for combinatorial biomarkers will be consequently limited to genes showing at least a 2^t-fold change of their expression in one of the samples in S_T.

3.3   Identification Process

The scaled training context is given by $K_T = (S_T, G_s, J)$, where G_s denotes the set of the scaled attributes and J is the obtained incidence relation. The biomarker identification utilizes the idea that all FCA classification approaches rely on and that is formulated for example in [11] in terms of positive and negative hypotheses. To avoid the splitting of K_T into a positive and negative subcontext as in [11] we simplify notations and consider so called homogeneous concept intents.

Definition: *Let (A, B) be a formal concept. B is called a **homogeneous concept intent** if all objects in A belong to the same class.*

Let $B \subseteq G_s$ be a homogeneous concept intent. Obviously, B identifies a subset of samples that belong to the same class. It is furthermore clear that if B_1 is a homogeneous concept intent and B_2 is a concept intent with $B_2 \supset B_1$, then B_2 is also a homogeneous concept intent. For the identification of combinatorial biomarkers we are especially interested in the homogeneous concept intents for the target class. Among these intents, the smallest ones, containing the least attributes, are the most general ones which apply to the largest subsets of target samples. Therefore, we only generate the smallest homogeneous concept intents, that is the ones which are closest to the top element of the concept lattice of K_T. If the extent corresponding to a homogeneous concept intent B contains all samples of the target class in the training context, then B contains all combinatorial biomarkers for that class. In fact, B is in this case the largest combinatorial biomarker for the target class in K_T. It is also possible that there is no single homogeneous concept intent for the target class. Then (with an appropriate scaling) the class is described by several homogeneous concept intents which can be interpreted as a separation into subclasses. In this case, we would treat each subclass separately and consequently construct combinatorial biomarkers for each of the subclasses.

The homogeneous concept intents can be computed in the following way:

1. Test whether $S_T \cap C_t$ is a concept extent. If so, return $(S_T \cap C_t)'$.
2. Otherwise, compute the iceberg lattice of the subcontext $((S_T \cap C_t)'', G_s, J)$ and use the homogeneity of a concept intent as a stopping criterion (instead of the usual support threshold used in iceberg lattices (see [14] for an introduction to iceberg lattices)). Return all found homogeneous concept intents.

This procedure can be implemented using any algorithm that computes the concept lattice in a top-down manner (for example we use Bordat's algorithm in the version described in [6]).

3.4   Postprocessing and Validation

Due to the imbalance in the gene expression data (there are thousands of genes describing only few samples) it is expected that the homogeneous concept intent B for the target class contains many more genes than are necessary to distinguish reliably between the target and background class. Combinatorial biomarkers that are to be used for diagnosis or prognosis of a disease should be both short and robust. If a biomarker contains too many genes its evaluation for a single patient is time consuming and tedious and therefore expensive. On the other hand, the marker must be highly robust against noise and identify its target class reliably. Having these requirements in mind, we extract from the found homogeneous concept intent subsets of genes that suffice to identify the target class, contain only few genes, and show a certain robustness within the training context. More precisely, we identify subsets $M \subset G_s$ with

a) $M' = B'$,
b) $|M| \leqslant k$, and   (7)
c) $\forall s \in S_T \setminus C_t$ there are at least r genes $g_1, \ldots, g_r \in M$
   with $\forall i \in \{1, \ldots, r\}\ g_i \notin s'$.

Here k is the maximum number of genes the biomarker is allowed to contain and r is the minimum number of genes in a sample from the background class that do not satisfy the conditions imposed by the biomarker. The values of these parameters must be selected according to the application at hand. To simplify notations, we set for $s \in S_T \setminus C_t$

$$\text{gap}(s) = |\{g \in M \; : \; g \notin s'\}|. \tag{8}$$

An exhaustive search for all combinations of at most k genes is clearly prohibitive due to its computational complexity. That is why we suggest a simple heuristic which is given in pseudo code in Table 1.

Table 1. Heuristic to extract candidate combinatorial biomarkers

1. set $M := \emptyset$, $L := S_T \setminus C_t$, `iter` $:= 0$, `maxiter` $:= 1000$
2. repeat
3.   find a sample $s \in L$ with $|s' \cap B|$ maximal
4. $a := r - \text{gap}(s)$
5. $D := \{g \in B \; : \; g \notin s'\}$
6.   randomly select genes $g_1, \ldots, g_a$ from D and set $M := M \cup \{g_1, \ldots, g_a\}$
7. $L := L \setminus \{s \in L \; : \; \text{gap}(s) \geqslant r\}$
8.   if $|M| > k$, then $M := \emptyset$, $L := S_T \setminus C_t$, `iter` $:=$ `iter`$+1$
9. until $(M' = S_T \cap C_t)$ and $(L = \emptyset)$ or (`iter` = `maxiter`)

Starting with an empty set of genes M, we successively collect genes from the homogeneous concept intent B into M until the conditions a) and c) from (7) are satisfied. For this purpose we keep a list L of the samples from the background class for which condition c) is not yet satisfied. In order for a sample $s \in S_T \setminus C_t$ to fulfill condition c), r genes from the set $D = \{g \in B \; : \; g \notin s'\}$ must be contained in the biomarker M. This is achieved by successively adding genes from D to M, starting with the most difficult samples, that is samples s for which $|s' \cap B|$ is maximal (lines 3. to 6.). When sample s is considered, M contains already $\text{gap}(s)$ genes whose threshold conditions are not satisfied by s. Therefore, only a genes need to be added to M to achieve the r genes required by condition c) for s (lines 4. and 6.). After each addition of genes to M, samples from the background set that satisfy condition c) are removed from L (line 7.). If during this procedure $|M|$ exceeds the preselected threshold k, then we restart the selection procedure (line 8.). To prevent a possible nonterminating loop, the selection procedure is run at most `maxiter` times. The randomness in the genes' selection produces a variety of different biomarker candidates when running the procedure several times.

3.5   Validation

After having produced a set of different candidate biomarkers, their validity has to be assessed. This is done by using the validation context $K_V = (S_V, G, W, I)$. Each biomarker candidate consists of a set of conditions of the form

$$\text{``the value of gene } g \lozenge t\text{''},$$

where $\diamond \in \{\leqslant, \geqslant, >\}$ and t is some threshold whose value depends on the scaling that was applied. We simply test for each sample $s \in S_V$ whether or not it satisfies the conditions specified in each of the biomarker candidates. A candidate biomarker whose conditions are satisfied by all samples $s \in S_V \cap C_t$ and by none of the background samples has proved its ability to recognize the target class correctly. This means that it is a valid combinatorial biomarker for class C_t with respect to our given expression data.

4   Application to Breast Cancer

We now apply the method described in Section 3 to a real world data set of gene expression data gained from biopsy samples of breast cancer tumors. The gene expression measurements were performed using Affymetrix GeneChip® technology ([9]). The data were produced in a study involving five clinical laboratories located in Cologne, Düsseldorf, Bonn, Berlin, and Munich. The data set contains 50 samples, 20 of which are metastases, 28 are primary tumors, and 2 are samples from healthy tissue. For each sample, the expression of 22.215 genes was measured, so we have a gene expression context $K = (S, G, W, I)$ with $|S| = 50$ and $|G| = 22.215$. Our aim is to find biomarkers that identify metastases. In addition to the noise problem that applies to all gene expression data, the data set from this study contains systematic biases because the biopsy samples were prepared by different laboratories. On the one hand this complicates the identification of a biomarker as classification procedures work best for homogeneous data sets, but on the other hand this is also a more realistic scenario as a biomarker used in clinical practice will of course have to work regardless of the laboratory that provides the sample.

From the 50 samples, 16 primary tumors and 13 metastases were randomly selected as the training context K_T. The validation context contains the remaining samples. The absolute expression values in both K_T and K_V are transformed into log ratios using the mean expression value of the 16 primary tumors in K_T as reference. Formally, the log ratio of gene g in sample s is computed as

$$l_{gs} = \log_2\left(\frac{e_{gs}}{\bar{e}}\right), \tag{9}$$

where $\bar{e}$ is the mean of the expression values of gene g across the 16 primary tumor samples in the training context. Scaling of K_T is performed according to the log ratio scaling method described in Subsection 3.2 with threshold $t = 1$. There is exactly one homogeneous concept intent B for the class of metastasis samples in K_T. This concept intent contains 42 genes. Using the heuristic from Table 1 with $k = 4$ and $r = 2$ we find 8 combinatorial biomarkers that pass the validation. The majority of the genes contained in these markers have functions related to the extracellular region, a finding that fits well with the fact that in order to migrate through the body, metastasis cells must adapt their outer structure and surrounding. Other genes are involved in processes that are specific to breast and brain tissue which also fits nicely because the samples in the data set stem from these locations (primary tumors from breast and metastases from brain). Three of the genes have already been reported in the literature to be associated to cancer progression and metastasis development.

To get an impression of how our FCA-based biomarker identification method compares with other classification methods, we used the scaled context of the breast cancer data set to infer decision trees with the function "treefit" supplied by the MATLAB® software ([13]). We used both our selection heuristic based on the homogeneous concept intent and the treefit function to generate 2000 biomarkers. The parameters for the selection method were $k = 4$ and $r = 2$. The treefit function was used with default parameters. As the decision tree builder is deterministic, we randomly ordered the columns of the training context in order to produce a variety of different decision trees. The results of the validation of the 2000 biomarkers are summarized in Table 2. We compare three parameters.

- **accuracy:** number of correctly classified samples divided by the number of all samples (averaged over all produced biomarkers).
- **valid:** number of valid biomarkers produced.
- **genes:** average number of genes used in the biomarkers.

Table 2. Comparison of the FCA method and the MATLAB® decision tree method for biomarker identification

	accuracy	valid	genes
FCA	85.3 %	29	3.8
treefit	81.0 %	0	1

One can see that the average accuracy of both methods is not dramatically different. However, the FCA method produces a quite satisfying number of valid biomarkers (that is the accuracy for these biomarkers is 100 %) while treefit produces not a single one. A reason for this can be seen in the third column of Table 2 which contains the average number of genes used in the biomarkers. All produced decision trees contain exactly one gene. This is because there are several single genes that are able to distinguish between primary tumor and metastasis in the training context, but none of them can make this distinction in the validation context.

5   Discussion and Outlook

The results from the breast cancer data are encouraging. They show that our method is able to find valid combinatorial biomarkers in real world gene expression data. Even in the difficult situation of working with a heterogeneous data set, the method produced satisfying results. The identified combinatorial biomarkers do not only show the desired behavior with respect to their ability of metastasis recognition but also include genes that make sense from a biological point of view. When comparing our method to a method using decision trees, we see that the FCA approach outperforms the decision tree algorithm. Given the rather small size of the study, we are still far from the introduction of our combinatorial biomarkers into clinical practice, but our results can already be used to investigate the processes involved in the development of metastases. Instead of the usual approach, that thrives to find single genes whose expression differs between metastases and primary tumors, our combinatorial biomarkers shed light on the interplay between genes that is characteristic for metastases. This meets the concerns of a recent direction of medical research which assumes that the development of metastases does not depend on single genes but is much more complex and requires a certain constellation of expression changes in order to occur.

Future research will center around the refinement of the method for combinatorial biomarker search and its application to other data sets. The basics of the method, namely to make use of the strength of FCA to find attribute combinations that identify given sets of objects uniquely, have already proven useful and will be kept. To speed up the process of the final biomarker selection, the addition of further criteria will be evaluated. Such criteria shall impose conditions on the maximum expression value of a gene to exclude genes that were measured near the detection limit or filter out genes that show a rather random behavior across the samples. Instead of stopping once a homogeneous concept intent is found, it might also prove interesting to unfold the structure of the concept lattice below this concept intent further to see whether there are subclasses and find gene combinations that describe them. Furthermore, a statistical analysis will be included to assess the significance of the identified combinatorial biomarkers and comparisons to further classification methods (such as artificial neural networks or support vector machines) will be made. To enter clinical practice, a much larger study is necessary to assess the applicability of FCA-based combinatorial biomarkers in the detail and soundness required by medical applications.

Acknowledgements

We thank Norbert Arnold (Division of Oncology, Department of Gynaecology and Obstetrics University Hospital Schleswig-Holstein, Kiel), Dieter Niederacher (Division of Molecular Genetics, Department of Gynaecology and Obstetrics Clinical Center University of Düsseldorf), Alfons Meindl (Department of Gynaecology and Obstetrics, Klinikum rechts der Isar at the Technical University Munich), and Susanne Seitz (Department of Tumor Genetics, Max Delbrueck Center for Molecular Medicine, Berlin) for providing biopsy samples, performing of the gene expression measurements, and sharing of their data. Finally, we would like to thank three anonymous referees whose suggestions helped to improve the paper. This work was in part supported by the rectorate of the University of Cologne and Köln Fortune via the MathEDatA project.

References

1. Alexe, G., et al.: Breast cancer prognosis by combinatorial analysis of gene expression data. Breast Cancer Research 8(4) (2006)
2. Alexe, G., et al.: Logical analysis of diffuse large b-cell lymphomas. Artificial Intelligence in Medicine 34(3), 235–267 (2005)

3. Alexe, G., et al.: Ovarian cancer detection by logical analysis of proteomic data. Proteomics 4(3), 766–783 (2004)
4. Antoniou, A., et al.: Average risks of breast and ovarian cancer associated with BRCA1 or BRCA2 mutations detected in case series unselected for family history: A combined analysis of 22 studies. American Journal of Human Genetic 72, 1117–1130 (2003)
5. Bastide, Y., et al.: Mining non-redundant association rules using frequent closed itemsets. In: Palamidessi, C., et al. (eds.) CL 2000. LNCS (LNAI), vol. 1861, pp. 972–986. Springer, Heidelberg (2000)
6. Berry, A., Bordat, J.P., Sigayret, A.: A local approach to cocept generation. Technical report, ISMIA/LIRMM (2004)
7. International Human Genome Sequencing Consortium. Finishing the euchromatic sequence of the human genome 431, 931–945 (2004)
8. Ganter, B., Wille, R.: Formal Concept Analysis, Mathematical Foundations. Springer, Heidelberg (1999)
9. GeneChip®, Affymetrix, Santa Clara, California
10. King, M.-C., Marks, J.H., Mandell, J.B.: For The New York Breast Cancer Study Group. Breast and ovarian cancer risks due to inherited mutations in BRCA1 and BRCA2 302, 643–646 (2003)
11. Kuznetsov, S.O.: Machine learning on the basis of formal concept analysis. Annotation and remote control 62(10), 1543–1564 (2001). Translated from Automatika i Telemekhanika, No. 10, 3–27 (2001)
12. Mann, G.J., et al.: The Kathleen Cuningham Concortium for Reasearch in Familial Breast Cancer. Analysis of cancer risk and BRCA1 and BRCA2 mutation prevalence in the kConFab familial breast cancer resource. Breast Cancer Research 8(1) (2006)
13. MATLAB®7(R14), The MathWorks, Natick, Massachusetts
14. Stumme, G., et al.: Computing iceberg concept lattices with TITANIC. Data Knowledge Engineering 42(2), 189–222 (2002)

Handling Spatial Relations
in Logical Concept Analysis
to Explore Geographical Data*

Olivier Bedel, Sébastien Ferré, and Olivier Ridoux

IRISA/Université de Rennes 1
Campus de Beaulieu
35042 Rennes Cedex, France
firstname.lastname@irisa.fr

Abstract. Because of the expansion of geo-positioning tools and the de-
mocratization of geographical information, the amount of geo-localized
data that is available around the world keeps increasing. So, the ability to
efficiently retrieve informations in function of their geographical facet is
an important issue. In addition to individual properties such as position
and shape, spatial relations between objects are an important criteria
for selecting and reaching objects of interest: e.g., given a set of touris-
tic points, selecting those having a nearby hotel or reaching the nearby
hotels. In this paper, we propose Logical Concept Analysis (LCA) and
its handling of relations for representing and reasoning on various kinds
of spatial relations: e.g., Euclidean distance, topological relations. Furthermore, we present an original way of navigating in geolocalized data,
and compare the benefits of our approach with traditional Geographical
Information Systems (GIS).

1 Introduction

In previous work [1], we have applied Logical Information Systems (LIS) to
explore of geographical data. LIS allow to query and browse a collection of
geographical objects using spatial properties, such as position and shape, and
non spatial ones, such as description and date. The geographical objects were
described in isolation, not taking into account their mutual organization. However, relations, and particularly spatial relations, play an important role when
describing and exploring geographical data. Indeed, geographical information
is tradionnaly split in thematic layers, and localization is often the only common property between these different layers. Quantitative and qualitative spatial
relations between geographical objects may be easily derived from objects localization. Spatial relations enhance the description of geographical data, improve
the expressivity of spatial querying and facilitate the combination of data dealing
with several thematics. For instance, one may want to find all bus stops under

---

* This work is funded by a scholarship from Région Bretagne.

R. Medina and S. Obiedkov (Eds.): ICFCA 2008, LNAI 4933, pp. 241–257, 2008.

some distance from some place; or to find appartments that are close to a public garden that contains a lake.

LIS are founded on Logical Concept Analysis (LCA) [2]. Like FCA, LCA allows to group objects into concepts on the basis of individual properties, but moreover LCA also enables to consider arbitrary relations between concepts [3]. FCA does not natively support relations, however several extensions have been proposed in this sense: *Power Context Families* [4] and *Relational Concept Analysis* (RCA) [5]. These extensions are discussed in Section 2, and the choice of LCA is motivated within the scope of describing spatial relations and exploring geographical data. Section 3 recalls the useful definitions of LCA.

There are various formalisms to represent and reason on spatial relations, from purely numeric relations [6] to purely symbolic relations [7]. We show LCA covers a wide range of spatial relations by applying it to two different kinds of spatial relations (Section 4). The first kind is a numeric spatial relation, the distance between objects, where relations are described by a precise value, and intervals of distance can be used in queries. The second kind is a symbolic spatial relation, topological relations such as "contains", "overlaps", or "touches". Furthermore, we show that these relations can be automatically derived from the position and shape of objects. So, having such relations costs nothing more to the application designer than the individual description of objects.

A prototype has been implemented, and is used to demonstrate the benefits of our approach for exploring geographical data (Section 5). The navigation facilities are especially emphasized as they prevent users from having to write complex queries. Finally, we compare our work with state of the art in Geographical Information Systems (GIS) (Section 6), and conclude with further works (Section 7).

2   Relations in Concept Analysis

To our knowledge, three main approaches have been proposed to take into account arbitrary relations between objects in FCA: Power Context Families, Relational Concept Analysis, and relations in Logical Concept Analysis. Each of them has its advantages and its drawbacks regarding the intended use. In this section, we discuss the oportunity of considering a formalism rather than another for the purpose of geographical data exploration in the FCA framework. In fact, we want to represent spatial relations between geographical objects and use them as a key for navigation and querying.

Power Context Families (PCF) extend standard FCA and enable to represent arbitrary n-ary relations between objects of a formal context. A PCF is a vector of formal contexts $(K_1, K_2, ..., K_n), n \geq 2$ with $K_i = (\mathcal{O}_i, \mathcal{A}_i, I_i)$, s.t. $\mathcal{O}_i \subseteq (\mathcal{O}_1)^i$. K_1 is the formal context that stores the description of objects and K_n describes the n-ary relations linking n of those objects. Dealing with n-ary relations may be interesting when representing complex interactions between objects, as it is the case in software engineering for instance. But, in the geographical domain, most of spatial relations that are used in GIS are expressed in terms of binary relations. Moreover, a drawback of PCF is that a different lattice is generated for each context. This means that one cannot use relational properties to navigate in the traditional concept lattice, i.e. the lattice derived from K_1. Yet, mixing object properties and relational properties in a navigation search is one goal we want to achieve.

Contrary to PCF, Relational Concept Analysis (RCA) and LCA enables to describe objects and their relations to each other in the same context, and so in the same lattice. RCA introduces into FCA abstractions of binary relations between formal concepts similar to role description ($\forall r.C, \exists r.C$) in Description Logics [5]. These abstractions result from a relational scaling on the object context, and appear in the corresponding concept lattice. However, RCA only deals with a flat set of relation names, and does not allow *a priori* to express valued relations or to represent a generalization ordering between relations. When dealing with spatial relations, this is a drawback as distance relations are valued relations, and topological relations can be organized in a taxonomy. LCA uses logic to describe objects and relations, as well as to express queries over those objects. Navigation between concepts is enabled by navigation links that take the form of $\exists r.f$, where r denotes the relation and f a description of the image through r. They can be nested or combined with Boolean operators to define complex queries. As shown in Section 4, logic enables to express and reason with valued properties, and also to consider a partial ordering over relational properties. This provides facilities to represent distance or topological relations. LCA has first been introduced to generalize FCA for the purpose of information retrieval [2]. In the same way as RCA extends FCA to support relations in concept analysis, the addition of relations in LCA [3] brings arbitrary relations in information retrieval with LCA. If RCA and LCA do not share a common goal, the same motivation in handling relations appears in both approaches.

An advantage of RCA is its ability to effectively build the concept lattice, which is useful for data-mining tasks. LCA supports the exploration of the concept lattice, where the focus is on the concept and its neighborhood. This is valuable when exploring large and dense concept lattices. And precisely, our aim is here information retrieval and among the previous approaches, LCA appears to be the most appropriate to deal with spatial relations for the purpose of exploring geographical objects with querying and navigation.

3   Relations in Logical Concept Analysis

In this section, we recall how LCA enables to take into account arbitrary binary relations between objects. We here limit our presentation to the parts that are relevant to querying and navigation mechanisms, i.e. we focus on the computation of extents and navigation links between concepts, and let out the computation of intents. For readers interested in the details of our approach, a complete introduction to relations in LCA can be found in [3].

3.1   Describing Objects and Relations

In LCA, the *object context* contains the description proper to each object individually, whereas the *relation context* describes the binary relations between objects of the *object context*.

Definition 1 (object context). *An* object context *is a triple* $K_1 = (\mathcal{O}, \mathcal{L}_1, d_1)$, *where:*

- $\mathcal{O}$ *is a set of objects.*
- $\mathcal{L}_1 = (L_1, \sqsubseteq_1)$ *is called a logic.* L_1 *is a language of formulas used to describe objects, i.e.,* L_1 *is composed of all words that can be used to build a formula, e.g. atoms or connectors.* $\sqsubseteq_1$ *is a partial ordering, called subsumption.*
- $d_1 : \mathcal{O} \to L_1$ *is a mapping from objects to their description.*

The extent of an L_1-formula q_1 is defined as the set of objects whose description is subsumed by q_1.

Definition 2 (object extent). *Let* $K_1 = (\mathcal{O}, \mathcal{L}_1, d_1)$ *be an object context and* $q_1 \in L_1$ *be a query. The* object extent *of* q_1 *in* K_1 *is defined by*

$$ext_1(q_1) =_{def} \{o \in \mathcal{O} \mid d_1(o) \sqsubseteq_1 q_1\}.$$

Example 1 (The film and artist context). Here we consider the context of Figure 1 about films and artists. The first row is to be read:

$$d_1(f_1) = \{title : Pulp\,Fiction\,, year : 1994\,, style : detective\}$$

In this example, $\mathcal{L}_1 = \mathcal{P}(\mathcal{A} \times \mathcal{L}_1^V)$ is a composite logic, where $a_1{:}v_1 \in \mathcal{A} \times \mathcal{L}_1^V$ is called a valued logical property and represents an attribute followed by its value. $\mathcal{A}$ and $\mathcal{L}_1^V$ are logics about attributes and values, $\times$ enables to define the product of two logics, and $\mathcal{P}$ enables to reason on sets of formulas of a logic. A composite logic $\mathcal{L}$ includes a composite language L and a composite subsumption relation $\sqsubseteq$. In this example, $\sqsubseteq_1$ correponds to a variant of set-theoretic inclusion where subsumption of values of the same attributes are taken into account. When querying, this enables to use patterns over values of properties. For instance, $\{year : 1994, style : Detective\} \sqsubseteq_1 \{year : \; >= 1990\}$. Other examples will be given in Section 4, but for a detailed explanation of the mechanism of logic composition, the reader should refer to [8].

The relation context is a kind of logical context, whose objects are pairs of objects of an object context, and there is an inverse operation for both objects and formulas.

Definition 3 (relation context). *Let* $K_1 = (\mathcal{O}, \mathcal{L}_1, d_1)$ *be an object context. A* relation context *is a triple* $K_2 = (\mathcal{R}, \mathcal{L}_2, d_2)$, *where:*

- $\mathcal{R}$ *is a set of pairs* $(o_1, o_2) \in \mathcal{O} \times \mathcal{O}$. *Each pair* (o_1, o_2) *represents an ordered binary relation from* o_1 *to* o_2. *Two mappings* start *and* end *are defined over* $\mathcal{R}$ *s.t.* $start((o, o')) =_{def} o$ *and* $end((o, o')) =_{def} o'$. *Furthermore,* $\mathcal{R}$ *is closed with an inverse operation* $^{-1}$, *i.e. for every pair* $r \in \mathcal{R}$, $start(r^{-1}) = end(r)$ *and* $end(r^{-1}) = start(r)$.

Object context								Relation context		
	type	title	year	style	name	age	sex		plays	directs
f_1	Film	Pulp Fiction	1994	detective				(p_1, f_1)	x	
f_2	Film	Planet Terror	2007	Action				(p_1, f_2)	x	
f_3	Film	Die Hard 4	2007	Action				(p_1, f_3)	x	
f_4	Film	Death Proof	2007	Action				(p_2, f_2)	x	x
p_1	Artist				Bruce Willis	52	M	(p_3, f_1)		x
p_2	Artist				Robert Rodriguez	39	M	(p_3, f_2)	x	
p_3	Artist				Quentin Tarantino	44	M	(p_3, f_4)	x	x

Fig. 1. An object context about artists and films, and the corresponding relation context. Relations $plays^{-1}$ and $directs^{-1}$ are not explicitly described, but can be automatically infered from the relation context.

- $\mathcal{L}_2 = (L_2, \sqsubseteq_2, .^{-1})$ is a logic of relations. L_2 is a language of formulas used to described relations. $\sqsubseteq_2$ is a subsumption relation, and $(.^{-1})$ denotes an inverse operation over formulas, considering the inverse of relations s.t. for every $f_2, g_2 \in L_2$, the following axioms are satisfied:
  - $(f_2^{-1})^{-1} \equiv_2 f_2$ (where $f_2 \equiv_2 g_2 =_{def} f_2 \sqsubseteq_2 g_2 \wedge g_2 \sqsubseteq_2 f_2$),
  - $f_2 \sqsubseteq_2 g_2 \Leftrightarrow g_2^{-1} \sqsubseteq_2 f_2^{-1}$.
- $d_2 : \mathcal{R} \to L_2$ is a mapping from pairs of objects to their description expressed as a logical formula. d_2 is compatible with the inverse relation, i.e. $\forall r \in \mathcal{R}$, $d_2(r^{-1}) \equiv_2 d_2(r)^{-1}$.

The extent of an L_2-formula q_2 over a relation context is the set of pairs of objects that are related by a relation subsumed by q_2.

Definition 4 (relation extent). *Let $K_2 = (\mathcal{R}, \mathcal{L}_2, d_2)$ be a relation context, and $q_2 \in L_2$ be a query. The relation extent of the query q_2 is defined by*

$$ext_2(q_2) =_{def} \{r \in \mathcal{R} \mid d_2(r) \sqsubseteq_2 q_2\}.$$

Example 2 (The playing and directing relation context). We now consider the relation context of Figure 1 indicating who plays in a film and who directs it. For instance, from this context, we can read:

$$d_2((p_3, f_1)) = \{directs\} \qquad \text{and} \qquad d_2((f_1, p_3)) = \{directs^{-1}\}$$
$$d_2((p_2, f_2)) = \{plays, directs\} \text{ and so } d_2((f_2, p_2)) = \{plays^{-1}, directs^{-1}\}$$

In this case, $\mathcal{L}_2 = \mathcal{P}(Ro)$ where $Ro = \{plays, plays^{-1}, directs, directs^{-1}\}$ is a logic of roles. Ro corresponds to the different roles an artist can have in a film, and vice versa. The subsumption relation $\sqsubseteq_2 = \supseteq$. $(.^{-1})$ maps a set of roles to the set of its inverse roles. For instance, $\{plays, directs\} \sqsubseteq_2 \{directs\}$ and $\{plays\}^{-1} = \{plays^{-1}\}$.

3.2 Querying

In LCA, queries include criteria on both objects and relations. So, for the purpose of querying and navigation, a new context $K = (K_1, K_2)$, combining the object and relation contexts, is considered. In the same way, a combined query language L is defined for representing expressive queries.

Definition 5 (combined context and language). *Let K_1 be an object context, and K_2 be a relation context. $K = (K_1, K_2)$ is the combined context, that gathers the individual descriptions of objects and their relationships to each others. The combined language L is defined as follows:*

$$L \to \top \mid \bot \mid L_1 \mid \exists L_2.L \mid \forall L_2.L \mid \neg L \mid L \sqcap L \mid L \sqcup L.$$

This language is the language of the description logic $\mathcal{ALC}$ [9], except atomic concepts are replaced by object formulas (coming from L_1), and atomic roles are replaced by relation formulas (coming from L_2). The previous definitions of object and relation extents are used as the basis for a combined extent that computes the extent of a combined query. The semantics behind its definition is the same as in description logics, except the closed world assumption is used instead of the open world assumption (e.g., the extent of the negation of a formula is always the complement of the extent of this formula).

Definition 6 (combined extent). *Let $K = (K_1, K_2)$ be a combined context, and $q \in L$ be a combined query. The combined extent of q is defined by the recursive set of definitions:*

- $ext(\top) =_{def} \mathcal{O}$
- $ext(\bot) =_{def} \emptyset$
- $ext(q_1) =_{def} ext_1(q_1)$, where $q_1 \in L_1$
- $ext(\exists q_2.q) =_{def} \{o \mid \exists r \in ext_2(q_2).(start(r) = o \wedge end(r) \in ext(q))\}$
- $ext(\forall q_2.q) =_{def} \{o \mid \forall r \in ext_2(q_2).(start(r) = o \Rightarrow end(r) \in ext(q))\}$
- $ext(\neg q) =_{def} \mathcal{O} \setminus ext(q)$
- $ext(q \sqcap q') =_{def} ext(q) \cap ext(q')$
- $ext(q \sqcup q') =_{def} ext(q) \cup ext(q')$

The formula $\exists q_2.q$ can be understood as *having at least one image through relation q_2 that satisfies the formula q*. Whereas $\forall q_2.q$ can be understood as *having all images through relation q_2 that satisfy q*.

Definition 7 (query reversal). *The query reversal is a query transformation defined as follows:*

$$rev(q' \sqcap \exists q_2.q'', \exists q_2.q'') = q'' \sqcap \exists q_2^{-1}.q' \text{ with } q', q'' \in L, q_2 \in L_2.$$

Query reversal allows to traverse backward relations already mentioned in a query. This allows to change the point of view by considering either one side of a relation or the other. This is useful because the query $\exists q_2^{-1}(\exists q_2 .q')$ is not always equivalent to the query q'.

3.3   Navigating

As seen before, query reversal already enables to navigate between queries. Moreover, in order to help users in building queries, a subset of navigation links can be computed for any query in order to refine it. These navigation links are taken in a finite vocabulary, whose elements are called *features*. We assume that the set of features for an object context K_1 (resp. relation context K_2) is given by a user-defined function, called $feat_1(K_1)$ (resp. $feat_2(K_2)$). In most cases,

$feat_1(K_1)$ (resp. $feat_2(K_2)$) contains at least the subset of L_1 (resp. L_2) that is used to describe objects, and the subset of patterns of L_1 (resp. L_2) that users have already entered in queries. Those two vocabularies need to be combined in order to provide a navigation vocabulary over a combined context. It has been proved [3] that the subset of the combined language useful to build navigation links can be restricted to:

Definition 8 (combined vocabulary). *Let $K = (K_1, K_2)$ be a combined context. The combined vocabulary is recursively defined as the set of features*

$$feat(K) =_{def} feat_1(K_1) \cup \{\exists x_2.x \mid x_2 \in feat_2(K_2), x \in feat(K)\}.$$

Features coming from K_1 are called object features, whereas those coming from K_2 are called relation features. As a query $\forall q_2.q$ can be rewritten as $\neg\exists q_2.\neg q$, there is no need for $\forall$ quantifier in relation features. Given a current query $q \in L$, only features occuring in the extent of q are presented to users as further query increments.

Definition 9 (query increments). *Let $K = (K_1, K_2)$ be a combined context, and a query $q \in L$. A feature $x \in feat(K)$ is a query increment if it has a common instance with the query q, i.e. $ext(q) \cap ext(x) \neq \emptyset$.*

Object features and relation features can both be used as query refinements, but relation features can also serve as paths to go through. A relation feature $\exists q_2.q'$ and may be used go from a query q to $q \sqcap \exists q_2.q'$ (refinement) or $q' \sqcap \exists q_2^{-1}.q$ (relation traversal). In order to facilitate the browsing of query increments, these navigation links are partially ordered according to subsumption. This requires to define subsumption between features by combining object and relation subsumptions.

Definition 10 (combined subsumption). *Let $K = (K_1, K_2)$ be a combined context. The combined subsumption $\sqsubseteq$ between features in $feat(K)$ is defined by*

$$x \sqsubseteq y =_{def} \begin{cases} x \sqsubseteq_1 y & \text{if } x, y \in L_1 \\ x_2 \sqsubseteq_2 y_2 \wedge x' \sqsubseteq y' & \text{if } x = \exists x_2.x', y = \exists y_2.y' \ (\text{for some } x', y' \in L) \\ false & \text{otherwise} \end{cases}$$

Example 3 (Query refinement, relation traversal and query reversal). We consider the *film* and *artist* context, and using the different ways of navigation, we progressively build a query. We start from the most general query $q_{t0} = \top$.

1. We start by refining q_{t0} with the object feature *Style:Action* to select action films: $q_{t1} = \top \sqcap style:Action = style:Action$.
2. Then, we use the relation feature $\exists plays^{-1}.(name:"Quentin Tarantino")$ to refine the query: $q_{t2} = style:Action \sqcap \exists plays^{-1}.(name:"Quentin Tarantino")$.
3. We choose then to traverse the relation $directs^{-1}$ using the relation feature $\exists directs^{-1}.(\top)$. Now $q_{t3} = \top \sqcap \exists directs^{-1}.(q_{t2})$ denotes the artists who direct an action film in which Quentin Tarantino plays.

4. We are then able to restrict to artists that are men: $q_{t4} = q_{t3} \sqcap sex{:}M$.
5. Last, we decide to come back to the selected films using the query reversal on relation $direct^{-1}$ traversed in q_{t3}. The new query q_{t5} is equal to:
   $style{:}Action \sqcap plays^{-1}.(name{:}"\text{Quentin Tarantino}") \sqcap \exists directs^{-1}.(sex{:}M)$.

4 LCA Applied to the Geographical Domain

In the sequel, we first present a logic $\mathcal{L}_1^g$ over geometries that enables to describe and reason about the spatial properties of geographical objects. Then, we give two examples of spatial relations defined as logical relations using LCA: a Euclidean distance relation defined as $\mathcal{L}_2^{[m]}$ and a set of topological relations defined as $\mathcal{L}_2^{Topo}$. Especially, we describe the logical reasoning over these kinds of spatial relations.

4.1 Spatial Properties

As introduced in Section 3, the object context contains the description of objects expressed as a conjunction of logical properties. To describe the spatial characteristics of a geographical object, we use a particular logical property, called the *geometry property*. The geometry property is a valued property whose name is **geometry** and whose value represents the shape of the geographical object. This value corresponds to a textual description of the geometry, expressed in the *Well Known Text* (WKT) format [10] defined by the Open Geospatial Consortium. Like most other geographical data formats, WKT encompasses the location in the shape description. In fact, the shape is defined as a sequence of absolute coordinates determining the spatial border of the geographical object. Examples of geometry properties are presented in Table 1. The domain of values of the geometry property is defined by a specialized logic over geometries: $\mathcal{L}_1^g = (L_1^g, \sqsubseteq_1^g)$.

Table 1. Three simple spatial descriptions illustrating point-wise, linear and area representation of geographical objects

object	geometry property
a subway station (position)	geometry:POINT(128.2 135.4)
a subway line	geometry:LINESTRING(102.3 99.4, 112.7 110.2, 120.9 123.0, 128.2 135.4, 129.6 155.3, 130.2 169.3, 150.4 168.2)
a building (covered area)	geometry:POLYGON((110.6 20.3, 110.6 22.1, 111.2 22.1, 111.2 20.3, 110.6 20.3))

L_1^g is equal to the WKT language, and $\sqsubseteq_1^g$ corresponds to the inclusion of geometries. According to $\mathcal{L}_1^g$, the value g =POLYGON((110.6 ... 20.3)) represents all geometries that are entirely inside g. For instance, POINT(0.0 0.0)$\sqsubseteq_1^g$ POLYGON((-1.0 -1.0,-1.0 1.0,1.0 1.0,1.0 -1.0,-1.0 -1.0)). When querying for geographical objects, a user has thus the ability to draw an area of interest on a map, and $\mathcal{L}_1^g$ can be used to retrieve all objects located inside the polygon corresponding to the drawn area. A more detailed explanation of the use of $\mathcal{L}_1^g$ can be found in [1].

4.2   Spatial Relations

Spatial relations are implicitly derived from to the geographical description of objects. The distance relation and the topological relations are function of the location and the shape of objects.

Distance Relation. Our first example of a spatial relation is the distance between objects. In the following, we consider only the case where coordinates of geographical objects are expressed in the same projected coordinate system, which corresponds to the common practice in GIS. We define the distance $dist$ between two points as the Euclidean distance. The distance $dist_g$ between two geometries g_1 and g_2 corresponds to the minimun distance between a point of g_1 and a point of g_2:

$$\forall g_1, g_2, \; dist_g(g_1, g_2) = \min\{dist(p_1, p_2) \mid p_1 \in g_1, p_2 \in g_2\}$$

With LCA, $dist_g$ can be represented with a logic of relations $\mathcal{L}_2^{[m]}$. Formulas take the form of valued attributes. The attribute name is `distance` and its values correpond to a distance expressed as a real number in the metric system. Intervals and several metric symbols, e.g. `m`, `cm` or `km` can be used as patterns. As a distance relation is symmetric, $f_2^{-1} = f_2$ for all $f_2 \in L_2^{[m]}$. We give some examples of formulas from $\mathcal{L}_2^{[m]}$, their inverse and their ordering w.r.t. $\sqsubseteq_2^{[m]}$:

distance:100m $\sqsubseteq_2^{[m]}$ distance: *(being distant of 100m is being distant)*
dist:1km $\sqsubseteq_2^{[m]}$ distance:1000m   *(1km is the same as 1000m)*
distance:1km $\sqsubseteq_2^{[m]}$ distance:in [10m,1km] $\sqsubseteq_2^{[m]}$ distance:>=100m
(distance:1km)$^{-1}$ $\equiv_2^{[m]}$ distance:1km

The logic $\mathcal{L}_2^{[m]}$ enables queries in the form $\exists(\texttt{distance:<=}d).f_1$, which selects all objects within a distance less or equal to d of objects described by f_1. This kind of queries corresponds to a fundamental functionality of GIS. This is a common way to define a *buffered area*, i.e. an area determined by both: a set of objects of interest (f_1) and a distance relation (`distance:<=`d). For instance, the following query q_1 selects fields close to a river:

$$q_1 = \texttt{type:field} \sqcap \exists(\texttt{distance:<=20m}).\texttt{type:river}$$

Furthermore, distance formulas can be combined and even nested to build complex queries. For instance consider the query q_2 dealing with apartments:

$$q_2 = (\texttt{apType:T3} \sqcup \texttt{apType:T4}) \sqcap \exists(\texttt{distance:<=500m}).\texttt{type:garden} \sqcap$$
$$\exists(\texttt{distance:<=200m}).(\texttt{type:busStop} \sqcap \exists(\texttt{travel\_time:<=10min}).$$
$$(\texttt{type:busStop} \sqcap \exists(\texttt{distance:<=200m}).\texttt{place:'IRISA'}))$$

Query q_2 selects 3- or 4-rooms apartments that are close (less than 500m) to a public garden, and that are near (less than 200m) a bus stop from where it takes less than 10min by bus to reach another bus stop that is close to IRISA (less than 20m).

Topological Relations. Topological relations are binary relations describing the spatial position of an object w.r.t another object. These relations are qualitative, they give information about the spatial organisation of objects independently from their size, shape or distance. The expression of topological relations is a domain that has already been widely investigated [7]. Several classifications have been proposed, some of which based on Galois lattices [11]. In the following, we consider a taxonomy of topological relations over regions [12], including the 8 base relations of the RCC8 model [13] and 7 intermediate relations (see Figure 2). Each organisation of 2 polygonal geometries in a 2-dimension space can be expressed as one of the 8 base relations. For the purpose of information retrieval, this classification has the advantage of being quite simple with only few relations (15) and understandable intermediate relations.

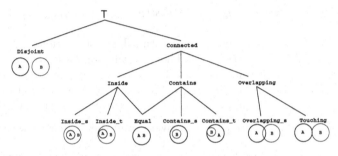

Fig. 2. A taxonomy of spatial relations over 2 regions proposed in [12]. Leafs of the taxonomy correspond to the 8 base relations of the RCC-8 model.

With LCA, we consider a logic of relations $\mathcal{L}_2^{Topo}$ to handle these topological relations. Formulas of L_2^{Topo} are the 15 spatial relations, and the ordering of $\sqsubseteq_2^{Topo}$ follows the taxonomy of Figure 2. For instance, **equal** $\sqsubseteq_2^{Topo}$ **connected**. The $\top$ stands for the general relation *being spatially related*, which we consider always true between two geographical objects. Notice also the symmetric properties between relations: $\texttt{inside}^{-1} = \texttt{contains}$, $\texttt{inside\_t}^{-1} = \texttt{contains\_t}$, $\texttt{inside\_s}^{-1} = \texttt{contains\_s}$, and $r^{-1} = r$ for all other relations. We now suppose that for each couple of area objects (o_1, o_2) of the object context, the 2 RCC-8 relations describing the position of o_1 w.r.t. o_2, and reciprocally, have been added in the relation context. At the moment, the computation of the relevant relation is done by an external GIS module. The logic $\mathcal{L}_2^{Topo}$ allows to build queries such as:

$$q_1 = \texttt{type:building} \sqcap \neg \exists\texttt{touching}.(\texttt{type:building})$$
$$q_2 = \texttt{type:cityBlock} \sqcap \forall\texttt{contains}.(\texttt{type:lodging})$$
$$q_3 = \texttt{type:road} \quad \sqcap \quad \exists\texttt{touching}.(\texttt{place:'my home'}) \quad \sqcap$$
$$\exists\texttt{connected}.(\texttt{type:garden} \sqcap \exists\texttt{contains}.(\texttt{type:lake}))$$

The query q_1 selects all non-terraced houses, q_2, residential areas, and q_3, roads next to my home and leading to a public garden having a lake.

5   Navigation Based on Spatial Relations

One asset of LCA is to combine querying and navigation in order to make information retrieval easier and faster. From any query, navigation is enabled through a set of navigation links that update the query, and eventually through query reversal. To facilitate the navigation, query increments are ordered in a *navigation tree*, according to the subsumption relation ($\sqsubseteq$). In previous works [1], we have already introduced a graphical interface that enables navigation inside geographical data, however it was not dealing with relational properties. This interface is composed of a *query box*, a *map area*, and a *navigation tree*. The query box recalls the current query, i.e. the navigation context and is manually editable by the user. By analogy with file systems, the current query is called the *working query* (*wq*). The map area is a cartographic representation of the explored context and enables to graphically select regions of interest as graphical query increments. The navigation tree is a visual representation of the partially ordered set of query increments, which are computed on demand and always relevant w.r.t. *wq*. In the following, we present a navigation tree allowing relational navigation through spatial relations. The query box has also been enhanced with the query reversal transformation. Non nested relation features appearing in *wq* are in fact hyperlinks that enable to traverse backward those relations.

To illustrate the use of navigation, we consider the following situation, where a traveller arrives in an unknown (fictive) city and looks for information about several points of interest, including lodging, lunch, recreation or transport offers. To help himself, he relies on a LIS device that let him query and navigate the city map context using a navigation tree and a query box. The set of objects $\mathcal{O}$ of the context corresponds to the points of interest, the streets and the position of our traveller. The individual spatial and non-spatial properties of the objects are described with a logic $\mathcal{L}_1^{city} = \mathcal{P}(\mathcal{A} \times (\mathcal{L}_1^V \cup \mathcal{L}_1^g))$. The distance relations and the topological relations the objects share are expressed with the logic $\mathcal{L}_2^{city} = \mathcal{P}(\mathcal{L}_2^{[m]} \cup \mathcal{L}_2^{Topo})$, allowing that descriptions of relations can include formulas from either $\mathcal{L}_2^{[m]}$ or $\mathcal{L}_2^{Topo}$. The description of the city map is presented in Figure 3a. Topological relations between all pairs of objects have been computed beforehand using a GIS. The disjoint relations, not relevant for this case of navigation, have not been expressed. However, objects disjoint from others can be accessed through the $\neg \exists connected. \top$ increment. Distance relations have also been computed beforehand between each point of interest and its neighbour streets. For instance, the "Bus stop 1" touches the "Street A", and the "Bar 2" is within a distance of 10 meters of "Street A".

5.1   The Navigation Tree

The navigation tree related to the citymap context is presented in Figure 3b. It displays information about the type of geographical objects (**type**), their name (**place_name**), and about the spatial relations they share (**distance** and **spatially_related** for topological relations). Query increments, also called

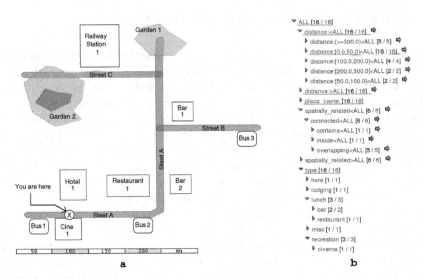

Fig. 3. Map of a city unfamiliar to our traveller (on the left), and the corresponding navigation tree (on the right)

features, may be of two kinds (see Section 3.3). Object features are properties shared by at least one object of wq. For instance, in the citymap navigation tree, there is one **lodging** object. Relation features denote a relation between some objects of the current query, with at least one other object of the context. Relation features correspond to $\exists x_2.x$ navigation links, but are expressed in the navigation tree with direction symbols > and <. For instance, in the citymap navigation tree, 6 objects are spatially related to another object. The feature **spatially_related>ALL** corresponds to $\exists spatially\_related.\top$, and **spatially_related<ALL** corresponds to $\exists spatially\_related^{-1}.\top$. In the tree of Figure 4, both features **distance:>ALL** and **distance:<ALL** are represented, altougth they are redundant because the distance relation is symmetric. This is also true for some topological relations. In the future, a special handling of symmetric relations is planed.

Each node of the tree represents a feature which can be used to change wq. When a node is expanded, this entails dynamically the computation of features that are specializations of the feature of this node. Then, the new features appear as its children in the tree. For instance, in Figure 3b, the taxonomy of points of interest is visible under the feature **type**, and distance intervals are visible under the **distance:<ALL** feature. The root of the tree is **ALL**, i.e., the most general formula. Each node of the tree is rendered with an icon, a label, two numbers, and possibly an arrow. The label is the formula representing the feature. The style of the label is informative: underlined labels correspond to formulas shared by all the objects in wq, whereas blue labels indicate properties that discriminate them. The two numbers indicate a proportion: the count of objects in wq that the feature leads to, i.e. the support, out of the count total of objects sharing the feature. In Figure 3b, the counts next to the feature **distance:[0.0,50.0]<ALL**

are both equal to 16, i.e. all geographical objects of the citymap context are within 50 meters of another object. Three actions are possible in the tree: (1) displaying more (resp. less) navigation links, i.e. expanding (resp. collapsing) a node by acting on its icon; (2) refining wq by selecting a label; (3) traversing a relation by selecting the arrow next to the node.

5.2   Example of Navigation

Just suppose that our traveller has left his luggage at the hotel and wants to have some recreation before having lunch. He looks at his LIS device, provides it with a new geographical object corresponding to its position (which we denote **here**). This entails the addition of distance relations (already displayed in the tree of Figure 3b) between the object **here** and the other points of interest of the context; the distances are computed on the basis of the shortest path, following streets. He then queries the system with:

$$wq = \texttt{recreation}$$

The LIS system provides him with the updated navigation tree of Figure 4a. The taxonomy of geographical objects (under feature **type**) has been reduced and now only contains the **recreation** feature. By expanding this feature, he can see 2 sub-features indicating 2 types of available recreation: **cinema** and **public_garden**. The first count of the feature **recreation** indicates that only 3 geographical objects correspond to recreational areas. The other features of the tree have also been updated. For instance, we can see that 2 recreation places are **connected** to "Street A".

Prefering being outside, our traveller decides to restrict his choice to public gardens not too close to his actual position. So, in the tree, he selects the feature **public_garden**, and manually modify wq so as to select public gardens located at more than 350 meters:

$$wq = \texttt{public\_garden} \sqcap \exists(\texttt{distance} :>= 350.0\texttt{m}).\texttt{here}$$

The navigation tree has been updated anew, and indicates him that two public gardens satisfy the query. By looking at the navigation tree, he can see the query increment **connected>ALL** denoting that some information about the topological organization of these gardens is available. Wanting more details about the environnement of the gardens, he traverses the **connected** relation by selecting the arrow next to the feature. This updates wq to:

$$wq = \exists\texttt{connected}^{-1}.(\texttt{public\_garden} \sqcap \exists(\texttt{distance} :>= 350.0\texttt{m}).\texttt{here})$$

The navigation tree now displays information concerning 3 geographical objects topologically linked with a public garden (see Figure 4b). By looking at the feature **type**, he can see that two of these objects are streets leading to a garden, and that the other one belongs to the miscellaneous category. Just by selecting the **misc** feature and looking at the feature **place_name**, our traveller can see that it corresponds to a lake which is in fact inside a public garden. To retrieve the corresponding garden, he then just has to follow backward the **connected**

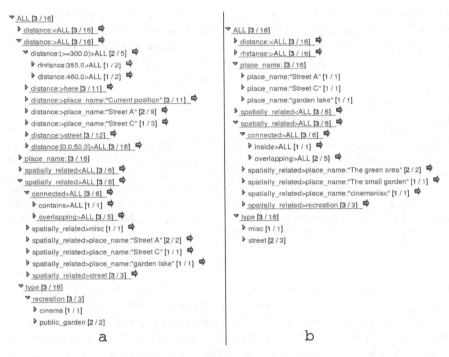

Fig. 4. The citymap navigation tree at two different steps. In tree a, $wq = \texttt{recreation}$, in tree b, $wq = \exists\texttt{connected}^{-1}.(\texttt{public\_garden} \sqcap \exists(\texttt{distance:>=350.0m}).\texttt{here})$.

relation (query reversal) by clicking on $\texttt{connected}^{-1}$ in the query box. Then wq becomes:

$$wq = \texttt{public\_garden} \sqcap \exists(\texttt{distance} :>= 350.0\text{m}).here \sqcap \exists\texttt{connected}.(\texttt{misc})$$

6   Benefits of Using LCA to Query Geographical Data

Regarding data organisation and retrieval, GIS have been widely influenced by Relational Database Management Systems (RDBMS) [14]. Most of the time, geographical information is structured into thematic layers, i.e., a layer for the streets, another for the points of interest, etc. Recently, XML-based formats for geographical data [15] allow to consider collections of heterogeneous geographical objects. But in every instance, retrieving data is done using querying interfaces and querying languages (XQuery-like or SQL-like) that integrate spatial predicates.

Our proposal of using LCA to manage geographical data not only enables to build queries similar to traditional GIS queries, it also offers a flexible organization of the data. Our data model is centered on the geographical object and enables to consider each collection of objects that share a common description expressed by a query q as a kind of flexible layer that can serve as a basis for further processing. In GIS, traditionally, querying for data disseminated into several collections implies making as many sub-queries as collections and merging the results. This service is automatically provided by LCA but requires the descriptions of similar objects to be comparable. Concerning the querying capabilities, firstly, the querying language available in LCA (see Section 4), combined with the spatial logics of distance relation and topological relations, already enables to express most of the queries traditionally used in a GIS, as shown in Section 4.2. For instance, concerning spatial queries, the formula $\exists(\texttt{distance:<=}d).q'$ enables to consider buffered areas of radius d around geographical objects described by q'. Moreover, like with GIS traditional languages, these buffered areas can be combined using Boolean operators and nested in other spatial predicates. Secondly, the expressivity of LCA relies for one part on the sub-logics used to reason on object descriptions. But once a new logic has been designed for a particular purpose, e.g. comparing areas derived from geometrical descriptions, it can be added to the system without reconsidering the whole theory. This makes the LCA querying language powerful as it can be easily extended, and customized to handle particular data. Currently, compared to SQL or XQuery, LCA lacks the ability to express agregates, although this is an interesting field we plan to investigate.

An advantage of LCA compared to traditional GIS is the navigation tree. Even in dedicated map search tools such as *the proximity business search* of Google Maps , the results are always delivered as a flat textual list with bullets on a map, where no indication is given on the structure of the answer and on how to refine it. In comparison, the navigation tree offers at least three assets. First, the navigation links give at a glance a summarized description of the currently selected dataset. Then, these navigation links provide a querying vocabulary which allows to build a query from scratch even with no knowledge about the data. Last but not least, the navigation links enable to refine the current query in a relevant way, i.e., ensuring the answer will be reduced but not empty. Giving aid in the building of queries is also a contribution of the VISCO system [12]. In VISCO, description logics are used to query a spatial database in a visual way. Like our prototype, VISCO enables to represent and to reason about spatial properties, e.g. area or perimeter, and topoligical relations of geographical objects. In addition to querying capabilities, VISCO also assists the user with query completion based on terminological default reasoning. However, contrary to our proposal, augmented queries may lead to empty answers because it does not take into account the content, but only the logic.

LCA also brings a new kind of navigation, *the relational navigation*. Considering a set of objects O that are instances of a query q and in relation r with other objects O' described by f, i.e. the link $\exists r.f$ is visible in the tree, one can directly jump to the related objects O'. In the geographical domain, this provides facilities when querying data organized in several thematic collections. Compared to traditional GIS practices, this kind of navigation prevents the user from building sub-queries corresponding to search criteria depending on several layers and then combining or nesting these sub-queries in the right way. With the relational navigation, the search process can be fully incremental, and does not impose any order on the building of a query.

7   Conclusion

In this paper, we first show that LCA is a framework that enables to easily model spatial relations between geographical objects. Valued spatial relations such as distance can be expressed in an intuitive manner thanks to the expressivity of logics. Furthermore, thanks to the partial ordering between logical relations, LCA naturally integrates taxonomies of relations, such as topological relations. Then, we present an original way to explore geographical data, using a navigation based on spatial and non-spatial properties of objects, as well as on the spatial relations between geographical objects. Especially, we illustrate the benefits of this paradigm of geographical data exploration provided by LCA, compared to traditonal GIS querying capabilities.

In the future, we plan to make the update of relations automatic and incremental when the description of geographical objects changes, e.g. their position. We also plan to work on a graphical representation of relation features in order to enhance the readability of the navigation tree and to assist the user in the building of queries involving spatial relations.

References

1. Bedel, O., et al.: Exploring a geographical dataset with GEOLIS. In: DEXA Workshop ACKE (2007)
2. Ferré, S., Ridoux, O.: An introduction to logical information systems. In: Information Processing & Management (2004)
3. Ferré, S., Ridoux, O., Sigonneau, B.: Arbitrary relations in formal concept analysis and logical information systems. In: Dau, F., Mugnier, M.-L., Stumme, G. (eds.) ICCS 2005. LNCS (LNAI), vol. 3596, pp. 166–180. Springer, Heidelberg (2005)
4. Wille, R.: Conceptual graphs and formal concept analysis. In: Delugach, H.S., Keeler, M.A., Searle, L., Lukose, D., Sowa, J.F. (eds.) ICCS 1997. LNCS, vol. 1257, pp. 290–303. Springer, Heidelberg (1997)
5. Rouane, M.H., et al.: A proposal for combining formal concept analysis and description logics for mining relational data. In: Kuznetsov, S.O., Schmidt, S. (eds.) ICFCA 2007. LNCS (LNAI), vol. 4390, pp. 51–65. Springer, Heidelberg (2007)
6. de Smith, M.J., Goodchild, M.F., Longley, P.A.: Geospatial Analysis: A Comprehensive Guide to Principles, Techniques and Software Tools. Troubador Publishing (2007)
7. Cohn, A.G.: Qualitative spatial representation and reasoning techniques. In: Brewka, G., Habel, C., Nebel, B. (eds.) KI 1997. LNCS, vol. 1303, Springer, Heidelberg (1997)
8. Ferré, S., Ridoux, O.: Logic functors: A toolbox of components for building customized and embeddable logics. Research Report RR-5871, Irisa (March 2006)
9. Donini, F.M., et al.: The complexity of concept languages. Information and Computation 134, 1–58 (1997)

10. Herring, J.R.: OpenGIS Implementation Specification for Geographic Information (06-103r3). Open Geospatial Consortium (OGC) (2006)
11. Napoli, A., Le Ber, F.: The Galois lattice as a hierarchical structure for topological relations. Ann. Math. Artif. Intell. 49(1-4), 171–190 (2007)
12. Wessel, M., Haarslev, V., Möller, R.: Visual spatial query languages: A semantics using description logic. In: Diagrammatic Representation and Reasoning, Springer, Heidelberg (2000)
13. Randell, D.A., Cui, Z., Cohn, A.: A Spatial Logic Based on Regions and Connection. In: Nebel, B., Rich, C., Swartout, W. (eds.) KR 1992, pp. 165–176. Morgan Kaufmann, San Francisco (1992)
14. Laurini, R., Thompson, D.: Fundamentals of Spatial Information Systems. Academic Press Limited, London (1992)
15. Cox, S., et al.: Geography Markup Language (GML) Encoding Spec. Open Geospatial Consortium (OGC) (2004)

Analysis of Social Communities with Iceberg and Stability-Based Concept Lattices

Nicolas Jay[1,2], François Kohler[2], and Amedeo Napoli[1]

[1] Laboratoire Lorrain de Recherche en Informatique et ses Applications,
Campus Scientifique - BP 239 - 54506 Vandoeuvre-lès-Nancy Cedex, France
jay@loria.fr
http://www.loria.fr/equipes/orpailleur
[2] Laboratoire SPI-EAO,
Faculté de Médecine - 9 Avenue de la Forêt de Haye -BP 184 54505 Vandoeuvre
Cedex, France

Abstract. In this paper, we presents a research work based on formal concept analysis and interest measures associated with formal concepts. This work focuses on the ability of concept lattices to discover and represent special groups of individuals, called social communities. Concept lattices are very useful for the task of knowledge discovery in databases, but they are hard to analyze when their size become too large. We rely on concept stability and support measures to reduce the size of large concept lattices. We propose an example from real medical use cases and we discuss the meaning and the interest of concept stability for extracting and explaining social communities within a healthcare network.

1 Introduction

Knowledge Discovery in Databases (KDD) is an iterative and interactive process for identifying valid, novel, and potentially useful patterns in data [1]. It is usually divided into three main steps: data preparation, data mining, and interpretation of the extracted units. Data mining is often considered as the central step in the KDD process. However, interpretation of data-mining results is also an important step within the KDD process. Indeed, one of the the success keys in KDD practice relies on the ability of easily producing units understandable as knowledge units. One way of achieving such a goal is to provide an adapted organization and representation of the extracted units, especially when the KDD system has to be used by novice users.

In parallel, Formal Concept Analysis (FCA) is a theory of data analysis introduced in [2], that is tightly connected with KDD [3,4], particularly regarding the search of frequent itemsets and the extraction of association rules [5]. Many algorithms relying on FCA central property of closure have been proposed to extract frequent closed itemsets: e.g. CLOSE [6], CLOSET [7], CHARM [8], TITANIC [9], and ZART [10]. The set of frequent closed itemsets may be used to determine the set of all frequent itemsets: closed itemsets are a loss less representation of frequent itemsets, while the set of closed itemsets can be orders of magnitude smaller than the set of all frequent itemsets.

R. Medina and S. Obiedkov (Eds.): ICFCA 2008, LNAI 4933, pp. 258–272, 2008.

FCA organizes information into a concept lattice representing inherent structures existing in data. A concept lattice can be visualized for analysis using graphical tools, e.g. TOSCANA [11], GALICIA [12]. FCA forms also the basis of a knowledge processing paradigm known as "Conceptual Landscapes" [13]. Furthermore, Stumme [9] has introduced the so-called iceberg lattices, which are concept lattices of frequent closed itemsets. Iceberg lattices serve as a support for visualization of association rules mined in large database. They can help analysts in selecting interesting patterns and organizing these patterns into understandable and reusable knowledge units. However, iceberg lattices may hide non frequent but still relevant concepts.

Following the idea of compact, reduced (loss less), and concise representation of extracted units (i.e. itemsets, association rules, or concepts), a number of numerical measures used for pruning itemsets, association rules, and in a certain sense concepts, have been proposed [14]. In this way, Kuznetsov has introduced stability as a new interest measure for concepts [15, 16]. Stability has been successfully used for pruning concept lattices, e.g. in the field of social networks [17–19]. Accordingly, in this article, we address the problem of exploring "social communities". By "social communities", we intend sets of agents or organizations whose members are linked by a common interest or objective [20]. One of our goals is to study the basis and to design a decision support system for assisting experts identifying social communities. The selection, organization, and discrimination of relevant units of knowledge, help to understand how agents interact in a social community and how they gather on specific topics. Moreover, we show in this paper that combining concept frequency together with concept stability provides a very efficient means for discovering and analyzing social communities.

The paper is organized as follows. Following the present first section, the second section introduces the definitions and the properties of FCA, of support and stability measures. The third section presents a qualitative discussion on stability and shows how stability enlighten concept with a high internal cohesion, i.e. stable and without exceptional individuals. Then, the fourth section gives details on an example of social community discovery within a healthcare network. A discussion on the example and on the knowledge units that can be extracted is proposed and precedes the conclusion of the paper.

2 Support and Stability: Interest Measures of Formal Concepts

2.1 Formal Concept Analysis

We describe here the FCA basics. FCA starts with a formal context $\mathbb{K} = (G, M, I)$ where G is a set of objects, M is a set of attributes, and the binary relation $I = G \times M$ specifies which objects have which attributes. Two operators, both denoted by $'$, connect the power sets of objects 2^G and attributes 2^M as follows:

$$' : 2^G \rightarrow 2^M, X' = \{m \in M | \forall g \in X, gIm\}$$

The operator $'$ is dually defined on attributes. The pair of $'$ operators induces a Galois connection between 2^G and 2^M. The composition operators $''$ are closure operators: they are idempotent, extensive and monotonous. For any $A \subseteq G$ and $B \subseteq M$, A'' and B'' are closed sets whenever $A = A''$ and $B = B''$.

A formal concept of the context $\mathbb{K} = (G, M, I)$ is a pair $(A, B) \subseteq G \times M$ where $A' = B$ and $B' = A$. A is called the *extent* and B is called the *intent*. A concept (A_1, B_1) is a *subconcept* of a concept (A_2, B_2) if $A_1 \subseteq A_2$ (which is equivalent to $B_2 \subseteq B_1$) and we write $(A_1, B_1) \leq (A_2, B_2)$. The set $\mathfrak{B}$ of all concepts of a formal context $\mathbb{K}$ together with the partial order relation $\leq$ forms a lattice and is called concept lattice of $\mathbb{K}$.

2.2 Iceberg Concept Lattices

This paragraph is based on [9] and introduces basics of iceberg lattices.

Definition 1. *Let* $B \subseteq M$. *The support count of the attribute set* B *in* $\mathbb{K}$ *is*

$$\sigma(B) = \frac{|B'|}{|G|} \qquad (1)$$

Let minsupp be a threshold $\in [0, 1]$, then B is said to be a frequent itemset if $\sigma(B) \geqslant$ minsupp.

A concept is called frequent concept if its intent is frequent.

Definition 2. *The set of all frequent concepts of a context* $\mathbb{K}$ *is called iceberg lattice of the context* $\mathbb{K}$.

The support function is monotonously decreasing: given two attribute sets B_1 and B_2, $B_1 \subseteq B_2 \Rightarrow \sigma(B_1) \geq \sigma(B_2)$. Thus an iceberg lattice is an order filter of the whole concept lattice and in general only a join-semi-lattice. Meanwhile, adding a bottom element makes it a lattice again.

Iceberg Lattices can be used to discover and visualize association rules. Within a formal context $K = (G, M, I)$, the task of mining association rules is to determine all pairs $X \rightarrow Y$ of M such that $\sigma(X \rightarrow Y) = \sigma(X \cup Y) \geqslant$ minsupp, and the *confidence* $\text{conf}(X \rightarrow Y) = \frac{\sigma(X \cup Y)}{\sigma(X)}$ is above a given threshold minconf $\in [0, 1]$.

Mining associations rules with FCA has two major advantages [21]. First, frequent closed itemsets are sufficient to deduce all frequent itemsets. Thus, algorithms can benefit from this property to reduce the search space. Second, iceberg lattices offer a reduced and lossless representation of association rules. They allow to directly read Luxenbourger basis for approximate association rules [22] from a line diagram.

2.3 Stability

Stability has been introduced (probably for the first time) in [15] and then revisited [16, 19]. Here, we rely on the definition given in [19].

Definition 3. *Let* (A, B) *a formal concept of* $\mathfrak{B}(\mathbb{K})$. *Stability of* (A, B) *is*

$$\gamma(A, B) = \frac{|\{C \subseteq A | C' = A' = B\}|}{2^{|A|}} \qquad (2)$$

The stability index of a concept indicates how much the concept intent depends on particular objects of the extent. Given a concept (A, B), the stability index measures the number of elements of G that are in the same *equivalence class* of A, where an equivalence class is defined as follows.

Definition 4. *Let* $X \subseteq G$, *we denote by* $\langle X \rangle$ *the* equivalence class *of* X *where:*

$$\langle X \rangle = \{Y \subseteq G | Y' = X'\} \qquad (3)$$

Note that when X is closed, any Y in $\langle X \rangle$ is a subset of X. Thus, considering a formal concept (A, B), definition 3 can be rewritten as:

$$\gamma(A, B) = \frac{|\langle A \rangle|}{2^{|A|}} \qquad (4)$$

Then, the larger the equivalence class of an extent is (wrt to extent size), the more stable the concept is. The idea behind stability is that a stable concept is likely to have a real world interpretation even if the description of some its objects (i.e. elements in the extent) is "noisy". Figure 1 shows an example of stability in a concept lattice. Each concept is labelled by its extent, intent and stability. For example, for the concept $(\{1, 5, 6\}, \{a\})$, we have:

$$\emptyset' = \{a, b, c, d\} \neq \{a\}$$
$$\{1\}' = \{a\}$$
$$\{5\}' = \{a\}$$
$$\{6\}' = \{a, b, c\} \neq \{a\}$$
$$\{1, 5\}' = \{a\}$$
$$\{1, 6\}' = \{a\}$$
$$\{5, 6\}' = \{a\}$$
$$\{1, 5, 6\}' = \{a\}$$

Thus $\gamma(\{1, 5, 6\}, \{a\}) = \frac{6}{8} = 0.75$. It can be noticed that stability is (by definition) always between 0 and 1. It can be still noticed that $\gamma(\bot) = 1$. This is always true, since for any subset X from the extent of $\bot$, X' is included in the intent of $\bot$.

Computing stability has been shown to be a #P-complete problem [16]. Meanwhile, once the concept lattice has been computed, a bottom-up traversal algorithm can efficiently compute stability [18]. Actually, a concept stability depends on the stability of its subconcepts. This can be shown as follows:

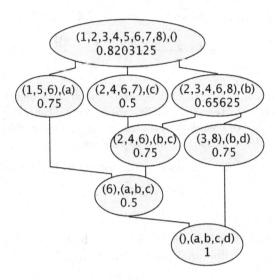

Fig. 1. Stability example

Proposition 1. *Let* (A, B) *a formal concept of* $\mathfrak{B}(\mathbb{K})$.

$$\gamma(A, B) = 1 - \sum_{X \subset A, X = X''} \gamma(X, X') 2^{|X| - |A|} \qquad (5)$$

Proof. For a formal concept (A, B), from (4), we have:

$$\gamma_{(A,B)} = \frac{|\langle A \rangle|}{2^{|A|}}$$

Let I_A be the set of subintents of A: $I_A = \{X \subseteq A | X = X''\}$. The set of equivalent classes $\{\langle X \rangle | X \in I_A\}$ forms a partition of 2^A. Thus $|2^A| = \sum_{X \in I_A} |\langle X \rangle|$, which gives:

$$|\langle A \rangle| = |2^A| - \sum_{X \in I_A, X \neq A} |\langle X \rangle|$$

Dividing by $|2^A|$ we obtain:

$$\frac{|\langle A \rangle|}{|2^A|} = 1 - \sum_{X \in I_A, X \neq A} \frac{|\langle X \rangle|}{|2^A|}$$

$$\gamma_{(A,B)} = 1 - \sum_{X \subset A, X = X''} \gamma(X, X') 2^{|X| - |A|}$$

3   A Qualitative Analysis of Stability

3.1   Stability and Cohesion

As stated in [19], a concept is stable if its intent does not depend much Âon each particular object of the extent. Stability is aimed at measuring how much a concept extent depends on some of its individual members. This may be useful in analyzing a dataset with a concept lattice, having a special attention to social communities. Here, a social community can be thought as a group of agents –human, software, or resource agents– sharing the same interests, or ideas, or needs [17]. For example, patients visiting the same hospitals with similar medical problems can be identified as a special social community. In the associated formal context, objects correspond to patients and hospital stays correspond to attributes (see hereafter). Always following the line of [19], an actual community has to be "internally cohesive" enough: a stable concept continues to be a concept even if a few members stop being members. This means also that a stable concept is resistant to noise and will not collapse when some members will be removed from its extent.

In this way, a stable concept is a meaningful concept, in the sense that it covers a group of objects, that considered together, have a high internal cohesion. The most stable concepts determine the most interesting groups of objects, that constitute their extents.

3.2   Stable Concepts Are of High Interest

FCA provides a powerful framework for identifying social communities [17, 23, 24]. Relations between agents and common interests can be modeled within a formal context. The associated concept lattice will allow to discover and identify which agents do share common interests and what are these interests. However, as the size of a formal context increases, the number of formal concepts in the lattice may grow dramatically. In this case, interest measures such as stability and support can reduce the complexity of the analysis of the concept lattice. Filtering concepts by support relies on the assumption that useful knowledge is represented by frequent patterns. But, this is not always true as pointed out in studies on rare itemsets, as in e.g. in [25], where it is shown that association rules with a low support but a high confidence may be of high interest.

Stability gives an alternative point of view on formal concepts. It indicates the probability of preserving a concept intent while removing some objects of its extent. Considering social communities, stability helps to identify groups of commons interest that dot not entirely depend on some specific agents. As stability is somewhat independent from support, it can be used to discriminate low-support concepts and detect small communities of strongly related agents. Moreover, stability also detects frequent concepts only depending on a small number of objects. For example, considering a lattice composed of the two following concepts $C_1 = (\{g_1, ..., g_{n-1}, g_n\}, \{m_1\})$ and $C_2 = (\{g_1, ..., g_{n-1}\}, \{m_1, m_2\})$ with n high, then it can be noticed that C_1 depends solely on the object g_n. Although C_1 and C_2 have both a support close to 1, stability of C_2 is 1 while stability of C_1 is $\frac{1}{2}$. In terms of social communities, the group of individuals $\{g_1, ..., g_n\}$ has not a sufficient "internal cohesion" or has not a "real existence".

Hence, stability, together with support, are a convenient means for identifying two types of concepts:

– *rare stable concepts*: concepts with a low support and a high stability,
– *frequent unstable concepts*: concepts with a high support and low stability.

In section 4, this point of view is discussed and illustrated within a real-world application aimed at detecting communities of patients, i.e. groups of patients being treated in the same groups of hospitals.

3.3   Stable Concepts Are Monothetic Rather Than Polythetic

As introduced above, stability can also be linked with exceptions, and, furthermore, with the so-called *monothetic* and *polythetic* characters of a class of individuals [26–28]. When building a concept lattice and analyzing groups of individuals through the extents of concepts, one problem is to recognize and explain *exceptions*. A subsequent question is to understand whether exceptions are linked to monothetic or polythetic classes.

A class of individuals C is said to be *monothetic* if and only if there exists a set of attributes Att that determines the membership of an individual to the class C (Att is a set of necessary and sufficient membership conditions). By contrast, given a set of attributes $Att = \{a_1, ..., a_n\}$, a class of individuals C is said to be *polythetic* if and only if:

– Every object that is an instance of the class C has an "important" –not necessarily fixed– number of attributes of Att.
– Every attribute of Att belongs to an "important" –not necessarily fixed– number of instances of C.
– There is not necessarily an attribute of Att belonging to every instance of C.

Relying on the fact that a stable concept has a high "internal cohesion", is resistant to noise, and does not collapse when some members stop being members of its extent, the more a concept is stable, the more it does not represent exceptional individuals, and, accordingly, the more the concept is able to represent cohesive groups of individuals, such as social communities. This means that stable concepts are rather monothetic and that unstable concepts are rather "exceptional" or polythetic, i.e. they include some exceptional character, shared by only a few individuals. For illustrating this view, let us consider the following example.

	A	B	C	D	E
1	x	x		x	x
2	x		x		
3	x	x	x		x
4		x	x		x
5	x	x	x		x

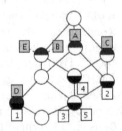

Here, attribute D can be considered as a "necessary and sufficient condition" for the membership of an individual to the class including individual 1. Indeed, the concept lattice includes the concept $(A, B) = (1, abde)$. However, this concept has a very low stability $(\frac{4}{16})$ and also a low support for its intent $(\frac{1}{5}$ when abde is considered as an itemset). This is in agreement with the view that a stable concept appears to be monothetic while an unstable concept tends to be polythetic.

4 Social Communities in a Regional Healthcare System

4.1 Motivations

Healthcare management and planning play a key role for improving the overall health level of the population. From a population point of view, even the best and state-of-the-art therapy is not effective if it cannot be delivered in the right conditions. Actually, many determinants influence the effective delivery of healthcare services: availability of trained personnel, availability of equipments, security constraints, costs, proximity... All of these should meet economics, demographics, and epidemiological needs in a given area. This issue is especially acute in the field of cancer care where many institutions and professionals must cooperate to deliver high level, long term, and costly care. Therefore, it is crucial for healthcare managers and decisions makers to be assisted by decision support systems that give strategic insights about the intrinsic behavior of the healthcare system.

On the one hand, healthcare systems can be considered as "data rich" as they produce massive amounts of data such as electronic medical records, clinical trial data, hospital records, administrative data, and so on. On the other hand, they can be regarded as "knowledge poor" as these data are rarely embedded into a strategic decision-support resource [29]. In France, the PMSI database is a national information system used to describe hospital activity with both an economical and medical point of view. In a previous work, we used this system together with iceberg lattices to discover how several institutions organize themselves into an implicit network to provide coordinated care at a regional level [30]. Our method has been used in real world by healthcare managers. It appeared that support based pruning had some limits, for example in analysing small institutions interactions.

4.2 The Difficulty of Choosing the Good Support Threshold

In this section we present an example of an iceberg lattice showing cooperations between hospitals in the field of cancer. We then discuss the choice of minsupp by studying concept support distribution. In our approach, we build a context in which objects are patients suffering from cancer and attributes are hospitals. A patient and a hospital are related if the hospital has delivered cancer care to this patient. In our experiment, the resulting context has 6036 patients and 170 hospitals. While the whole concept lattice holds 865 concepts, an iceberg lattice built with a minsupp of 0.0033 (20 patients) gives 93 frequent concepts. A small excerpt of this iceberg is shown in figure 2.

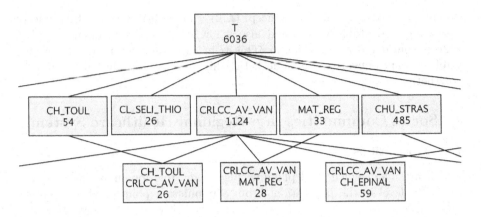

Fig. 2. Iceberg of cancer treatment cooperations

For the sake of clarity, ⊥ was removed. Although its right and leftmost parts are not drawn, the iceberg lattice is much more wide than deep because the context is sparse and data are poorly correlated. This means that cooperations are most of the time tightly partitioned, and that patients are rarely hospitalized in more than two hospitals. The intent of co-atoms, i.e. immediate descendants of ⊤, is always a singleton, indicating that a hospital never shares all of its patients with another one, or if it is so, less than 20 patients are involved in the interaction. The intent of atoms, i.e. the immediate ascendant of ⊥, is always a pair. The extent of atoms gives an idea of the strength of the cooperation between the two hospitals lying in the intent: the larger is the cardinal of the extent, the higher is the strength of the cooperation (i.e. the more patients are shared between the two hospitals). The examination of the iceberg brings different types of knowledge:

- some concepts that are both atoms and co-atoms (for example : CL-SELI-THIO). They represent institutions that share a few patients with others. This is that either they treat a few patients, or they work in a relative autonomy, or cooperation is split with many other hospitals.
- other concepts have at least a sub-concept (different from ⊥). They represent a hospital receiving a significant number of patients, and having collaborations with at least another establishment. For example MAT-REG and CLCC-AV-VAN share 28 patients.
- The concept representing the CLCC-AV-VAN hospital has a high support and many sub-concepts. This hospital is a specialized anti-cancer center. It employs highly skilled and specialized personnel. Treatments given there rely on state-of-art technology. Furthermore, It actively participates in anti-cancer research programs and thus can be considered as a reference institution.

The choice of the minimum support strongly influences the interpretation of the iceberg. It must be sufficiently low to convey meaningful knowledge and sufficiently high to keep this knowledge readable for a human expert. Here, the

Table 1. Concept support count and intent size

Intent size / Number of patients	0	1	2	3	> 4
< 10	0	40	285	297	76
[10, 20[	0	13	52	7	0
[20, 30[	0	10	20	0	0
[30, 40[	0	6	2	0	0
> 40	1	34	20	0	0

whole lattice holds 865 concepts. Table 1 shows the distribution of their support according to the size of their intent. By choosing a *minsupp* of corresponding to 20 patients, we see we can miss interesting knowledge:

- 13 hospitals treat at least 10 patients,
- 52 cooperations of two hospitals involve at least 10 patients,
- 7 concepts represent cooperations between three hospitals and involve at least 10 patients.

In our case, support is a weak mean for discriminating "not-so-frequent" concepts. Some concepts not appearing in the iceberg may be of interest for different reasons:

- They illustrate a cooperation of one hospital sharing almost all of its patients with another one.
- They concern a hospital treating few patients but not sharing them with any other one.
- They concern a 3-hospitals interaction.

Lowering the support threshold can let these concepts appear in the iceberg, but at the expense of readability. Moreover support measure is not specific enough to discriminate concepts with the above characteristics.

4.3 Stability Analysis

In this section, we study stability of concepts within the whole concept lattice. Figure 3 shows the histogram of concept stability.

Most of the concepts hold only one object in their extent. As ⊥ extent is empty, they have a stability of 0.5. The next important group consists of concepts having stability very close to 1. It corresponds generally to the most frequent concepts. Indeed, 93% of the concepts having support greater than 10 have stability greater than 0.99. Figure 4 is a scatter-plot of stability and support, also featuring intent size.

Values are presented on a log-scale for better visualization of both support and stability ranges. Moreover, we prefer to use the stability odds defined as follows:

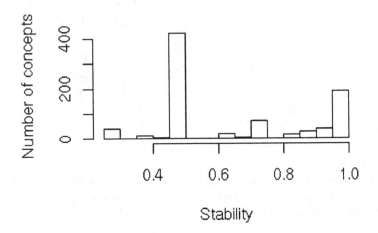

Fig. 3. Concept stability histogram

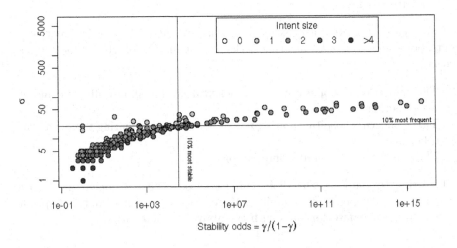

Fig. 4. Concept stability and support

Definition 5. *Given a concept* $(A, B) \neq \bot$, *we call stability odds of* (A, B):

$$o_\gamma(A, B) = \frac{\gamma(A, B)}{1 - \gamma(A, B)} \qquad (6)$$

Stability odds are the ratio between the number of subsets of an extent A which belong to the equivalence class of A, and of those which do not.

Stability odds illustrate more clearly the distribution of stability for values close to 1 and show that stability has a better discriminant power than support. The whole set of concepts has 87 distinct support values and 152 distinct stability values. For concept having support between 10 and 19, we observe 57 distinct stability values. Figure 4 displays thresholds for the 10% most frequent concepts and 10% most stable concepts. In the next sections we discuss the differences between those two thresholds.

4.4 Frequent Unstable Concepts

The upper left quarter of figure 4 shows the 9 concepts having support greater than the 10% most frequent threshold and stability less than the 10% most stable threshold, i.e. frequent unstable concepts. They represent institutions playing a significant role in cancer care as secondary or tertiary care[1] centers. They share most of their patients with other hospitals. This gives rise to cooperations of 2 kinds.

Figure 5 shows the order ideals of two frequent unstable concepts. Two hospitals have tight interaction each with another tertiary care center, i.e. MAT-REG-NANCY with CRLCC-AV-VAN, CL-ARCENCIEL with CH-EPINAL. These two institutions are secondary care centers usually delivering surgery and referring patients to a tertiary care center for radiotherapy. Their activity can be almost entirely explained by an exclusive cooperation. This induces a form of dependency with the tertiary care center. Patient not concerned by the cooperation can be considered as exceptions, i.e. patients not following the usual care pathway for some reason (for example because they suffer from a very specific pathology). After pruning according to stability, only the concepts in grey on figure 5 will remain.

One center, as shown on Figure 6, is a highly specialized anti-cancer institution. IGR-PARIS is an international well-known anti-cancer center located in Paris. Many Lorraine local hospitals refer patients to IGR for rare tumors. If we apply stability pruning here, the whole sub-lattice will disappear, which may be desirable for readability. Meanwhile, its suggests that searching for this type of unstable frequent concept can also be an interesting knowledge mining task.

4.5 Rare Stable Concepts

Three concepts are located in the right lower corner of figure 4.

Two have an intent of size 2 and illustrate thus cooperations between two hospitals sharing 18 patients. These cooperations differ from others of the same support in that they are not split themselves in cooperations involving a third institution. Thus, similarity of concerned patients is entirely explained by those cooperations. The last concept has a size 3 intent and illustrates the cooperation between three large specialized hospitals located in the same city of Nancy. This is the most frequent and most stable concept of that type.

Stability allows to distinguish rare concepts that cannot be separated from others by support. Rare stable concepts differ from other rare concepts in that their attributes suffice to explain the similarity of their objects.

---

[1] Secondary care is delivered by a broadly skilled specialist (e.g. a general surgeon, a general internist, or an obstetrician). Tertiary care is provided by a sub-specialist(e.g. an orthopedic surgeon, a neurologist, or neonatologist).

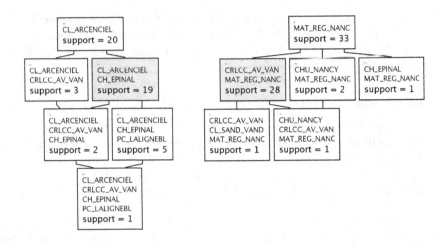

Fig. 5. Frequent unstable concepts: exclusive cooperations

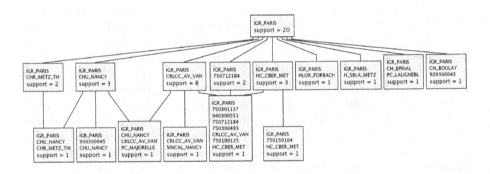

Fig. 6. Frequent unstable concepts : **IGR-PARIS** as a national referred care center

5   Synthesis

Our system is used in real world by healthcare managers. First, at the individual level, it helps healthcare professionals to assess their activity in the regional landscape. While physicians are able to cite the names of people they are used to cooperate with, they cannot measure the strength of these cooperations. And it is even harder for hospital managers to count patients shared with other institutions due to the gaps and lacks of adapted processes in information systems. Second, at the regional level, it provides for the administrative staff a decision support to reorganize care resources according to the implicit behavior of the healthcare system. Actually, French law establishes activity thresholds in the field of cancer: e.g. an hospital must treat at least 30 patients a year to be authorized for digestive cancer surgery. Our system has allowed to enlighten and accordingly to promote cooperations between institutions that could not reach the thresholds alone. It has also demonstrated how administrative decisions could impact the social healthcare network, given the existence of many dependencies between structures.

6 Conclusion

Our main objective is to build a system allowing for visualization of a social healthcare network in the field of cancer care. This system is used today by healthcare managers. Things must be kept simple while conveying enough information for assisting strategic decisions. The use of concept interest measures has a strong impact both on readability and semantics of discovered knowledge. Together with support, stability can successfully identify two kind of concepts: frequent unstable concepts and rare stable concepts. In our experiment, the formal context is sparse and we need to mine concepts with very low support. Stability brings additional knowledge that helps to discover interesting rare concepts that can not be discriminated by support. We believe that it could have significant implications in the field of rare itemsets mining [25]. Furthermore, stability enhances lattice visualization when pruning frequent unstable concepts. Besides, frequent unstable concepts may also be a subject of interest.

References

1. Fayyad, U., Piatetsky-Shapiro, G., Smyth, P.: The kdd process for extracting useful knowledge from volumes of data. Communication of the ACM 29(11), 27–34 (1996)
2. Wille, R.: Restructuring lattice theory: an approach based on hierarchies of concepts. In: Rival, I. (ed.) Ordered Sets, Reidel (1982)
3. Wille, R.: Why can concept lattices support knowledge discovery in databases? J. Exp. Theor. Artif. Intell. 14(2-3), 81–92 (2002)
4. Valtchev, P., Missaoui, R., Godin, R.: Formal concept analysis for knowledge discovery and data mining: The new challenges. In: Eklund, P.W. (ed.) ICFCA 2004. LNCS (LNAI), vol. 2961, pp. 352–371. Springer, Heidelberg (2004)
5. Agrawal, R., Imielski, T., Swami, A.: Mining association rules between sets of items in large databases. In: Proceedings of the ACM SIGMOD Int'l Conference on Management of Data, pp. 207–216 (1993)
6. Pasquier, N., et al.: Efficient mining of association rules using closed itemset lattices. Journal of Information Systems 24, 25–46 (1999)
7. Pei, J., Han, J., Mao, R.: Closet: An efficient algorithm for mining frequent closed itemsets. In: Proceedings ACM SIGMOD Workshop DMKD 2000, Dallas, Texas, USA, May 2000, pp. 21–30 (2000)
8. Zaki, M.J., Hsiao, C.J.: Charm: An efficient algorithm for closed itemset mining. In: Grossman, R.L., et al. (eds.) SDM, SIAM, Philadelphia (2002)
9. Stumme, G.: Efficient Data Mining Based on Formal Concept Analysis. In: Hameurlain, A., Cicchetti, R., Traunmüller, R. (eds.) DEXA 2002. LNCS, vol. 2453, Springer, Heidelberg (2002)
10. Szathmary, L., Napoli, A., Kuznetsov, S.O.: ZART: A Multifunctional Itemset Mining Algorithm. In: Proc. of the 5th Intl. Conf. on Concept Lattices and Their Applications (CLA 2007), Montpellier, France (October 2007)

11. Vogt, F., Wille, R.: Toscana - a graphical tool for analyzing and exploring data. In: Tamassia, R., Tollis, I(Y.) G. (eds.) GD 1994. LNCS, vol. 894, pp. 226–233. Springer, Heidelberg (1995)
12. Valtchev, P., et al.: Galicia: an open platform for lattices. in using conceptual structures. In: Contributions to the 11th Intl. Conference on Conceptual Structures (ICCS 2003), pp. 241–254. Shaker Verlag (2003)
13. Wille, R.: Conceptual landscapes of knowledge: a pragmatic paradigm for knowledge processing. In: Gaul, W., Locarek-Junge (eds.) Classification in the Information Age, pp. 344–356. Springer, Heidelberg (1999)
14. Guillet, F., Hamilton, H.J.: Quality Measures in Data Mining (Studies in Computational Intelligence). Springer, Secaucus, NJ, USA (2007)
15. Kuznetsov, S.O.: Stability as an estimate of the degree of substantiation of hypotheses derived on the basis of operational similarity. Nauchn. Tekh. Inf., Ser.2 (Automat. Document. Math. Linguist.) 12, 21–29 (1990)
16. Kuznetsov, S.O.: On stability of a formal concept. Annals of Mathematics and Artificial Intelligence 49, 101–115 (2007)
17. Roth, C.: Co-evolution in epistemic networks – reconstructing social complex systems. Structure and Dynamics: eJournal of Anthropological and Related Sciences 1(3) (2006)
18. Roth, C., Obiedkov, S., Kourie, D.G.: Towards concise representation for taxonomies of epistemic communities. In: Yahia, S.B., Nguifo, E.M. (eds.) CLA 4th International Conference on Concept Lattices and their Applications, Tunis, Faculté des Sciences de Tunis, pp. 205–218 (2006)
19. Kuznetsov, S., Obiedkov, S., Roth, C.: Reducing the representation complexity of lattice-based taxonomies. In: Priss, U., Polovina, S., Hill, R. (eds.) ICCS 2007. LNCS (LNAI), vol. 4604, pp. 241–254. Springer, Heidelberg (2007)
20. Ahuja, M.K., Carley, K.M.: Network structure in virtual organizations. Journal of Computer-Mediated Communication 3(4) (1998)
21. Lakhal, L., Stumme, G.: Efficient mining of association rules based on formal concept analysis. In: Ganter, B., Stumme, G., Wille, R. (eds.) Formal Concept Analysis. LNCS (LNAI), vol. 3626, pp. 180–195. Springer, Heidelberg (2005)
22. Luxenbouger, M.: Implications partielles dans un contexte. Mathmatiques, Informatique et Sciences Humaines 29(113), 35–55 (1991)
23. Freeman, L.C., White, D.R.: Using galois lattices to represent network data. Sociological methodology 23, 127–146 (1993)
24. Duquenne, V.: Latticial structures in data analysis. Theoretical Computer Science 217, 407–436 (1999)
25. Szathmary, L., Napoli, A., Valtchev, P.: Towards rare itemset mining. In: Proceedings of the IEEE International Conference on Tools with Artificial Intelligence (ICTAI), Patras, Greece, IEEE Computer Society Press, Los Alamitos (2007)
26. Sokal, R.: Phenetic taxonomy: Theory and methods. Annual Review of Ecology and Systematics 17, 423–442 (1968)
27. Bailey, K.: Monothetic and polythetic typologies and their relation to conceptualization, measurement and scaling. American Sociological Review 38(1), 18–33 (1973)
28. Lerman, I.: Les bases de la classification automatique. Gauthier-Villars diteurs, Paris (1970)
29. Abidi, S.S.: Knowledge management in healthcare: towards 'knowledge-driven' decision-support services. Int. J. Med. Inform. 63(1-2), 5–18 (2001)
30. Jay, N., Kohler, F., Napoli, A.: Using formal concept analysis for mining and interpreting patient flows within a healthcare network. In: Proceedings of CLA 2006 (2006)

Formal Concept Analysis Enhances Fault Localization in Software

Peggy Cellier[1,2], Mireille Ducassé[2], Sébastien Ferré[1], and Olivier Ridoux[1]

[1] IRISA/University of Rennes 1,
[2] IRISA/INSA de Rennes
firstname.lastname@irisa.fr
http://www.irisa.fr/LIS/

Abstract. Recent work in fault localization crosschecks traces of correct and failing execution traces. The implicit underlying technique is to search for *association rules* which indicate that executing a particular source line will cause the whole execution to fail. This technique, however, has limitations. In this article, we first propose to consider more expressive association rules where several lines imply failure. We then propose to use Formal Concept Analysis (FCA) to analyze the resulting numerous rules in order to improve the readability of the information contained in the rules. The main contribution of this article is to show that applying two data mining techniques, association rules and FCA, produces better results than existing fault localization techniques.

1 Introduction

The execution of a program in a testing environment generates a set of data about the execution, called a *trace* of the execution. Traces allow the program to be monitored and permit the program to be debugged when some executions fail, namely produce unexpected results. A trace can contain different kinds of information, for example the executed lines, and the *verdict* of the execution (*FAIL* or *PASS*). Fault localization often investigates the contents of traces to find the reasons of failures. There exist several approaches to crosscheck traces. Some are based on the differences between a passed execution and a failed execution [RR03, CZ05]. Others use statistical indicators in order to rank lines of the program [JHS02, LNZ+05, LYF+05]. In particular, Jones *et al.* [JHS02] propose to measure a kind of correlation between executing a given line and failing a test. Denmat *et al.* [DDR05] show that this is similar to search for a restricted form of *association rules* [AIS93, AS94] and that the restriction leads to limitations.

Searching for association rules is a well-known data mining task with a well-documented rationale. The knowledge context is represented by a set of *transactions (objects)* described by a set of *items (attributes)*. Searching that context for association rules consists in searching for implications where the premise and the conclusion are sets of attributes. In order to measure the relevance of the computed rules, some statistical indicators are used, such as *support, confidence,* or *lift.* In the framework of association rules, the method of Jones *et al.* consists

R. Medina and S. Obiedkov (Eds.): ICFCA 2008, LNAI 4933, pp. 273–288, 2008.

Table 1. Mutants of the Trityp program

Mutant	# passed exec.	# failed exec.	Faulty line
1	288	112	`[84] if ((trityp == 3) && (i+k > j))`
2	384	16	`[79] trityp = 0 ;`
3	308	92	`[64] trityp = i+1 ;`
4	280	120	`[87] if ((trityp != 3) && (j+k > i))`
5	305	95	`[65] if (i >= k)`

in searching for rules with only one line in their premise and only the attribute *FAIL* in conclusion. Note that the general association rule framework allows for several attributes in the premise.

Formal Concept Analysis (FCA) [GW99] has already been used for several software engineering tasks: to understand the complex structure of programs, in order to "refactor" class hierarchies for example [Sne05]; to design the class hierarchy of object-oriented software from specifications [GV05, AFHN06]; to find causal dependencies[Pfa06]. Tilley *et al.* [TCBE05] presented a survey on applications of FCA for software engineering activities : e.g. architectural design, software maintenance. FCA finds interesting clusters, called *concepts*, in data sets. The input of FCA is a *formal context*, i.e. a binary relation describing elements of a set of objects by subsets of properties (*attributes*). A *formal concept* is defined by a pair *(extent, intent)*, where *extent* is the maximal set of objects that have in their description all attributes of *intent*, and *intent* is the maximal set of attributes common to the description of all objects of *extent*. The concepts of the context can be represented by a lattice where each concept is labelled by its intent and extent.

The main contribution of this article is to show that applying two data mining techniques, association rules and FCA, produces better results than existing fault localization techniques. This is discussed in detail Section 6. Another contribution is to propose to build upon the intuition of existing methods: the difference or correlation between execution traces contains significant clues about the fault. We combine the expressiveness of association rules to search for possible causes of failure, and the power of FCA to explore the results of this analysis. The kind of association rules that we use allows some limitations of other methods to be alleviated. The goal of our method is no longer to highlight the faulty line but to produce an explanation of the failure thanks to the lattice.

In the sequel, Section 2 describes the running example used to illustrate the method. Section 3 presents the two contexts used by the method. Section 4 shows how to interpret the rule lattice in terms of fault localization. Section 5 discusses the statistical indicators. Section 6 presents the main contribution, the benefits of our method compared to other methods. Section 7 discusses further work.

2   Running Example

Throughout this article, we use the Trityp program given in Figure 1 to illustrate our method. It classifies sets of three segment lengths into four categories:

```
public int Trityp(){
[57] int trityp ;
[58] if ((i==0) || (j==0) ||
        (k == 0))
[59]    trityp = 4 ;
[60] else
[61]   {
[62]     trityp = 0 ;
[63]     if ( i == j)
[64]        trityp = trityp + 1 ;
[65]     if ( i == k)
[66]        trityp = trityp + 2 ;
[67]     if ( j == k )
[68]        trityp = trityp + 3 ;
[69]     if (trityp == 0)
[70]       {
[71]         if ((i+j <= k) ||
              (j+k <= i) ||
              (i+k <= j))
[72]           trityp = 4 ;
[73]         else
[74]           trityp = 1 ;
[75]       }
[76]     else
[77]       {
[78]         if (trityp > 3)
[79]           trityp = 3 ;
[80]         else
```

```
[81]              if ((trityp == 1)
                    && (i+j > k))
[82]                trityp = 2 ;
[83]              else
[84]                if ((trityp == 2)
                      && (i+k > j))
[85]                  trityp = 2 ;
[86]                else
[87]                  if((trityp == 3)
                        && (j+k > i))
[88]                    trityp = 2 ;
[89]                  else
[90]                    trityp = 4 ;
[91]       }
[92]     }
[93]     return(trityp) ;}
static public
string conversiontrityp(int i){
[97]   switch (i){
[98]     case 1:
[99]         return "scalen";
[100]    case 2:
[101]        return "isosceles";
[102]    case 3:
[103]        return "equilateral";
[104]    default:
[105]        return "not a ";}}
```

Fig. 1. Source code of the Trityp program

scalene, isosceles, equilateral, not a triangle. The program contains one class with 130 lines of code. It is a classical benchmark for test generation methods. Such a benchmark aims at evaluating the ability of a test generation method to detect errors by causing failure. To this purpose slight variants, *mutants*, of the benchmark programs are created. The mutants can be found on the web[1], and we use them for evaluating our localization method. For the Trityp program, 400 test cases have been generated with the *Uniform Selection of Feasible Paths* method of Petit and Gotlieb [PG07]. Thanks to that method, all feasible execution paths are uniformly covered.

Table 1 presents the five mutants of the Trityp program that are used in this article. The first mutant is used to explain in details the method. For mutant 1, one fault has been introduced at Line 84. The condition (trityp == 2) is replaced by (trityp == 3). That fault implies a failure in two cases. The first case is when trityp is equal to 2. That case is not taken into account as a particular case and thus treated as a default case, at Lines 89 and 90. The second case is

---

[1] http://www.irisa.fr/lande/gotlieb/resources/Java exp/trityp/

Table 2. Example of the trace context for mutant 1 of the Trityp program

	Line 66	Line 68	Line 81	Line 84	Line 85	Line 87	Line 90	Line 93	⋯	PASS	FAIL
$exec_1$	×	×						×		×	
$exec_4$	×		×	×		×	×	×		×	
$exec_{69}$		×						×		×	
$exec_{108}$		×	×	×	×			×			×
$exec_{113}$	×		×	×		×	×	×			×
$exec_{114}$		×	×	×	×			×		×	
...									×		
$exec_{400}$								×		×	

when `trityp` is equal to 3. That case should lead to the test Line 87, but due to the fault it is first tested at line 84. Indeed, if the condition (i+k>j) holds, `trityp` is assigned to 2. However, (i+k>j) does not always entail (j+k>i), which is the real condition to test when `trityp` is equal to 3. Therefore, `trityp` is assigned to 2 whereas 4 is expected.

The fault of mutants 2 and 3 are on assignments. The fault of mutants 4 and 5 are on conditions.

3   Two Formal Contexts for Fault Localization

This section presents the information on which the localization process is based. Firstly, a context is built from execution traces, the *trace context*. Secondly, particular association rules are used to crosscheck the trace context. Thirdly, a second context is introduced in order to reason on the numerous rules, the *rule context*. How to interpret the *rule lattice*, associated to the rule context, is presented in Section 4.

The Trace Context. In order to reason about program executions we use traces of these executions. There are many types of trace information and discussing them is outside the scope of this article (see for example [HRS+00]). Let us only assume that each trace contains at least the executed lines and the verdict of the execution, $PASS$ if the execution produces the expected results and $FAIL$ otherwise. This is a common assumption in fault localization research. This forms the *trace context*. The objects of the trace context are the execution traces. The attributes are all the lines of the program and the two verdicts. Each trace is described by the executed lines and the verdict of the execution.

Table 2 gives a part of the resulting trace context for mutant 1. For instance, during the first execution, the program executes lines 66, 68, ... and passes[2].

Association Rules. In order to understand the causes of the failed executions, we use a data mining algorithm [CFRD07] which searches for association rules. In

---

[2] Complete context: http://www.irisa.fr/LIS/cellier/icfca08/trace_context.txt

Table 3. Example of the rule context for mutant 1 of the Trityp program with $minlift = 1.25$ and $minsup = 1$

	Line 81	Line 84	Line 87	Line 90	Line 105	Line 66	Line 78	Line 112	Line 113	$\cdots$	Line 17	Line 58	Line 93
r_1	×	×	×	×	×	×	×	×	×		×	×	×
r_2	×	×	×	×			×	×	×		×	×	×
										$\cdots$			
r_8	×						×	×	×		×	×	×
r_9							×	×	×		×	×	×

addition, to reduce the number of association rules, we focus on association rules based on *closed itemsets* [PBTL99]. Namely, we search for association rules where 1) the premises are the intents of concepts whose extent mostly contains failed execution traces and few passed execution traces (according to the statistical indicators) and 2) the conclusion is the attribute $FAIL$ (Definition 1). This corresponds to the selection of the concepts that are in relation with the concept labelled by $FAIL$. Note that those concepts can be in relation with the concept labelled by $PASS$, too.

Definition 1 (association rules for fault localization). *The computed association rules for fault localization have the form: $L \rightarrow FAIL$ where L is a set of executed lines such that $L \cup \{FAIL\}$ is the intent of a concept in the trace context and $L \cap \{FAIL\} = \emptyset$.*

Only the association rules that satisfy the minimum thresholds of the selected statistical indicators are generated. We have chosen the *support* and *lift* indicators. The support indicates the frequency at which the rule appears. In our application, it measures how frequently the lines that form the premise of a given rule are executed among the lines in a failure. The lift indicates if the occurrence of the premise increases the probability to observe the conclusion. Relevant rules are thus filtered with respect to a minimum support, *minsup*, and a minimum lift, *minlift*. The threshold *minsup* can be very low, for instance to cope with failures that are difficult to detect (for details see Section 5). The threshold *minlift* is always greater or equal to 1 because otherwise the lift indicates that the premises *decrease* the probability to observe the conclusion.

The Rule Context. The computation of association rules generates a lot of rules, and especially rules with large premises. Understanding the links that exist between the rules, for example if a rule is more specific than another, is difficult to do by hand. The computed association rules, however, correspond to concepts of the trace context. They are partially ordered according to their premises; indeed $L_1 \rightarrow FAIL$ is more specific than $L_2 \rightarrow FAIL$ when L_1 and L_2 are sets of lines such that $L_2 \subset L_1$. Therefore, in order to help analyze the rules, we propose to build a new context, the *rule context*. The objects are the association rules; the attributes are lines. Each association rules is described by the lines of its premise.

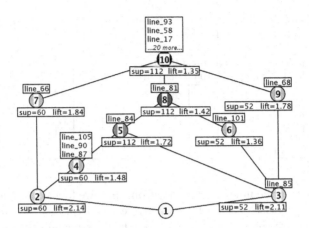

Fig. 2. Rule lattice for mutant 1 associated to the rule context of Table 3

Table 3 shows a part of the rule context for mutant 1 of the Trityp program with the support threshold, *minsup*, equal to 1 object and the lift threshold *minlift* equal to 1.25. The premise of $rule_1$ contains : line 81, line 84, line 87, line 90, ...[3] In addition, line 78, line 112, line 113, ..., line 17, line 58, line 93 are present in all rules which means that they are executed by all failed executions.

4   A Rule Lattice for Fault Localization

The *rule lattice* is the concept lattice associated with the rule context. It allows association rules to be structured in a way that highlights the partial ordering which exists between them. Figure 2 displays the rule lattice associated with the rule context of Table 3[4]. The remaining of this section presents a description of the rule lattice and then gives an interpretation of it.

4.1   Description of the Rule Lattice

The rule lattice has the property that each concept can be labelled at most by one object. Indeed, two different rules cannot have the same premise. If, in the trace context, several executions have exactly the same description they are abstracted in a single rule in the rule context.

The rule lattice is presented in a reduced labelling. In that representation each attribute and each object is written only once. Namely, each concept is labelled by the attributes and the objects that are specific to it. It is the most widespread representation. As a consequence, the premise of a rule *r* can be computed by collecting the attributes labelling all the concepts above the concept that is labelled by *r* [GW99]. For example, on Figure 2 the premise of the rule which

---

[3] Complete rule context: http://www.irisa.fr/LIS/cellier/icfca08/rules_context.rl

[4] The lattice was generated with the ToscanaJ tool (http://toscanaj.sourceforge.net/)

labels Concept 3 is line 85, line 84, line 68, line 101, line 81, line 93, line 58, line 17, plus the other 20 attributes (lines) that label the top concept.

In the rule lattice, the more specific rules are close to the bottom. When the support threshold for searching for association rules is one object and the lift threshold is close to 1, each most specific rule represents a single execution path in the program that leads to a failure. For example, on Figure 2 there are two very specific rules, the two concepts closest to the bottom: Concept 2 and Concept 3. The rule which labels Concept 2 contains in its premise the lines 66, 105, 90, 87, 84, 81 and the label of the top concept. It corresponds to the failure case when `trityp` is equal to 2 (see Section 2). The rule which labels Concept 3 contains in its premise the lines 85, 84, 68, 101, 81 and the label of the top concept. It corresponds to the failure case when `trityp` is equal to 3 (see Section 2). By looking at the support value of each rule, we note that three rules relate to 60 failed executions (Concepts 2, 4, 7), in fact failed executions when `trityp` is equal to 2; three rules relate to 52 failed executions (Concepts 3, 6, 9), in fact failed executions when `trityp` is equal to 3; and three rules relate to 112 failed executions (Concepts 5, 8, 10), namely all failed executions.

4.2   Interpretation for Fault Localization

Navigating in the rule lattice bottom up first displays rules that are in general too specific to explain the error. It then displays rules that are more general and maybe more informative, and finally displays the top of the lattice which is labelled by the attributes (line numbers) that are common to all failed executions.

The bottom concept of the rule lattice in Figure 2 has no attribute in its labelling. During the debugging session two paths are proposed to follow. The leftmost path from the bottom concept, Concept 2, corresponds to the case where variable `trityp` is equal to 3 and condition (i+k>j) holds whereas the condition (j+k>i) does not hold. It leads to two concepts. The first concept is Concept 7 labelled by line 66, it is the statement which initializes `trityp` to 2. The second concept is Concept 4 labelled by three line numbers: 105, 90, 87. These lines correspond to the case when the variable `trityp` is equal to 2 and `trityp` is assigned to 4 when 2 is expected, i.e. the triangle is labelled as not a triangle instead of isosceles. Those two concepts are too specific but by looking at the rule of the concept upwards, the faulty line is localized. Concept 5 covers the greatest number of failed executions (support=112) and has the greatest lift among rules which have support equal to 112. The same reasoning can be done with the rightmost concept, Concept 3. It also leads to line 85. It corresponds to the *then branch* of the faulty conditional, i.e. the line where variable `trityp` is assigned to 2 when 4 is expected. The rule of that concept is too specific to understand the fault, it covers 52 failed executions. Following this path, three paths open upwards: two concepts whose rules have the same support as the rule of the concept that is labelled by line 85, Concept 6, 9; and a concept which is labelled by line 84, the faulty line, Concept 5, whose rule covers the most number of failed executions (support=112).

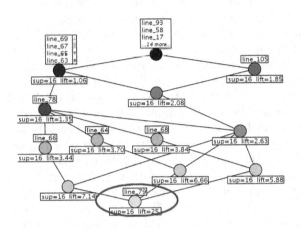

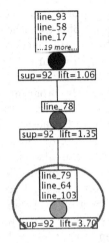

Fig. 3. Rule lattice from program Trityp with mutant 2 (faulty line 79) with minlift=1 and minsup=1

Fig. 4. Rule lattice from program Trityp with mutant 3 (faulty line 64) with minlift=1 and minsup=1

This example shows that the rule lattice gives relevant clues for exploring the program. The faulty line is not highlighted immediately but exploring the lattice bottom up guides the user in its task to understand the fault.

4.3   More Examples

One failed execution path. Figure 3 gives the rule lattice for mutant 2 when association rules are computed with $minlift=1$ and $minsup=1$. Figure 4 gives the rule lattice for mutant 3 when association rules are computed with $minlift=1$ and $minsup=1$. On those two examples the fault is a bad assignment. Only one execution path leads to a failure. We see in those cases that the faulty lines are immediatly highlighted in the label of the bottom concept. For example, in Figure 3 we see the faulty line 79 at the bottom. We remark that the dependencies between lines appear in the labelling. Indeed, in Figure 4 we see that the bottom concept is labelled not only by the faulty line 64 but also by line 79 and line 103. It is explained by two facts. Firstly, the execution of line 64 in a faulty way always implies the execution of line 79 and 103. Secondly, few executions that imply lines 64, 79 and 103 together pass. Note that all association rules of these examples have the same support. They cover all failed execution traces.

Several failed execution paths. The execution of mutant 1 and 4 can fail with different execution paths. Mutant 1 was detailed in the previous section. Figure 5 gives the rule lattice for mutant 4 when association rules are computed with $minlift = 1$ and $minsup = 1$. The fault of mutant 4 is at line 87. In the rule lattice of Figure 5, line 87 labels the concept which is labelled by a rule with the greatest support. In addition, among rules with the greatest support, this one

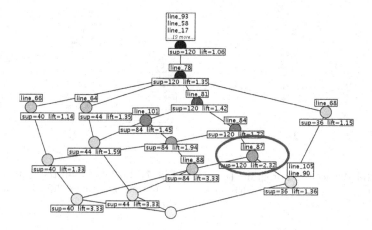

Fig. 5. Rule lattice from program Trityp with mutant 4 (faulty line 87) with minlift=1 and minsup=1

has the greatest lift. In other words, line 87 is one of the common lines of all failed executions and with the least relation with passed executions.

Borderline case. Figure 6 gives the rule lattice for mutant 5 when association rules are computed with *minlift*=1 and *minsup*=1. In the rule lattice, two failed execution paths are highlighted. One covers 33 of the failed executions. The other covers 62 of the failed executions. Looking at the common lines of those two execution paths, we find line 66. The associated rule of line 66 has the greatest lift among rules with support equal to 95. But the faulty line is 65. Line 65 labels the top concept of the rule lattice. The faulty line is thus not highlighted. It is due to the fact that line 65 is an if-statement executed by all executions. The number of failed executions that execute line 65 is thus very low with respect to the number of passed executions that execute line 65. However, line 66 is the *then branch* of the condition of line 65. The method does not highlight the faulty line but gives clues to find it.

5   Statistical Indicators

In order to compute and evaluate the relevance of association rules, several statistical indicators have been proposed, for example *support, confidence,* and *lift* [BMUT97]. The choice of statistical indicators and their thresholds is an important part of the computation of association rules. Depending on the selected thresholds, the rule lattice may significantly vary and so does its interpretation.

Typically, the support is the first filtering criterion for the extraction of association rules. It measures the frequency of a rule. For the fault localization problem, the value of *minsup* indicates the minimum number of failed executions that should be covered by a concept of the trace context to be selected.

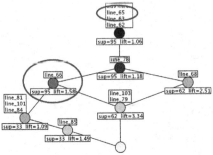

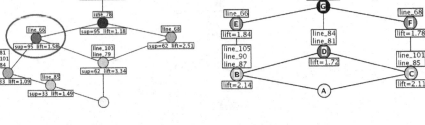

Fig. 6. Rule lattice from program Trityp with mutant 5 (faulty line 65) with min-lift=1 and minsup=1

Fig. 7. Lattice from the rule context of mutant 1 with minlift=1.5 and min-sup=1

Choosing a very high threshold, only the most frequent execution paths are represented in the set of association rules. Choosing a very low threshold, *minsup* equals to one object, all execution paths that are stressed by the test cases are represented in the set of association rules. In our experiments, we choose *minsup* equals to one object. It is equivalent to search for all rules that cover at least one failed execution. The threshold can be greater in order to select less rules, but some failed execution paths would be lost, although in fault localization rare paths might be more relevant than common ones.

The other well-known indicator that we use is lift. In our approach, the lift indicates how the execution of a set of lines improves the probability to have a failed execution. The lift threshold can be seen as a *resolution*[5] cursor. On the one hand, if the threshold is high, few rules are computed, therefore few concepts appear in the lattice. It implies that the label of each concept contains a lot of lines, but each concept is very relevant. On the other hand, selecting a low threshold implies that more rules are computed. Those rules are less relevant and the number of concepts in the lattice increases. The number of attributes per label is thus reduced. For example, let us consider mutant 1 of program Trityp. Figure 2, already described in Section 4, shows the result lattice when the lift threshold is equal to 1.25. Figure 7 shows the result lattice when the lift threshold is equal to 1.5. Concept D of Figure 7 merges Concepts 5 and 8 of Figure 2. Concepts D and 5 actually correspond to the same rule. Concept 8 correspond to a more general rule with a lower lift. The same applies to Concepts C and 3, 6.

In addition, the concepts of the rule lattice have two properties thanks to the statistical indicators related to the trace lattice. Property 1 states that the support of rules that label the concepts of the rule lattice decreases when exploring the lattice top down. In the following, $label(c)=r$ means that rule r labels concept c of the rule lattice. The extent of a set of attributes, X, is written $ext(X)$, and $\|Y\|$ denotes the cardinal of set Y.

---

[5] The word "resolution" is used here as in "image display resolution".

Property 1. Let c_1 and c_2 be two concepts of the rule lattice such that $label(c_1)=r_1$ and $label(c_2)=r_2$. If $c_2 < c_1$ then $sup(r_2) \leq sup(r_1)$.

Proof. The fact that $c_2 < c_1$ implies that the intent of c_2 contains the intent of c_1. The intent of concept c_1 (resp. c_2) is the premise, p_1 (resp. p_2), of rule r_1 (resp. r_2), thus $p_1 \subset p_2$. Conversely $ext(p_2) \subset ext(p_1)$. As the definition of the support of a rule $r = p \rightarrow FAIL$ is $sup(r) = \|ext(p) \cap ext(FAIL)\|$, $sup(r_2) \leq sup(r_1)$ holds.

Property 2 is about the lift value. If two ordered concepts in the rule lattice are labelled by rules with the same support value, the lift value of the rule which labels the more specific concept is greater. That is why the rule lattice is explored bottom up (see Section 4.2).

Property 2. If $c_2 < c_1$ and $sup(r_2) = sup(r_1)$ then $lift(r_2) > lift(r_1)$.

Proof. In the previous proof we have seen that $c_2 < c_1$ implies that $ext(p_2) \subset ext(p_1)$. The definition of the lift of a rule $r = p \rightarrow FAIL$ is $lift(r) = \frac{sup(r)}{\|ext(p)\| \|ext(FAIL)\|} * \|\mathcal{O}\|$. As $sup(r_2) = sup(r_1)$, and $\|ext(p_2)\| < \|ext(p_1)\|$, $lift(r_2) > lift(r_1)$ holds.

6 Benefits of Using Association Rules and FCA

The contexts and lattice structures introduced in the previous sections allow programmers to see all the differences between execution traces as well as all the differences between association rules. There exists other methods which compute differences between execution traces. Section 6.1 shows that the information about trace differences provided by our first context (and the corresponding lattice) is already more relevant than the information provided by four other methods proposed by Renieris and Reiss [RR03], as well as Zeller *et al.* [CZ05]. Section 6.2 shows that explicitly using association rules with several lines in the premise alleviate many of the limitations of Jones *et al.*'s method mentioned in the introduction [JHS02]. Section 6.3 shows that reasoning on the partial ordering given by the proposed rule lattice is more relevant than reasoning on total order rankings [JHS02, LNZ$^+$05, LYF$^+$05].

6.1 The Trace Context Structures Execution Traces

The first context that we have introduced, the trace context, contains the whole information about execution traces (see Section 3). In particular, the associated lattice, the *trace lattice*, allows programmers to see in one pass all differences between traces. Figure 8 shows the trace lattice of mutant 1.

There exists several fault localization methods based on the differences of execution traces. They all assume a single failed execution and several passed executions. We rephrase them in terms of search in a lattice to highlight their advantages, their hidden hypothesis and limitations.

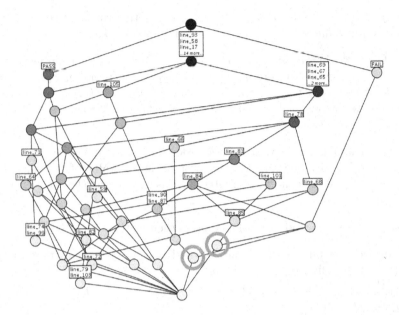

Fig. 8. Lattice from the trace context of mutant 1 of the Trityp program

The *union model*, proposed by Renieris and Reiss [RR03], aims at finding features that are specific to the failed execution. The method is based on trace differences between the failed execution f and a set of passed executions S: $f - \bigcup_{s \in S} s$. The underlying intuition is that the failure is caused by lines that are executed only in the failed execution. Formalized in FCA terms, the concept of interest is the one whose label contains $FAIL$, and the computed information is the lines contained in the label. For example, in Figure 8 this corresponds to the upper concept on the right-hand side which contains no line in its label, namely the information provided by the union model is empty. The trace lattice presented in the figure is slightly different from the lattice that would be computed for the union model, because it represents more than one failed execution. Nevertheless, the union model often computes an empty information, namely each time the faulty line belongs to failed and passed execution traces. For example, a fault in a condition has a very slight chance to be localized. Our approach is based on the same intuition. However, as illustrated by Figures 8 and 2, the lattices that we propose do not lose information and help to navigate in order to localize the faults, even when the faulty line belongs to failed and passed execution traces.

The union model helps localize bug when executing the faulty statement always implies an error, for example the bad assignment of a variable that is the result of the program. In that case, our lattice does also help, the faulty statement label the same concept as $FAIL$.

The *intersection model* [RR03] is the complementary of the previous model. It computes the features whose absence is discriminant of the failed execution:

$\bigcap_{s \in S} s - f$. Replacing $FAIL$ by $PASS$ in the above discussion is relevant to discuss the intersection model and leads to the same conclusions.

The *nearest neighbor* approach [RR03] computes the Ulam's *distance* metrics between the failed execution trace and a set of passed execution traces. The computed trace difference involves the failed execution trace, f, and only one passed execution trace, the nearest one, p: $f - p$. That difference is meant to be the part of the code to explore. The approach can be formalized in FCA. Given a concept C_f whose intent contains $FAIL$, the nearest neighbor method search for a concept C_p whose intent contains $PASS$, such that the intent of C_p shares as many lines as possible with the intent of C_f. On Figure 8 for example, the two circled concepts are "near", their share all their line attributes except the attributes $FAIL$ and $PASS$, therefore $f = p$ and $f - p = \emptyset$. The rightmost concept fails whereas the leftmost one passes. As for the previous methods, it is a good approach when the execution of the faulty statement always involves an error. But as we see on the example, when the faulty statement can lead to both a passed and a failed execution, the method is not sufficient. In addition, we remark that there are possibly many concepts of interest, namely all the nearest neighbors of the concept which is labelled by $FAIL$. With a lattice that kind of behavior can be observed directly.

Note that in the trace lattice, the executions that execute the same lines are clustered in the label of a single concept. Executions that are near share a large part of their executed lines and label concepts that are neighbors in the lattice. There is therefore no reason to restrict the comparison to a single pass execution. Furthermore, all the nearest neighbors are naturally in the lattice.

Delta debugging, proposed by Zeller *et al.* [CZ05], reasons on the values of variables during executions rather than on executed lines. The trace information, and therefore the trace context, contains different types of attributes. Note that our approach does not depend on the type of attributes and would apply on traces containing other attributes than executed lines.

Delta debugging computes in a memory graph the differences between the failed execution trace and a single passed execution trace. By injecting the values of variables of the failed execution into variables of the passed execution, the method tries to determine a small set of suspicious variables. One of the purpose of that method is to find a passed execution relatively similar to the failed execution. It has the same drawbacks as the nearest neighbor method.

6.2   Association Rules Select a Part of the Trace Context

As it was presented in the introduction, Jones *et al.* [JHS02] compute association rules with only one line in the premises. Denmat *et al.* show that the method has limitations, because it assumes that an error has a single faulty statement origin, and lines are independent. In addition, they demonstrate that the *ad hoc* indicator which is used by Jones *et al.* is close to the lift indicator.

By using association rules with more expressive premises than the Jones *et al.* method (namely with several lines), the limitations mentioned above are alleviated. Firstly, the fault can be not only a single line, but the execution of several lines together. Secondly, the dependency between lines is taken into account. Indeed, dependent lines are clustered or ordered together.

The part of the trace context which is important to search for localizing a fault is the set of concepts that are around the concept labelled by $FAIL$; i.e. those that have a non-empty intersection with the concept labelled $FAIL$. Computing association rules with $FAIL$ as a conclusion computes exactly those concepts, modulo the $minsup$ and $minlift$ filtering. In other words the focus is done on the part of the lattice around the concept labelled by $FAIL$. For example, in the trace lattice of the Trityp program presented in Figure 8, the rule lattice when $minlift$ is very low (yet still attractive, i.e. $minlift>1$), is drawn in bold lines.

6.3   The Rule Lattice Structures Association Rules

Jones *et al.*'s method presents the result of their analysis to the user as a coloring of the source code. A red-green gradient indicates the correlation with failure. Lines that are highly correlated with failure are colored in red, whereas lines that are highly not correlated are colored in green. Red lines typically represents more than 10% of the lines of the program, whithout identified links between them. Some other statistical methods [LNZ$^+$05, LYF$^+$05] also try to rank lines in a total ordering. It can be seen as ordering the concepts of the rule lattice by the lift value of the rule in their label. From the concept ordering the lines in the label of those concepts can be ordered.

For example, on the rule lattice of Figure 2, the obtained ranking would be: line 85, line 66, line 68, line 84, ... No link would be established between the execution of line 85 and line 68 for example.

The user who has to localize a fault in a program has a background knowledge about the program, and can use it to explore the rule lattice. The reading of the lattice gives a context of the fault and not just a sequence of independent lines to be examined, and reduces the number of lines to be examined at each step (concept) by structuring them.

7   Further Work

The process proposed in this article for fault localization is already usable at the end of the debugging process. When the programmer has a rough idea of the location of the faults and that only a small part of the execution has to be traced, the current techniques to visualize lattices can be directly used.

However, we conjecture that the technique, with some extensions, can be used to analyze large executions. At present, the information is presented in terms of lines whereas it is not always the most relevant information granularity. For example, given a basic block (i.e. a sequence of instructions with neither branching nor conditional), all its lines always appear in the same label. Displaying the location of the basic block would be more relevant. This will help keeping concept labels to a readable size. We are currently working on the presentation of the results to reduce their size and to give more semantics to them so that they are more tractable for users.

Another further work is the computation of the concepts of the rule context directly from the trace context. Indeed, the concepts of the rule lattice belong to the trace lattice. However, gathering the two computations, association rules from the trace context and concepts from the rule context, is not a trivial task due to the statistical indicators which are not monotonous, for example the lift.

8   Conclusion

In this article we have proposed a new application of FCA: namely, fault localization. The process combines association rules and formal concept lattices to give a relevant way to navigate into the source code of faulty programs. Compared to existing methods, using a trace lattice has three advantages. Firstly, thanks to the capabilities of FCA to crosscheck data, our approach computes in one pass more information about execution differences than all the cited approaches together. In particular, the information computed by two methods, the union model and the intersection model, can be directly found in the trace lattice. Secondly, the lattice structure gives a better reasoning basis than any particular distance metrics to find similar execution traces. Indeed, all differences between the sets of executed lines of passed and failed executions are represented in the trace lattice. A given distance between two executions can be computed by dedicated reasoning on specific attributes. Thirdly, our approach treats several failed executions together, whereas the presented methods can analyze only one failed execution.

Moreover, the generality of the FCA framework makes it possible to handle traces that contain other information than lines numbers. In summary, casting the fault localization problem in the FCA framework helps analyze existing approaches, as well as alleviate their limitations.

References

[AFHN06]   Arévalo, G., et al.: Building abstractions in class models: Formal concept analysis in a model-driven approach. In: Nierstrasz, O., et al. (eds.) MoD-ELS 2006. LNCS, vol. 4199, pp. 513–527. Springer, Heidelberg (2006)

[AIS93]   Agrawal, R., Imielinski, T., Swami, A.N.: Mining association rules between sets of items in large databases. In: Buneman, P., Jajodia, S. (eds.) Int. Conf. on Management of Data, ACM Press, New York (1993)

[AS94]   Agrawal, R., Srikant, R.: Fast algorithms for mining association rules. In: Bocca, J.B., Jarke, M., Zaniolo, C. (eds.) Int. Conf. Very Large Data Bases, pp. 487–499. Morgan Kaufmann, San Francisco (1994)

[BMUT97]   Brin, S., et al.: Dynamic itemset counting and implication rules for market basket data. In: Peckham, J. (ed.) Proc. Int. Conf. on Management of Data, pp. 255–264. ACM Press, New York (1997)

[CFRD07]   Cellier, P., et al.: A parameterized algorithm for exploring concept lattices. In: Kuznetsov, S.O., Schmidt, S. (eds.) ICFCA 2007. LNCS (LNAI), vol. 4390, Springer, Heidelberg (2007)

[CZ05]   Cleve, H., Zeller, A.: Locating causes of program failures. In: Int. Conf. on Software Engineering, pp. 342–351. ACM Press, New York (2005)

[DDR05]   Denmat, T., Ducassé, M., Ridoux, O.: Data mining and cross-checking of execution traces: a re-interpretation of jones, harrold and stasko test information. In: Int. Conf. on Automated Software Engineering, ACM, New York (2005)

[GV05]   Godin, R., Valtchev, P.: Formal concept analysis-based class hierarchy design in object-oriented software development. In: Ganter, B., Stumme, G., Wille, R. (eds.) Formal Concept Analysis. LNCS (LNAI), vol. 3626, pp. 304–323. Springer, Heidelberg (2005)

[GW99]   Ganter, B., Wille, R.: Formal Concept Analysis: Mathematical Foundations. Springer, Heidelberg (1999)

[HRS+00]   Harrold, M.J., et al.: An empirical investigation of the relationship between spectra differences and regression faults. Softw. Test., Verif. Reliab. 10(3), 171–194 (2000)

[JHS02]   Jones, J.A., Harrold, M.J., Stasko, J.T.: Visualization of test information to assist fault localization. In: Int. Conf. on Software Engineering, pp. 467–477. ACM, New York (2002)

[LNZ+05]   Liblit, B., et al.: Scalable statistical bug isolation. In: Conf. on Programming Language Design and Implementation, pp. 15–26. ACM Press, New York (2005)

[LYF+05]   Liu, C., et al.: Sober: statistical model-based bug localization. In: European Software Engineering Conf. held jointly with Int. Symp. on Foundations of Software Engineering, ACM Press, New York (2005)

[PBTL99]   Pasquier, N., et al.: Discovering frequent closed itemsets for association rules. In: Int. Conf. on Database Theory, pp. 398–416. Springer, Heidelberg (1999)

[Pfa06]   Pfaltz, J.L.: Using concept lattices to uncover causal dependencies in software. In: Missaoui, R., Schmidt, J. (eds.) Formal Concept Analysis. LNCS (LNAI), vol. 3874, pp. 233–247. Springer, Heidelberg (2006)

[PG07]   Petit, M., Gotlieb, A.: Uniform selection of feasible paths as a stochastic constraint problem. In: Int. Conf. on Quality Software, October 2007, IEEE, Los Alamitos (2007)

[RR03]   Renieris, M., Reiss, S.P.: Fault localization with nearest neighbor queries. In: Int. Conf. on Automated Software Engineering, IEEE, Los Alamitos (2003)

[Sne05]   Snelting, G.: Concept lattices in software analysis. In: Ganter, B., Stumme, G., Wille, R. (eds.) Formal Concept Analysis. LNCS (LNAI), vol. 3626, pp. 272–287. Springer, Heidelberg (2005)

[TCBE05]   Tilley, T., et al.: A survey of formal concept analysis support for software engineering activities. In: Ganter, B., Stumme, G., Wille, R. (eds.) Formal Concept Analysis. LNCS (LNAI), vol. 3626, Springer, Heidelberg (2005)

Refactorings of Design Defects Using Relational Concept Analysis

Naouel Moha[1], Amine Mohamed Rouane Hacene[2], Petko Valtchev[3],
and Yann-Gaël Guéhéneuc[1]

[1] DIRO, University of Montréal, CP 6128, Montréal, H3C 3J7, Canada
[2] LORIA, BP 239 - 54506 Vandoeuvre-lès-Nancy, Cedex France
[3] LATECE, Université du Québec à Montréal, CP 8888, Montréal, H3C 3P8, Canada

Abstract. Software engineers often need to identify and correct design defects, *i.e.*, recurring design problems that hinder development and maintenance by making programs harder to comprehend and/or evolve. While detection of design defects is an actively researched area, their correction — mainly a manual and time-consuming activity — is yet to be extensively investigated for automation. In this paper, we propose an automated approach for suggesting defect-correcting refactorings using relational concept analysis (RCA). The added value of RCA consists in exploiting the links between formal objects which abound in a software re-engineering context. We validated our approach on instances of the *Blob* design defect taken from four different open-source programs.

Keywords: Design Defects, Refactoring, Relational Concept Analysis.

1 Introduction

Design defects are "bad" solutions to recurring design problems that generate negative consequences on the quality characteristics of object-oriented (OO) software systems, such as evolvability and maintainability, and therefore increase the cost of software development [5,16]. Design defects, such as antipatterns [28] (e.g., the Blob addressed below), are distinguished from low-level defects, such as code smells [5] (e.g., long methods and large classes). Automatic detection and correction of design defects are thus keys for the improvement of software quality.

We proposed a systematic method to specify design defects consistently and precisely and to generate detection algorithms from their specifications automatically [17]. We specified a language based on rules that allows to define these specifications with structural, semantic, and measurable properties that characterize a design defect. This method was a first step towards the systematic detection of design defects. Yet both detection and correction of such defects are time-consuming and error-prone activities hence leaving room for automated techniques and tools. On the one hand, approaches exist for detecting design defects, for instance, using metrics [15,21], coupled with visualisation tools [13,14] and/or structural data [9]. On the other hand, to the best of our knowledge,

R. Medina and S. Obiedkov (Eds.): ICFCA 2008, LNAI 4933, pp. 289–304, 2008.
© Springer-Verlag Berlin Heidelberg 2008

none of them attempts to correct discovered defects in a semi- or fully automated manner.

Thus, design defects are still dealt with manually through tedious code analyses and transformations, which divides into three main steps, possibly repeated through trials and errors: (1) Identification of the modifications to correct the design defects, (2) Application of the modifications on the program, (3) Evaluation of the resulting modified program. Step two of correction has been made easier by the recent introduction of *refactorings* [5], *i.e.*, changes performed on the source code of a program to improve its internal structure without changing its external behaviour. Thus, possible transformations are now well understood and documented and the emphasis lies on step one, *i.e.*, the decision of which modifications (or refactorings) to apply.

In the literature, Trifu *et al.* [27] proposed correction strategies mapping design defects to possible solutions. However, a solution is only an example of how the program *should have been implemented* to avoid a defect rather than a list of steps that a software engineer could follow to correct the defect. Huchard and Leblanc [11] used formal concept analysis (FCA) to suggest class hierarchy restructuring so as to maximise the sharing of specifications and code and to remove code smells (see [6] for a broader discussion on the restructuring of class hierarchies through FCA). In summary, both approaches address important issues with design defects but none attempts *to suggest refactorings to correct them*.

We propose to apply RCA, that extends FCA with the processing of individuals with links, on a suitable representation of a program to help identify appropriate refactorings for specific design defects. In particular, we examine the benefits of RCA for the correction of a very common design defect, the Blob [28, p. 73–83], also known as *God Class* [22]. The Blob reveals a procedural design (and thinking) implemented with an OO programming language. It manifests itself through a large controller class that plays a God-like role in the program by monopolizing the computation, and which is surrounded by a number of smaller data classes providing many attributes but few or no methods.

Blobs are common and RCA is particularly well-suited to suggest refactorings to correct them. Indeed, correcting a Blob amounts to splitting the Blob class into smaller cohesive sets by grouping class members that collaborate to realize a specific responsibility of the Blob class. In our context, cohesive sets are identified using formal concepts whose intents involve both proper characteristics and inter-member links, such as calls between methods. Our enhanced approach is illustrated using a running example of a library management system, which includes a Blob.

The present work builds upon a previous study described in [18] that relied on standard FCA. Its contribution is three-fold. First, a more powerful approach is adopted based on finer and richer modeling of the problem through RCA. Then, a set of enhanced rules for candidate class extraction out of the concept sets is designed, each rule is provided with an effective algorithm. Third, a mechanism to automatically interpret the results is introduced to suggest the refactorings to apply. A validation thereof involving Blobs from four different open-source programs is also presented. The results show that RCA suggests a high rate of relevant refactorings and we briefly discuss the application of our method on further design defects.

The paper starts by a short presentation of design defects correction (Section 2). Follow a summary on RCA (Section 3) and the description of our approach (Section 4). Section 5 presents the results of a preliminary empirical study of the approach validity. Related work is summarised in Section 6 while future research directions are given in Section 7.

2   The Defect Correction Problem

In the following, we relate design defects to general quality criteria for OO designs using an occurrence of the Blob as running example. The defects are shown to decrease scores on these criteria. The improvement brought by the FCA-based refactorings is discussed in later sections.

2.1   Quality Criteria

Design defects are the results of *bad* practices that transgress *good* OO principles. Thus, we use the degree of satisfaction of those principles before and after the correction as a measure of improvement. We rely on quantification of *coupling* and *cohesion*, which are among the most widely acknowledged software quality characteristics, keys for maintainability [2].

The cohesion of a class reflects *how closely the methods are related to the instance variables in the class* [4] and is typically measured by the LCOM metric (Lack of COhesion Metric: between 0 and 1) which uses *the number of disjoint sets of methods* [4]. A low LCOM score characterises a cohesive class whereas a value close to 1 indicates a lack of cohesion and suggests the class might better be split into cohesive sets. The coupling of a class to the rest of a program is defined as the degree of its reliance on services provided by other classes [4]. It is measured by the CBO metric (Coupling Between Objects) [3] that counts the classes to which a class is coupled. A well-designed program exhibits *high* average cohesion and *low* average coupling, but it is widely known that these criteria are antinomic hence a trade-off is usually sought.

2.2   Further Design Defects

We choose to illustrate our approach with the Blob because it impacts negatively both cohesion and coupling: blobs show low cohesion and high coupling. Moreover, it is a frequent defect in OO programs. For example, a previous study revealed 1,146 Blobs in the Eclipse IDE [19] even though it is recognised for the quality of its design.

Yet, we found that a good number of other design defects are infected by low cohesion and high coupling, e.g., Divergent Change [5, page 79], Feature

Envy [5, page 80], Inappropriate Intimacy [5, page 85], Lazy Class [5, page 83], Shotgun Surgery [5, page 80], or Swiss Army Knife [28, page 197]. Therefore, our approach could be adapted to these defects.

2.3 Running Example

Our running example (see Figure 1) was inspired by a simple library management system, which includes a Blob (described in [28]). The large controller class is the class Library_Main_Control that accesses to data of the two surrounding data classes Book and Catalog.

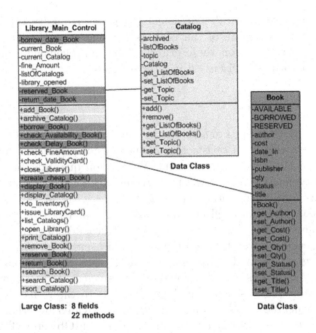

Fig. 1. Library Blob class diagram

Refactoring a Blob consists in moving class members away from the large controller class to its surrounding classes or to new specifically designed classes. For the class Library_Main_Control, we notice that all methods and fields related to Book or Catalog could be moved to their respective data classes. As a result, data classes gain more behaviour while the large class becomes less complex. However, the process of choosing and applying refactoring is long and tedious: software engineers need to go through all methods and fields of the large class to identify the subsets thereof that form consistent cohesive sets. Yet it is a necessary pain because the result of the process may substantially improve the quality of the program.

3   Relational Concept Analysis

FCA offers a framework to derive conceptual hierarchies from sets of individuals based on the properties that these individuals share[1].

3.1   Formal Concept Analysis

FCA describes (formal) concepts both extensionally and intentionally, *i.e.*, as sets of individuals and sets of shared properties, and organizes them hierarchically—according to a generality relation—into a complete lattice, called the concept lattice. The lattice structure allows easy navigation and search as well as optimal representation of information comparable to the classical OO requirement of maximal factorisation (each property/individual is canonically represented by a unique concept). For instance, the table on the left-hand side of Fig. 2 illustrates a binary context derived from the class `Library_Main_Control` where individuals are the Blob methods while properties are methods and accessed fields[2]. Fig. 3 depicts a simplified (reduced) labeling of the concept lattice derived from this context, yet enriched by additional properties as described later in this section.

Left: Context of methods

	'add_Book()'	'borrow_Book()'	'check_Availability_Book()'	'check_FineAmount()'	'close_Library()'	'issue_LibraryCard()'	'open_Library()'	'remove_Book()'	'reserve_Book()'	'return_Book()'	'search_Book()'	'sort_Catalog()'	R-current_book	R-fine_amount	W-borrow_date_book	W-library_opened	W-reserved_book	W-return_date_book
add_Book()	X												X					
borrow_Book()		X											X		X			X
check_Availability_Book()			X										X					
check_FineAmount()				X										X				
close_Library()					X											X		
issue_LibraryCard()						X												
open_Library()							X									X		
remove_Book()								X										
reserve_Book()									X								X	
return_Book()										X					X			X
search_Book()											X							
sort_Catalog()												X						

Right: Binary relation 'call' between methods

	add_Book()	check_Availability_Book()	check_FineAmount()	remove_Book()
add_Book()				
borrow_Book()	X			
check_Availability_Book()				
check_FineAmount()				
close_Library()				
issue_LibraryCard()			X	
open_Library()				
remove_Book()				
reserve_Book()	X			
return_Book()				
search_Book()				
sort_Catalog()	X			X

Fig. 2. Left: Context of methods. **Right:** Binary relation 'call' between methods.

Formal concepts naturally endow "cohesiveness" because their extents comprise members sharing *all* the properties in the respective intents. Conversely,

---

[1] We use *individuals* for *objects* and *properties* for *attributes* to avoid confusion with OO objects and attributes.

[2] The prefixes `R-` and `W-` that appear in the field names specify the access mode, *i.e.*, read and write, respectively.

concept extents are *maximal* sets for the respective intent. In order to identify highly cohesive sets that could jointly replace the Blob class and hence
improve the overall quality, a suitable formalization consists in using class methods as individuals and instance variables as properties. For example, the concept
({open_Library(), close_Library()},{W-library_opened}) (concept $c9$ in Fig. 3)
could generate a smaller, hence more cohesive, class.

Furthermore, in an attempt to reduce coupling in the resulting OO code, we
consider the links between class members such as method calls (see Fig. 2, on
the right). For instance, both methods borrow_Book() and reserve_Book() call
check_Availability_Book(). Assigning the first two methods to the same class
inevitably decreases the class coupling in the OO code. Therefore, we would like
to define an approach that allows grouping these methods. We use RCA to do
so because grouping individuals based on the links they share, *i.e.*, the calls of
same or comparable methods, is beyond the scope of classical FCA.

3.2 Bringing Relations to Concept Intents

Relational concept analysis (RCA) is an approach for extracting formal concepts from sets of individuals described by properties, called also *local properties*, and links. RCA comes up with formal concepts that are connected in the
same way that description logics concepts are connected, *i.e.*, by means of role
restrictions involving logical quantifiers. RCA input data is organized within a
structure called *relational context family* (RCF) that comprises a set of binary
contexts $\mathcal{K}_i = (O_i, A_i, I_i)$ and set of binary relations $r_k \subseteq O_i \times O_j$, where O_i
and O_j are the individual sets of $\mathcal{K}_i$ (domain) and $\mathcal{K}_j$ (range), respectively. For
instance, the context encoding the access of fields by methods and the binary
relation *call* that links methods of the Blob with one another form a sample RCF
(see Fig. 2).

A scaling mechanism is used to translate links into context properties: relations are interpreted as features whose values are sets of individuals, hence the
target properties are predicates describing these sets. The predicates are derived
from the available concept lattice on the underlying context. Thus, for a given
relation seen as a function $r : O_i \rightarrow 2^{O_j}$, new properties, called *relational*, of the
form $qr{:}c$, are added to $\mathcal{K}_i$, where c is concept on $\mathcal{K}_j$ and q a scaling operator
(comparable to role restriction connectors in description logics). An individual
$o \in O_i$ gets a property $qr{:}c$ depending on the relationship between its link set
$r(o)$ and the extent of $c = (X, Y)$. The relationship can be either inclusion,
i.e., $r(o) \subseteq X$ (called *universal* scaling schema, q is $\forall$), or non-empty intersection, *i.e.*, $r(o) \cap X$ (called *existential* scaling schema, q is $\exists$). Formally, given a
context $\mathcal{K}_i=(O_i, A_i, I_i)$, a relation $r \subseteq O_i \times O_j$ and the lattice $\mathcal{L}_j$ of $\mathcal{K}_j$, the
image of $\mathcal{K}_i$ for the existential scaling operator is: $sc_\exists(\mathcal{K}_i) = (O_i, A_i^+, I_i^+)$, where
$A_i^+ = A_i \cup \{\exists r : c | c \in \mathcal{L}_j\}$ and $I_i^+ = I_i \cup \{(o, \exists r : c) | o \in O_i, c = (X, Y) \in
\mathcal{L}_j, r(o) \cap X \neq \emptyset\}$. In the present study, as in the vast majority of software engi-
neering applications of RCA, current or anticipated, only the existential scaling
is suitable. Hence we shall be systematically omitting the $\exists$ sign in attribute
names to keep notations simple.

Table 1. Scaling of the Blob context along the relation `call`. For space limitation, individuals that are not affected by relational scaling are omitted.

	call:c0	call:c2	call:c4	call:c5	call:c6	call:c11
`borrow_Book()`		×	×	×		
`issue_LibraryCard()`				×	×	
`reserve_Book()`		×	×	×		
`sort_Catalog()`	×	×		×		×

For example, assume methods are scaled along relation *call* regarding the lattice of the context in the left hand side of Fig. 2, which is composed of the concepts $\{c0, c2, c4, c5, c6, c11\}$ and the respective precedence links illustrated in Fig. 3. Since the method `sort_Catalog()` calls the method `add_Book()` which appears in the extent of concepts $c0$, $c2$ and $c5$ and calls the method `remove_-Book()` which belong to the extent of concepts $c11$ and $c5$, the Blob context is extended by the relational properties *call:c0*, *call:c2*, *call:c5* and *call:c11*. Table 1 presents the integration of the relation *call* to the Blob context.

The scaling mechanism is only one step in the global analysis process which, given a RCF, yields a set of lattices, one per context, called *relational lattice family* (RLF). The RLF is defined as the set of lattices whose concepts jointly reflect *all* the shared properties and links among individuals of the RCF. Its construction is an iterative process because the scaling mechanism modifies contexts and thereby the corresponding lattices, which in turn may require a new scaling to reflect the newly formed concepts and the link sharing they provoke. Iterations stop whenever a fixed point is reached, *i.e.*, further scaling leaves all the lattices in the RLF unchanged.

Lattice evolution is illustrated though the analysis of the Blob RCF in Fig. 2: RCA yields the concept lattice illustrated in Fig. 3. The final lattice of the Blob is different from the initial one due to the relational information inserted into the scaled version of the Blob context. Indeed, the individuals are assigned relational properties that lead to the sharing of more properties among these individuals. By factoring out the new properties into concept intents, links between individuals are lifted up to the concept level, yielding relations between concepts[3]. Thus, in Fig. 3, previously existing concepts obtain new properties while completely new concepts emerge. For example, the concept $c16$ that represents the method `sort_catalog()` has been assigned the relational properties `call:c0` and `call:c11`, which means that `sort_catalog()` calls methods in the extent of concept $c0$ and $c11$, namely `add_book()` and `remove_book()`. Furthermore, methods `borrow_Book()` (concept $c3$) and `reserve_Book()` (concept $c12$) have top concept as immediate successor in the initial lattice. Their link with the method `check_Availability_-Book()` (concept $c4$) has been revealed through scaling. They form a new concept $c19$ (see Fig. 3) that represents the set of methods that call `check_Availability_-Book()`.

---

[3] Observe that for compactness reasons, only non-redundant relational properties are visualized in concept intents, *i.e.*, those referring to the most specific concepts.

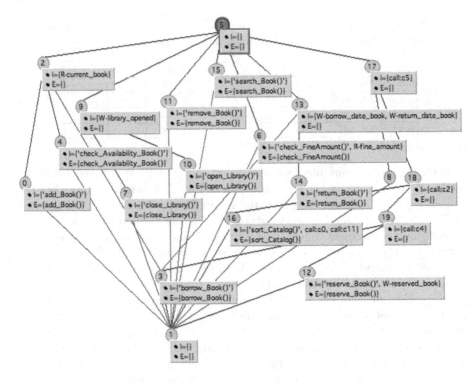

Fig. 3. The lattice of the context of methods shown in Fig. 2

4 Correction of Design Defects Using RCA

Our intuition is that design defects resulting in high coupling and low cohesion could be improved by redistributing class members among existing or new classes to increase cohesion and/or decrease coupling. RCA provides a particularly suitable framework for the redistribution because it can discover strongly related sets of individuals with respect to shared properties and inter-individual links and hence supports the search of cohesive subsets of class members. Fig. 4 depicts our approach for the identification of refactorings to correct design defects in general and the Blob in particular. It shows the tasks of detection of design defects and of correction of user-validated defects.

4.1 Overall Process

We define a three-step RCA-based correction process that follows a two-step defect detection process. First, we build a model of the program that is simpler to manipulate than the raw source code and therefore eases the subsequent activities of detection and correction. The model is instantiated from a meta-model to describe OO programs. Next, we apply well-known algorithms based on metrics and–or structural data on this model to single out suspicious classes having potential design defects [17]. For each suspicious class, we automatically extract a RCF that encodes relationships among class members from the model of the program. Then, the obtained RCF is fed into a RCA engine that derives the corresponding concept lattices. Finally, the discovered concepts are explored using some simple algorithms, which apply a set of refactoring rules that allow the identification of cohesive sets of fields and methods. The approach suggests a set of refactorings that jointly amount to splitting the Blob into as many classes as there are cohesive sets and merge the content of the surrounding classes with the new classes whenever appropriate.

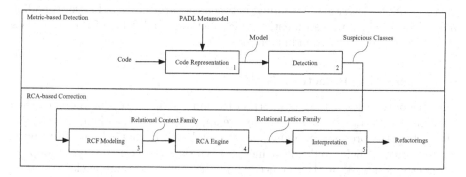

Fig. 4. RCA-based Workflow for the Detection and Correction of Design Defects

4.2   RCF Extraction

To correct design defects, we need to identify cohesive sets of methods with respect to the mode of usage of fields, *i.e.*, read or write, and call between methods. Hence, the individuals are methods of the large class and properties are its fields. The incidence relation represent the access of fields in read/write mode. In order to differentiate between the two access modes, the prefixes R- and W- are added to the name of the fields as illustrated in Fig. 2. Method invocations within the large class are encoded by a dedicated inter-individuals relation denoted *call* (see the table in the right hand side of Fig. 2).

The formal attributes were derived from names of methods and added to the method context. These attributes allow the emergence of a single concept for each method, called *method concept*[4], in the corresponding lattice. Beside listing the entire set of properties of a given method, the concept method helps preserving one-to-one invocation between methods. These details can be lost during the scaling step that aims at integrating the relation *call* into the context of the large class by substituting one-to-many invocations for those of type one-to-one.

---

[4] The smallest extent in the lattice containing this method.

4.3 Deriving the Lattice

Fig. 3 represents the concept lattice obtained by the RCF engine from the context given in Fig. 2. The concepts of the lattice represent the refactoring opportunities of the design defect. Indeed, concepts such as $c9$ exhibit group of methods using the same sets of fields and fields used by cohesive sets of methods. These concepts are considered as class candidates because they are cohesive. In addition, concepts such as $c3$ and $c12$ highlight subsets of cohesive methods, because methods calling the same set of other methods are highly cohesive. A third category of concepts such as $c9$ and $c13$ represent the *use-relationship* between methods of the large class and the surrounding data classes. The study of these concepts allow to assess the coupling between the large class and its surrounding data classes. Thus, we can identify which methods and fields of the large class should be moved to surrounding classes.

4.4 Suggesting Refactorings

The RLF of the Blob is used to interpret the inner structure of the Blob and then suggest refactorings. More specifically, we apply algorithms looking for concepts that reflect the presence of highly cohesive and weakly coupled sets. Intuitively, shared usages of fields and calls of methods is a sign of cohesion whereas coupling is directly expressed by the reliance of a method on a surrounding class (method and-or field). Following these design guidelines, we correct the Blob in two ways. First, we move disjoint and cohesive sets of methods and-or fields that are related to a data class in that data class. Two refactorings describe such migration between classes: *Move Method* [5, p.142] and *Move Field* [5, p.146]. Second, we organise cohesive subsets that are not related to data classes in separate classes. In addition to the two previous refactorings, we use the refactoring *Extract Class* [5, p.149], which consists in creating a new class and moving the chosen fields and methods from the old class to the new class using the two first previous refactorings.

We have specified three refactoring rules to build incrementally cohesive sets by visiting the concept lattice of methods. These rules are applied in sequence, *i.e.*, we apply the two first rules that deal with the access of fields by methods in read-write mode and then the rule that handle method calls.

Rule 1. Methods accessing in write mode the same set of fields are gathered in a single cohesive set.

Rule 2. Methods accessing in read mode the same set of fields are gathered in a single cohesive set if the number of common fields that they access is higher than the number of fields they access separately.

These two rules are inspired from the object identification approach described in [23] where grouping of methods is based on the accessed fields, with respect to the number of fields they access separately. The obtained cohesive sets are merged according to the following rule:

Rule 3. Methods that call the same set of methods are put in a single cohesive set if the number of jointly called methods is higher than the number of methods called separately.

For example, by applying the three previous rules on the running example of the library Blob class, we obtain several cohesive sets as illustrated in Fig. 5, on the left. The cohesive sets that should be migrated in the data classes are shown in Fig. 5 on the right. This last step is currently performed manually but planned to be automated.

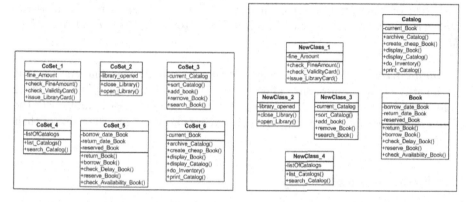

Fig. 5. Left: The cohesive sets obtained from class `Library_Main_Control` depicted in Fig. 1. **Right:** Moving these cohesive Sets to existing data-classes or new-classes.

We provide the implementation details of these rules in the following.

Implementing Rule 1. We iterate the lattice and record all concepts related to fields with the prefix W-. We mark all these concepts as visited. We sort this list in reverse order by the number of fields with W-. Thus, fields that are accessed in write mode by a high number of methods are processed first. For example, the concept $c3$ in Fig. 3 is processed first because of the related concept $c13$. For each concept of the list, we create a new cohesive set and apply the method APPLYRULEWRITE(). This method consists in moving the current(s) field(s) (`borrow_date_book` et `return_date_book` in concept $c13$) with the refactoring *Move Field* and for each method in the intent of the current concept (`borrow_-Book()`) that has not yet been included in a set (*i.e.*, not yet visited), we move it to the current cohesive set using the refactoring *Move Method*. Then, recursively, we check the parents of the current concept and the children of a parent if interesting to explore. The children of a parent are interesting to explore if the parent contains at least one W- field also contained in the current concept. For example, only the children of the parent $c13$ of the concept $c3$ are interesting to explore. We reapply the rule APPLYRULEWRITE() on the children.

Implementing Rule 2. This rule consists in finding the best cohesive set of methods that access to a common set of fields in read mode. For each concept related to common fields in read mode and not yet visited *i.e.*, not processed when applying Rule 1, and thus not included in a set, we calculate a ratio. The ratio corresponds to the number of fields in common with their total number of fields. We calculate the mean of all the ratios corresponding to each concept and retain only groups of concepts that have a mean higher than 0.5, *i.e.*, concepts whose methods accessing a common number of fields is higher than their own number of fields in average. We obtain thus a list of candidate sets of concepts that we sort in reverse order to process first concepts with a greater ratio. For each sets of concepts, we create a new cohesive set by moving the methods and fields with the respective appropriate refactorings (*Move Field* and *Move Method*).

Implementing Rule 3. This rule is similar to rule 2. The difference is that we identify common methods called by one or several methods of the resulting cohesive sets built from Rules 1 and 2. We calculate also a ratio and select the best candidates, and then merge the cohesive sets according to the value of their ratio.

5   Experimental Study

We use PADL [10] to model source code and GALICIA v.2.1 [20], to construct and visualize the concept lattices. PADL is the meta-model at the heart of the PTIDEJ tool suite (*Pattern Trace Identification, Detection, and Enhancement in Java*) [8]. GALICIA is a multi-tool open-source platform for creating, visualizing, and storing concept lattices [20]. Both tools communicate by means of XML files describing data and results. Thus, an add-on to PTIDEJ generates contexts in the XML format of GALICIA, which are then transformed by the tool into lattices and shown on screen for exploration.

In order to validate the proposed approach for the detection and correction of Blob design defects, we consider four different open-source programs. We use *freely available* programs to ease comparisons and replications of our experiments. We provide some information on these programs in Table 2.

In Azureus, we found 41 Blobs by applying our detection algorithms. We notice that the underlying classes are difficult to understand, maintain, and reuse because they have a large number of fields and methods. For example, the class `DHTTransportUDPImpl` in the package `com.aelitis.azureus.core.-dht.transport.udp.impl`, which implements a distributed sloppy hash table (DHT) for storing peer contact information over UDP, has an atypically large size. It declares 42 fields and 66 methods for 2,049 lines of code. It has a medium-to-high cohesion of 0.542 and a high coupling of 81 (8th highest value among 1,626 classes). The data classes that surround this large class are: `Average`, `HashWrapper` in package `org.gudy.azureus2.core3.util` and `IpFilterManagerFactory` in package `org.gudy.azureus2.core3.ipfilter`.

Table 2. List of Programs

Name	Version	Lines of Code	Number of Classes	Number of Interfaces
Azureus	2.3.0.6	191,963	1,449	546
A peer-to-peer client implementing the BitTorrent protocol				
Log4J	1.2.1	10,224	189	14
A logging Java package				
Lucene	1.4	10,614	154	14
A full-featured text-search Java engine				
Nutch	0.7.1	19,123	207	40
An open-source web search engine, based on Lucene				

Table 3. Blob Classes in Four Different Programs and the Number of Cohesive Sets

System	Blob Class	Size (number of fields and methods)	Lines of code	Cohesion	Coupling	Number of fields and methods moved	Number of cohesive sets	Number of real cohesive sets
Azureus v2.3.0.6	DHTTransportUDPImpl	(42+66) 108	2,049	0.542	81	(27+32) 59	10	7
	DHTControlImpl	(47+80) 127	1,868	0.52	67	(35+62) 97	19	11
	TRTrackerBTAnnouncerImpl	(36+47) 83	1,393	0.948	54	(24+33) 57	16	5
Log4j v1.2.1	LogBrokerMonitor	(29+105) 134	1,591	0.479	86	(23+85) 108	31	17
	Category	(9+53) 62	1,042	0.831	46	(8+44) 52	18	9
Lucene v1.4	IndexReader	(7+52) 59	593	0.661	68	(5+30) 35	4	2
	QueryParser	(36+48) 84	1,085	0.3	26	(24+37) 61	13	10
Nutch v0.7.1	FSNamesystem	(24+35) 59	1,211	0.908	23	(17+25) 42	18	9
	JobTracker	(22+31) 53	910	0.938	21	(17+18) 35	11	8

Table 3 provides the results of applying our rules on three different Blobs classes detected in Azureus and on two Blobs classes in the three other programs. It is noteworthy that the results provided by our method have been assessed manually: Among the set of all cohesive sets in the output we identified those whose semantics could be clearly established and it confirmed their cohesiveness. A measure for the precision of our method is the ratio of the real cohesive sets to the total number of sets output by the method. As Table 3 indicates, the precision may vary within a wide range (from 30 to 70 % of correct guesses). The cohesive sets suggested by our approach include an important number of small cohesive sets, which include generally at most one field and one or two methods. This explains why we did not get a good precision. The other concise sets gather between 10 and 20 fields/methods and are good candidates for the creation of new classes because they define a specific responsibility or semantics.

To increase the robustness of our approach, we need to define additional rules related to the access of fields and methods by methods not only within one class but also located in other associated classes. Moreover, our analysis is purely static. Thus, we need to enhance our method with a dynamic analysis to preserve the behavior of the program. Finally, the restructuring should be semi-supervised by an expert because only experts could assess the relevance of grouping elements. The method should be seen as a support for restructuring huge number of data. Thus, we share Snelting's opinion that an interactive restructuring performed by the software engineer is more appropriate [24].

6   Related Work

Few studies have explored the semi-automatic correction of design defects. Thus, we only list work related to design defects and to the use of FCA in software maintenance.

Sahraoui et *al.* in [23] proposed an approach for identifying objects in *procedural* code, a problem that is similar to the split of a Blob (in this case the Blob corresponds to the entire application or to a module thereof). The approach combines metrics calculation with several FCA-based analysis steps in class identification and further graph-based reasoning to detect associations among newly identified classes.

Snelting and Tip [24] proposed a FCA-based method for adapting a class hierarchy to a specific usage thereof. It comprises a study of the way class members are used in the client code of a set of applications. The study enables the identification of anomalies in the design of class hierarchies, *e.g.*, class members that are redundant or that can be moved into a derived class. In contrast, we detect design defects at a higher level as specified in the literature. Moreover, beyond pure hierarchies, we are interested in classes with associations.

Godin and Mili [7] used concept lattices for class hierarchy redesign based on classes signatures. Yet like [11], they find useful hierarchy restructuring and member redistribution but ignore any possible relationships among the members of a class.

Marinescu [16] presented an approach based on *detection strategies* which applies metrics computation. Combinations of metrics through filtering and composition are used to capture deviations from good design principles and heuristics. Yet the method is inherently limited as design flaws admit no easy detection exclusively by metrics: the *structure* of a design matters and it is impossible to capture in numbers. In contrast, our approach relies on a combination of metrics for the detection of design defects with a clustering and visualisation technique, FCA, that allows the design structure to be fully comprehended.

The work of Kirk et *al.* [12] comes close to ours. Yet they use attribute slicing to refactor large classes, *i.e.*, they slice the variable set of the class into subsets based on the usage of variables by methods. The approach was designed to deal with the Large Class code smell and hence has a scope of a single class whereas Blob involves multiple classes. Conversely, they use an intra-method slicing techniques that allows the precise set of instruction manipulating a instance variable to be detected and the isolated. The practical validation of the approach is yet to be done.

Tonella and Antoniol used FCA to infer recurring patterns in program models [26]. Their study yielded impressive results in terms of groups of classes having common structural relations. However, their approach seems of limited interest for us because it detects only structural relations, whereas design defects are often characterised by measurable properties (*e.g.*, a large class has *a large number* of fields and methods). FCA is not devised to deal with numerical measurement, hence it could benefit from metrics-based techniques.

Arévalo *et al.* [1] applied FCA to identify implicit dependencies among classes in program models (extracted from source code). A set of views at different levels of abstraction are built: At the class level, views show the access of methods to variables and the patterns of calls among methods in a class, hence they help to assess the class cohesion. At the class hierarchy level, views highlight common and irregular forms of hierarchies so as to deduce possible refactorings. At the program level, they refined and extended the approach of Tonella *et al.* to any (recurring) regularities such as design patterns, architectural constraints, idioms, etc. Our approach is similar in that it detects flaws, but our choices of the elements and properties to be analysed are guided by the descriptions of the defects.

7 Conclusion

We proposed an approach that uses RCA to suggest appropriate refactorings to correct certain design defects. In particular, we showed how our approach can help refactoring programs with Blob design defects. Unlike other FCA-based restructuring approaches, we worked on whole lattice regions rather than on separate concepts because candidate refactoring are inferred from several concepts in the lattice. We illustrated our approach using an example of a Library management system and validated it on Azureus v2.3.0.6 and three other programs. We showed that using RCA, our approach could suggest relevant refactorings to improve the program. The generalisation of our results to other design defects is briefly discussed and will be developed in future work. Future work will also include assessing more programs via our approach and discussing the proposed refactorings with their developers and apply them. We also plan to performed quantitative studies on the trade-off between cohesion and coupling.

References

1. Arévalo, G.: High Level Views in Object Oriented Systems using Formal Concept Analysis. PhD thesis, University of Berne (January 2005)
2. Du Bois, J.V.B., Demeyer, S.: Refactoring - improving coupling and cohesion of existing code. In: Proceedings of WCRE, pp. 144–151 (2004)
3. Chidamber, S.R., Kemerer, C.F.: A metrics suite for object oriented design. IEEE Transactions on Software Engineering 20(6), 476–493 (1994)
4. Fenton, N., Pfleeger, S.L.: Software metrics: a rigorous and practical approach, 2nd edn. PWS Publishing Co. (1997)
5. Fowler, M.: Refactoring – Improving the Design of Existing Code, 1st edn. Addison-Wesley, Reading (1999)
6. Godin, R., Valtchev, P.: Formal concept analysis-based hierarchy design in oo software development. In: Ganter, B., Stumme, G., Wille, R. (eds.) Formal Concept Analysis. LNCS (LNAI), vol. 3626, pp. 304–323. Springer, Heidelberg (2005)

7. Godin, R., Mili, H.: Building and maintaining analysis-level class hierarchies using galois lattices. In: Proceedings of OOPSLA, pp. 394–410 (1993)
8. Guéhéneuc, Y.-G.: A reverse engineering tool for precise class diagrams. In: Proceedings of CASCON, pp. 28–41 (2004)
9. Guéhéneuc, Y.-G., Albin-Amiot, H.: Using design patterns and constraints to automate the detection and correction of inter-class design defects. In: Proceedings of TOOLS, pp. 296–305 (2001)
10. Guéhéneuc, Y.-G., Albin-Amiot, H.: Recovering binary class relationships: Putting icing on the UML cake. In: Proceedings of OOPSLA, pp. 301–314 (2004)
11. Huchard, M., Leblanc, H.: Computing interfaces in java. In: Proceedings of ASE, pp. 317–320 (2000)
12. Kirk, D., Roper, M., Walkinshaw, N.: Using attribute slicing to refactor large classes. In: Seminar Proceedings of Beyond Program Slicing, vol. 05451 (2006)
13. Lanza, M.: CodeCrawler—Lessons learned in building a software visualization tool. In: Proceedings of CSMR, pp. 409–418 (2003)
14. Lanza, M.: Object-Oriented Reverse Engineering – Coarse-grained, Fine-grained, and Evolutionary Software Visualization. PhD thesis, Institute of Computer Science and Applied Mathematics (May 2003)
15. Marinescu, R.: Measurement and Quality in Object-Oriented Design. PhD thesis, Politehnica University of Timisoara (October 2002)
16. Marinescu, R.: Detection strategies: Metrics-based rules for detecting design flaws. In: Proceedings of ICSM, pp. 350–359 (2004)
17. Moha, N., Guéhéneuc, Y.-G., Leduc, P.: Automatic generation of detection algorithms for design defects. In: Proceedings of ASE (2006)
18. Moha, N., et al.: Using FCA to suggest refactorings to correct design defects. In: Proceedings of CLA (2006)
19. Object Technology International / IBM. Eclipse platform – A universal tool platform (2001)
20. Galicia (September 2005), http://sourceforge.net/projects/galicia/
21. Raţiu, D., et al.: Using history information to improve design flaws detection. In: Proceedings of CSMR, pp. 223–232 (2004)
22. Riel, A.J.: Object-Oriented Design Heuristics. Addison-Wesley, Reading (1996)
23. Sahraoui, H.A., et al.: A concept formation based approach to object identification in procedural code. Automated Software Eng. 6(4), 387–410 (1999)
24. Snelting, G., Tip, F.: Understanding class hierarchies using concept analysis. ACM TOPLAS 22(3), 540–582 (2000)
25. Azureus (June 2003), http://azureus.sourceforge.net/
26. Tonella, P., Antoniol, G.: Object oriented design pattern inference. In: Proceedings of ICSM, pp. 230–240 (1999)
27. Trifu, A., Dragos, I.: Strategy-based elimination of design flaws in object-oriented systems. In: Proceedings of WOOR (2003)
28. Brown, W., et al.: AntiPatterns Refactoring Software, Architectures, and Projects in Crisis. Robert Ipsen (1998)

Contingency Structures and Concept Analysis

Alex Pogel and David Ozonoff

Physical Science Labortory, New Mexico State University, Las Cruces, NM 88003,
USA
Department of Environmental Health, Boston University School of Public Health,
Boston, MA 02118, USA
Alex.Pogel@psl.nmsu.edu, dozonoff@bu.edu

Abstract. Formal Concept Analysis has found many uses in knowledge representation and data mining, but its penetration into established data-based research disciplines has been slower. Marrying application motivations, structures, and methods from epidemiology and the mathematical formalisms of FCA, we define *Generalized Contingency Structures* and *Tagged Contingency Structures*, two new objects that generalize the contingency table, an ad hoc data summarization device in epidemiology, to a mathematical object with well-understood structure. We have extended the FCA repertoire by adding to the Formal Context an associated structure that we call a Tag Context, which formally incorporates important kinds of background knowledge. We illustrate the motivation and use of these new ideas, formats, and objects with some brief examples.

The relationship between mathematics and its applications is most productive when each enriches the other. A mathematical formalism may be grafted entirely onto an application without modification of either the application or the mathematics - a conceptual isometry - but this is unusual and provides little new to either partner in the complex dance that is applied mathematics. We describe here an interpretation of mathematical structures from Formal Concept Analysis into epidemiology through which we expect the ideas and repertoire of each discipline will be broadened and taken in new directions.

1 Epidemiology

Epidemiology is a core discipline in public health. But as Rothman and Greenland observed in one of the major texts in the field [RG], there seem to be even more definitions of epidemiology than epidemiologists. One widely used definition is the study of the patterns and determinants of disease in populations.

Describing the *patterns* of disease in populations is an epidemiological subproject often referred to as *descriptive epidemiology*. Patterns are descriptions of how particular features in the data relate to each other. Features in descriptive epidemiology are given in terms of attributes of an individual (person), such as an individual's location (place), or when the individual's feature is observed

R. Medina and S. Obiedkov (Eds.): ICFCA 2008, LNAI 4933, pp. 305–320, 2008.
© Springer-Verlag Berlin Heidelberg 2008

(time or position in a sequence), or summary measures of these over the entire population (e.g., median age). The patterns of health states by person, place, and time are further described by how they co-occur (Who is getting the disease? When? Where?) or how some health states are functionally related to others (e.g. how a person's blood pressure is related to their weight, or where cancer occurs on a map). These data are much used for administrative purposes, for example to support decisions on the need for facilities, to estimate the cost of services, or to plan for targeting high risk groups, like pregnant women.

Thus in many respects there is a straightforward relationship between conventional descriptive epidemiology (the co-occurrence of variable levels) and typical applications of Formal Concept Analysis (FCA) and association rule mining; only here the question is focused on which epidemiological features are associated. On the other hand, epidemiology has its own data structures and conventions, due to its particular set of motivating principles and corresponding historical development. One of these is the widespread use of contingency tables to prepare and analyze data, and there are many methods that flow from this format. Later in this paper, we describe how the concept lattice can be viewed as a more general version of a contingency table, and also introduce some analytic procedures that reflect epidemiologists' use of contingency tables.

Description is not the only goal in epidemiology, nor even the main goal. As Rothman and Greenland [RG] suggest, the main goal is usually to understand the *determinants* of the patterns:

> If the subject of epidemiologic inquiry is taken to be the occurrence of disease and other health outcomes, it is reasonable to infer that the ultimate goal of most epidemiologic research is the elaboration of causes that can explain patterns of disease occurrence.

Fixing the determinants of the pattern is the subject of *analytic epidemiology*, Rothman's "elaborations of causes." This, too, is in the language of co-occurrence or functional relationships, but now with directionality, from cause to effect (never the other way around). Once the mysterious notion of causality enters the picture with its directionality requirement, we need to employ theoretical guesses or assertions about "how the world works." We are not only carving out certain features of the world (in descriptive epidemiology, health status according to person, place, and time) but also employing background knowledge such as tagging some of these features as explanatory variables or outcome variables (causes or effects, respectively), or ancillary variables that are not of primary interest but needed to understand the relationship of cause and effect (covariates; age and gender are prime examples). Along with identifying these categories of variables, we are in addition asserting that only specific forms of associations meet the directionality requirement.

Causality [or as philosophers refer to it, causal necessity] is a slippery philosophical concept. Most scientists do not fret over it unduly, believing that like pornography, "they'll know it when they see it." But causality is a species of association, i.e., a regular conjunction of certain *types* of events. We know many such associations are not causal in nature. There are two traditions of scientific thinking regarding the hallmarks of causal associations [Spir], one traceable to the Bernoullis and the other to Bacon and J. S. Mill. The first identifies causality with dependence (or lack of independence): one thing causes another if the second is somehow dependent upon the first. This is the core of the statistical approach. The second tradition is consistent with the first but adopts a more empirical perspective: X causes Y if a change in X is always accompanied by a change in Y, all other things being equal. The origin of this notion rests in control. If you change the cause variable, all other things being equal, you expect to see a change in the effect variable. The hallmark causal investigation in this tradition is The Experiment. The investigator controls the independent variable and observes the effects on other variables, all other things kept the same. Of course all other things are rarely equal, so additional techniques to take into account uncontrolled differences between the "cause-on" and "cause-off" experiments are often required. When these other factors are unknown or not easy to control, the crucial maneuver is to randomly assign the causal variable to two different groups. This allows the use of well-recognized statistical techniques to evaluate whether differences in outcome, possibly arising upon application of the posited causal factor, are due to the factor or could be due to some uncontrolled differences between the two groups.

Important branches of epidemiology employ experimental designs that allow reasonably precise statements about the relationship between cause and effect. For example, randomized clinical trials are used to determine the effect of various treatments on disease states. More often, however, we cannot control the treatment ("cause") variable for ethical or practical reasons and we are left to observe real world events. In essence, we are looking for natural experiments, such as comparing the health experience of asbestos workers with the health experience of the general population. While we did not control who got exposed and who didn't, we are able to observe what happened to those who worked with asbestos compared to those who didn't. We then arrange our observations to be as useful as possible, often using mathematical tools such as statistics to make inferences about how the world works.

At the heart of these causal judgments is a comparison of various combinations of variable states. One group has the causal factor while the other does not, but in the real world there are usually many other differences as well. Conventional statistical techniques employed in epidemiology compare the relative frequencies of the different combinations of explanatory variables (causes) and response variables (effects). Another approach would be to employ techniques whereby comparison and co-occurrence are natural ideas and supported by relational structure, and Order Theory and FCA [DP, GW] provide well-developed theories to build upon. We present in this paper an interpretation of epidemiologic constructions in the concept lattice and discuss ways of introducing distinctions between cause, effect, and covariates to allow conceptual exploration of the dataset in pursuit of the main epidemiological project of discerning the determinants of patterns.

2   Contingency Tables in Epidemiology

We enter the epidemiological arena at the point where data on a population of individuals have been collected and arrayed in a table giving the value of a health status indicator for each. In FCA this is a Formal Context, the result of conceptual scaling of multi-valued data, and the starting point from which concept lattices are constructed for various purposes, such as conceptual exploration, knowledge discovery in data mining, and association rule mining (ARM). In epidemiology further processing is usually done, so that the data is displayed with less resolution but more clarity (for epidemiologists), in the form of a contingency table. Contingency tables are widely used in epidemiology to prepare categorical data for further exploration. The underlying idea is to display the co-occurrence of health status of individuals in a study or target population, as the first step in looking for or describing patterns of health status indicators by person, place, or time, or for discerning causal associations between them.

A two-way contingency table classifies each object (in epidemiology the objects are almost always individual subjects) by two different health status variables. The variables are partitioned into levels or factors (by the process of scaling, in FCA terms) - for example, male and female are levels or factors of the variable *sex*, while 0, $1 - 10$, $11 - 20$, $21 - 100$ are levels or factors of the variable *cigarettes smoked per day* (abbreviated by CD in the sequel). The content of a table cell is the cardinality of the set of objects that satisfy the conjunction of the particular attribute levels of the variables that specify the cell in question. If there are three variables of interest we have a three-way contingency table, a "cubical" table, while four-way and higher tables can only be shown on paper by slicing them into two- or three-way tables and displaying the slices separately, or displaying them recursively, by nesting tables within cells.

While this is the first step in an epidemiologic exploration of the data, several study design commitments have already been made prior to construction of the table. The n subjects have been gathered together; health status variables of interest have been selected; the variables for each subject have been assigned a value by measurement; and the variables (features or attributes) have been scaled by partitioning them into nominal or ordinal categories (known as conceptual scaling in FCA). Probability models assume various processes for generating the data. If the final population of n subjects is arrived at by counting the occurrence of a particular health status measurement during a fixed time period, we often speak of Poisson sampling. If the number of n randomly selected subjects has been fixed ahead of time, we speak of multinomial sampling. If the marginal totals of either the rows or columns are fixed ahead of time we talk of independent or product bi/multinomial sampling, while if the margins of both rows and columns are fixed, we have hypergeometric sampling. The sampling scheme is not signaled by the form of the table. To identify it, you also need to know how the subjects were obtained. Nor does the form of the table reveal any information about the accuracy or reliability of the measurement, that is, whether the assigned value of a variable level to an individual is both correct and repeatable.

Another aspect of contingency tables that is not obvious at first is that each dimension of the table is a single variable and its levels are a partition of possible outcomes of measuring the variable in an individual. Because a partition is mutually exclusive and exhaustive, the margins of the tables - the sum of all the cells for each level of a variable - have a special meaning and can be used for further analysis. For example, a common statistical question is whether the rows and columns are independent of each other in the sense that a particular level that specifies a row does or does not indicate a high likelihood that those cases will fall into a particular column. Pearson's chi-square statistic is a typical statistical test used for this purpose, and it was observed in [SW] and elsewhere that Pearson's statistic can be used in the concept analysis context to determine dependence between two scales. Some common measures of association in epidemiology, such as the odds ratio (OR), can be used appropriately for any sampling scheme, while others, such as the relative risk (RR), only with some. The OR is also symmetric with regard to directionality while the RR is not. For many common situations, the OR and RR are very close to each other.

We now step back from the initial epidemiological maneuver, the construction of a low dimensional contingency table whose variables are controlled by convention, hypothesis, or habit, and go back to the formal context, the epidemiologist's raw data.

3   Generalized and Tagged Contingency Structures

Standard statistical tests for contingency tables assume that the levels partition the set of values of the variables. An immediate generalization is to relax that requirement, allowing the factors or levels of each variable to overlap or nest, as is common in various ordinal scales. In usual epidemiological practice, a single contingency table, whether one-way (which is just a variable frequency table), two-way, or multiway (cross-tabulations), is still a partial view of some combinations or patterns among features in the dataset, so another generalization is to allow all possible conjunctions of attributes. Overall, we lose nothing useful in contingency table analysis while gaining generality and the possibility of using the language and theory of conceptual scaling within FCA. We now recall two generalizations of the contingency table presented in [OPH], and follow these with some new definitions. We write $\mathbb{N}$ for the set of non-negative integers, $\mathbb{K} = (G, M, I)$ for a standard formal context, and $Int(\mathbb{K})$ for the set of all concept intents of $\mathbb{K}$.

Definition: The *generalized contingency table* (GCT) of $\mathbb{K}$ is the function $t : \mathcal{P}(M) \to \mathbb{N}$, where $t(A) = |A'|$, for all $A \in \mathcal{P}(M)$; the *closed set contingency table* (CSCT) of $\mathbb{K}$ is the function $T : Int(\mathbb{K}) \to \mathbb{N}$ where $T(A) = |A'|$, for all $A \in Int(\mathbb{K})$.

Like the conventional contingency table, the GCT and CSCT include only the cardinality of the set of subjects that share a conjunction of levels of one or more variables. The GCT represents a superset of the cells of any other conventional contingency table and is a rich mathematical object in its own right, unlike a single conventional contingency table, which is only a fragment of the GCT with no particular properties of its own. In the CSCT we clearly seek the same practical gains from the reduction to closed attribute sets that makes FCA so attractive in various knowledge representation and association rule mining algorithm developments [Pea, ZO]. In terms of expressive power, we can naturally extend the GCT and CSCT by replacing the count of A cases with A' itself, the extent of the concept generated by A. This yields modifications of the definitions above, where we change "table" to "structure" to reflect the use of extents, in place of counts:

Definition: The *generalized contingency structure* of a formal context (G, M, I) is the function $t : \mathcal{P}(M) \to \mathcal{P}(G)$ defined for each $A \in \mathcal{P}(M)$ by $t(A) = A'$.

Recall a semiconcept is a pair (A, B), with $A \subseteq G$, $B \subseteq M$, such that $A \subseteq B'$ or $B \subseteq A'$ [GW]. Thus, in FCA terms, an alternative description is that the generalized contingency structure (GCS) is the set of all semiconcepts of (G, M, I) of the form (B', B), where $B \subseteq M$. Like the GCT, the GCS provides a global view of co-occurrences of variables and their levels in the data, but also includes a record of the cases that witnessed those co-occurrences. While there is an obvious extension of the CSCT, to a definition of a *closed set contingency structure*, this is entirely unnecessary, as this structure *is precisely the concept lattice*, the set of ordered pairs (A', A), where $A \in Int(\mathbb{K})$.

Thus far, by directly generalizing (GCT), extending (GCS), and reducing (from GCS to the concept lattice), we have found that the concept lattice is a generalization of the contingency table, a major format for data reduction, display, and analysis in epidemiology. Because we include the actual object sets (the individual elements in the power set) and not just their cardinality, the GCS and concept lattice include all the information in the formal context.

However, we have still not captured the background knowledge encoded in the dimensions of a conventional contingency table. To provide a true generalization of the contingency table, we must not only include support for additional structural features (such as object sets), but also include all that is tacitly implied to an epidemiologist with a conventional contingency table. This means we must incorporate into the GCS and the concept lattice those summaries and analytical actions an epidemiologist would make with a conventional contingency table, including the directionality of associations, via recognition of the status of constituents as cause, effect, or covariate. To remedy this deficiency, and support later rule measure definitions that will capture the directionality requirement for causation, we introduce tag contexts.

Definition: a *tag context* of (G, M, I) is a formal context (M, C, J).

The intention behind this definition is that M is the set of attributes of (G, M, I), C is a set of classes (or *tags*) to which the attributes in M may belong, and J indicates class membership. With this tag context we may easily identify subcontexts of (G, M, I) in terms of any intent D of (M, C, J), by setting $(G, M, I)_D = (G, M \cap D^J, I|_{G \times D})$; $(G, M, I)_D$ is the restriction of (G, M, I) to the set D^J of attributes in the extent of concept (D^J, D) of (M, C, J). Many other works have proposed similar augmentations to the base formal context. The appended tag context is similar to the *taxonomy* employed in [CFRD], used there to avoid redundancy in intents, except where the taxonomy is a relation on M, the tag context indicates classes (usually disjoint from M) that reflect background knowledge regarding the attributes. Since the tag context shares the set M with (G, M, I), the addition of this auxiliary structure is closely related to the RCA proposal in [RHNV], except that there the family of contexts is augmented with a set of binary relations between the object sets.

As first described in [HKB] and cited in [RS], preliminary data from a health survey in the Netherlands found an association between keeping pet birds and increased risk of lung cancer. To test the association, epidemiologists compared the frequency of bird keeping among 49 lung cancer patients from four hospitals in the Netherlands with bird keeping in 98 urban residents who didn't have lung cancer. They also collected information about age, sex, and cigarette smoking from all the subjects. The goal was to see if bird keeping was more common among lung cancer patients than the general population after accounting for other factors that might affect lung cancer risk. Table 1 shows an example of a tag context (M, C, J) for a formal context (G, M, I) for which $|M| = 20$ ([URL] presents the full Birdkeeping-LungCancer context), and the attribute class set is $C = \{\textbf{Cause, Effect, Covariate}\}$.

Table 1. A tag context for the Birdkeeping-LungCancer context posted at [URL]

	Cause	Covariate	Effect
Is A Birdkeeper (BK)	X	-	-
Is not a Birdkeeper (¬BK)	X	-	-
Cigarettes per day (CD) ≤ 20	-	X	-
Cigarettes per day (CD) ≤ 10	-	X	-
Cigarettes per day (CD) ≥ 20	-	X	-
Cigarettes per day (CD) ≥ 10	-	X	-
Years Smoking (YR) ≤ 20	-	X	-
Years Smoking (YR) ≤ 10	-	X	-
Years Smoking (YR) ≥ 20	-	X	-
Years Smoking (YR) ≥ 10	-	X	-
Has Lung Cancer (LC)	-	-	X
Does Not Have Lung Cancer (¬LC)	-	-	X
Socio-economic Status High (SS)	-	X	-
Socio-economic Status Low (¬SS)	-	X	-
Age ≤ 60	-	X	-
Age ≤ 50	-	X	-
Age ≥ 60	-	X	-
Age ≥ 50	-	X	-
Sex = Female	-	X	-
Sex = Male	-	X	-

Now, to address the fact that background knowledge is encoded in the dimensions of the contingency table, the causality aims of the study, and the choice of effect measures (measures of strength of association), we augment the previously defined structures with a tag context.

Definition: A *tagged generalized contingency structure* of a formal context (G, M, I) is the GCS of (G, M, I) with a specified tag structure (M, C, J), and a *tagged core contingency structure* of a formal context $\mathbb{K} = (G, M, I)$ is the concept lattice $\mathfrak{B}(\mathbb{K})$ with a specified tag structure (M, C, J).

With either of these *tagged structures*, we have sufficient expressive power to capture standard epidemiological constructions over all the possible contingency tables that can be formed from a given formal context. Later sections of this paper explain how we use these structures to extend standard epidemiological practice to these objects, and we are currently engaged in further work to complete this vision. In [OPH] we connect the concept lattice diagram with the conventional 2×2 contingency table that cross-classifies any two combinations of variable levels. The rows and columns of each 2×2 table present a partition of some variables X and Y, into attributes A and B, respectively, along with their complements $\neg A$ and $\neg B$, respectively. We focus on four particular cells: the upper left interior cell corresponds to $|A' \cap B'|$, the upper right and lower left cells are the marginals $|A'|$ and $|B'|$, respectively, and the bottom right corner (the grand total) is the size $|G|$ of the universal set G of subjects. Consequently, all cells of such a contingency table are determined by the values of only these four of its nine cells (the nine cells are the four interior cells, the four marginal totals on the rows and columns, and the grand total). If one draws a Hasse diagram of the order relations of these four cells, using the standard concept lattice convention of ordering attribute sets by reverse inclusion, this contingency table has a naturally associated subdiagram (a diamond consisting of (G, G'), (A', A''), (B', B''), and $((A \cup B)', (A \cup B)'')$ order embedded in the lattice diagram. An example is shown in Figure 1 below.

Thus, every meet, $(A', A'') \wedge (B', B'')$, in the concept lattice is a particular two-way contingency table that cross-classifies with respect to attributes A and

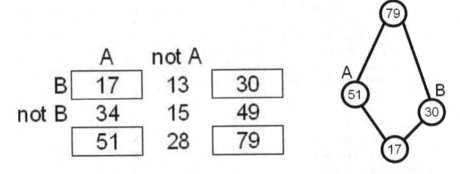

Fig. 1. A 2×2 contingency table and the corresponding Hasse diagram

B, and so *each concept D is an equivalence class of two-way contingency tables* (representing the different meets of pairs of concepts that yield D). Furthermore, every two-way contingency table is (with an appropriate relation, regarding equality of sets of cells) an equivalence class of rules, as, for example, $A \to B$ and $A \to \neg B$ can be viewed as having the same two-way table).

All of the *structures* defined above are new objects of study for epidemiology. Thus FCA has enlarged that discipline by directing its attention to new constructions that have a mathematical and conceptual richness, and various implementations and directions of algorithmic development. The concept lattice provides a minimal global description of the dataset, with no loss of information, and also provides a description of the hierarchical structure of the data, as visualized in the lattice diagram. Moreover, specific subposets of the lattice diagram are connected to the more familiar contingency tables of epidemiology. We now use some of the maneuvers of epidemiology to increase the analytic capabilities of the concept lattice in the epidemiological domain, and also in other domains focused on determination of risk factors [HLK].

4 Epidemiological Association Rule Measures

Association rules are a centerpiece of data mining, and have well-known computational links with the concept lattice, but they are a relatively weak form of knowledge and their very abundance in even small datasets produces a serious problem in applications [Im]. Epidemiologists often find little value in ARM, either because the rules produced are trivial or redundant, or because they express associations of no particular interest for the project at hand. Conversely, there are often specific kinds of associations of particular interest, for example, those relating putative causes with putative effects. We would like to discover associations with particular forms and to test whether they are simply expected expressions of internal combinatorial complexity rather than some external causal necessity.

Epidemiologists work from the bottom up, first examining simple associations between a few variables using low dimensional contingency tables or perhaps using a multivariate model such as logistic regression to incorporate more variables. Even with multivariate methods, however, the number of interaction terms is usually small, and the overall result is model-dependent. Only a highly selected portion of the data is used and this frequently has additional structure imposed upon it for ease of analysis (for example that some relationships between variables are linear). The lattice diagram, on the other hand, has all the richness of the dataset but overwhelms the analyst with too much information and does not incorporate subject-specific content or assumptions to help navigate around features of little interest.

We now describe some ways the concept lattice diagram and association rule lists can be made more useful for epidemiologists. Given the epidemiological interpretation of the concept lattice discussed above, it is important that the viewing of lattice diagrams incorporates two central epidemiological effect measures used in contingency table analysis, namely, the Odds Ratio (OR) and a measure of its precision, the confidence interval. After recalling these two measures, we turn to some new rule measures based upon causation's directionality requirement. The new measures are computed through reference to the tag context that is part of each tagged structure, as defined in the previous section. In the next section we describe how these measures may be used in a highlighting scheme to help view a concept lattice diagram, and finally we display examples of these methods in a software implementation.

Support and confidence both offer natural pruning capabilities in association rule generation and have often been used to define the central problem of ARM (the overwhelming amount of information it generates), despite the limited utility offered by these measures in most applications. Many other association measures can be computed for each conjunctive association rule $A \to B$, including lift, prevalence ratio, conviction, and leverage. Here we recall the definition of two rule measures that are very useful for practicing epidemiologists:

- *Odds Ratio*: odds of B given A divided by the odds of B given $\neg A$, i.e.

$$\frac{\frac{|A' \cap B'|}{|A' \cap (G \setminus B')|}}{\frac{|(G \setminus A') \cap B'|}{|(G \setminus A') \cap (G \setminus B')|}}$$

- (Test-based) *Confidence Interval of Odds Ratio*: The range of values of the odds ratio that would give a p-value greater than a stipulated nominal value (e.g., 5%), assuming only random error, here expressed as [OR95-, OR95+].

Now we use the tag context within any tagged structure defined over (G, M, I) to define two rule measures that capture, to varying degrees of specificity, the directionality requirement for causation:

Definition: Association rule $A \to B$ of (G, M, I) has

- $PremisePurity(A \to B) = \frac{Max_{c \in C}(c^J \cap A)}{|A|}$,
- $ConclusionPurity(A \to B) = \frac{Max_{c \in C}(c^J \cap B)}{|B|}$, and
- $Purity(A \to B) = PremisePurity(A \to B) * ConclusionPurity(A \to B)$.

Definition: If (M, C, J) is the tag context in a tagged structure of (G, M, I), and $C \to D$ is a conjunctive association rule of (M, C, J), we say that conjunctive rule $A \to B$ of (G, M, I) *has the form $C \to D$* provided that $A \subseteq C'$ and $B \subseteq D'$.

Example: The Tag Context in Table 1 is sufficient to express that association rules from (G, M, I) have forms such as Cause→Effect, Covariate→Effect, or Covariate→Cause.

Of course, rules of the simple Forms in this example also have Purity= 1. Nothing in the definition of Form requires such simplicity - any concept intents of

(M, C, J) can be used to express the premise and conclusion constraints for Form. We note that Purity and Form are related measures: those association rules of (G, M, I) with tag context (M, C, J) that satisfy Purity= 1 are precisely those rules of the Form $c \rightarrow d$, for some attributes $c, d \in C$. Also, since Purity is $[0, 1]$-valued, it can be used to express conditions such as Purity$\geq v$ for some fixed value $v \in [0, 1]$, in order to filter a set of association rules to those of sufficiently high Purity.

5   Highlighting Concepts Via Rule Measure Conditions

[Bec] discussed the implementation within ToscanaJ of the annotation of lattice diagrams with results of numerical computations. We have a similar aim in this section, since we want to describe how to highlight particular concepts that witness association rules satisfying specific constraints based on rule measures, and display a simple example. We view concepts as equivalence classes of association rules that are expressed as bipartitions of key sets, and we then discuss how to use highlighting to communicate which concepts and rules in a potentially large lattice diagram satisfy rule measures of interest. For examples of larger lattices, which would be space prohibitive for this paper, see [URL].

We recall that each association rule $A \rightarrow B$ has its support witnessed by the extent of the concept $(A', A'') \wedge (B', B'')$. Of course, multiple association rules are witnessed by any one concept with more than one element in its intent; for a simple example, suppose $A \neq B$, and consider $A \rightarrow B$ and $B \rightarrow A$, and for any $b \in B$, $A \cup b \rightarrow B \setminus b$; all these rules are witnessed by the same concept $((A \cup B)', (A \cup B)'')$. With this observation as a backdrop, we note that our method for generating rules from a concept D is to determine each bipartition $K_1 \sqcup K_2$ of each $K \in \mathcal{K}(D)$, where $\mathcal{K}(D)$ is the set of key (attribute) sets of concept D, and then check whether the rule $K_1 \rightarrow K_2$ satisfies each of a list of rule-based criteria, such as OR ≥ 2, Lift > 1.1 and Purity $= 1$; if any such rule passes, then concept D will maintain its usual color, else it will be colored light gray (thereby highlighting the concepts that pass the set of criteria). Furthermore, for any concepts that remain normally colored (i.e. that are highlighted), a mouse-click on the concept indicates on the lattice diagram (via the standard concept analysis convention of writing attribute labels above the associated concept node) a key set rule that passed the rule measure filter, and at another window, it shows all the rules of this same form that passed the rule measure filter. An example highlighting this method is shown in the last part of this section.

At the current level of development of rule mining in Seqer, the user is provided the ability to select which attributes from the current subcontext are potential premises and which are potential conclusions. For example, in the upper left part of the screen shown below in Figure 2, we have chosen all attributes except LC and ¬LC as Premise attributes, and only LC and ¬LC as Conclusion attributes. This selection screen allows the user to restrict attention to rules that have a particular form, e.g. Cause→Effect. Rules are generated by computing a concept lattice from the premise subcontext (the restriction of the attribute set of the current subcontext to those checked as premise attributes) and similarly computing a lattice for the conclusion subcontext. Only rules of the form $I_p \rightarrow I_c$ are generated and then filtered, where I_p is an intent of a concept in the premise lattice and I_c is an intent of a concept in the conclusion lattice.

If we choose the entire attribute set of the Lung Cancer and Bird Keeping formal context [URL], for both the Premise and Conclusion subcontexts, and we use the filter [Support $\geq$ 0.02, Confidence $\geq$ 0.1, OR 95-$\geq$ 1.1 , OR$\geq$ 1.5, OR95+$\geq$ 2.0, Premise Depth $<=$ 4 and Conclusion Depth $= 1^1$] then the AR algorithm just described returns (502) rules; however, if we add the condition Purity $= 1$ to the filter list, this total of (502) is reduced to (121) generated rules. Not only are there less rules to contend with, but they gain interpretability through the application of purity.

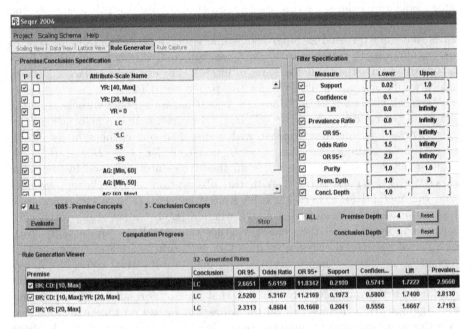

Fig. 2. Form is applied through choice of premise and conclusion attributes, and then the filters applied at right have allowed the selection of the top three OR95- rules from the list of (32)

Similarly, if we enforce form as described above, by removing Lung Cancer from the premises and setting Lung Cancer as the only possible conclusion, and if we do not include Purity $= 1$ (but keep the same filters described above), the ARM algorithm yields (83) rules, whereas adding Purity $= 1$ in this case

---

[1] Here "depth" refers to the cardinality of the set of attributes in the premise or attribute conjunctions in the rule.

reduces the set to only (32) rules. In this case, the reason is that various Cause and Covariate attributes are combining in premises, and these combinations are denied once the Purity = 1 condition is applied.

Seqer allows the set of attributes in any selected set of rules to determine the attributes of a subcontext of the current context, whenever the user requests this additional focus. In the latter table of (32) rules, if we isolate our attention to the attributes involved in the top three rules, namely, the four attributes YR$\geq$ 20, CD $\geq$ 10, BK, and LC, we get a simplified lattice diagram with (16) concepts; Figure 3 indicates how the highlighting scheme functions in this example. Here we have applied rule filter criteria that will highlight only a small number of the concepts in the diagram, including OR95- $\geq$ 1.5, and OR $\geq$ 2 and OR95+ $\geq$ 3, and a few other criteria. In the diagram in Figure 3, Seqer has responded to a user's mouse-click on a highlighted concept by revealing the rule(s) that caused the concept to be highlighted in the first place; here the rule is BK$\rightarrow$LC, following a mouse-click on the dark concept near bottom.

Notice that of the (15) concepts in the lattice diagram, there are roughly (8) concept clusters, after an application of the spring-based lattice drawing algorithm in [HP]. The cluster at the bottom of the lattice includes the universal lower bound and two other related concepts; that universal lower bound is obviously the meet of all four attributes, but it is interesting to note that it has

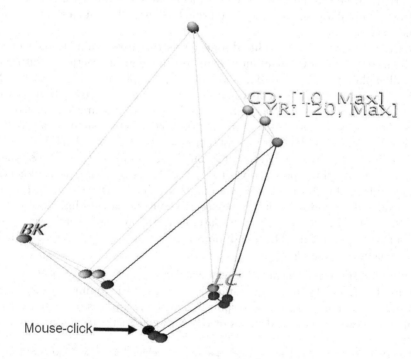

Fig. 3. The lattice of the subcontext determined by the three highest OR95- rules; with the highlighting scheme, we see the rule BK$\rightarrow$LC

support 19.7%, versus the common value of 0% for the bottom of the lattice. In this lattice diagram, since concept height is determined by support of concept, as in [PHM], we witness the approximate numerical values of all the cell values underlying all the 2 × 2 contingency tables that can be formed from the set of attributes in the three highest OR95- rules from the original set of rules. Not only are the rules available (via mouse-click on concepts, as shown above), but the origination of their effect measure values are also directly apparent, through identification of the appropriate diamond subposet that corresponds to the underlying contingency table (as in Figure 1 and [OPH]).

Diagram-based variants of the zooming method seen in this example (wherein a lattice was derived from rules that passed a filter) are also available within Seqer. For example, from a large, complex lattice derived from the Birdkeeping-LungCancer formal context, the user can request the subdiagram consisting of only those concepts below the concept (LC', LC'') whose support exceeds $x\%$ of the support of (LC', LC'') (say, with $x = 80$), and then select in a subcontext only those attributes related to the filtered set of concepts, and then generate ARs from this selected set of attributes, filter the results using rule measure conditions, and again restrict attention to the lattice generated by some smaller set of attributes determined by filtering the rules. See [URL] for examples of such conceptual exploration sequences. In general, Seqer has been designed to support iterative interactions between subcontext selection, a variety of lattice diagram manipulations, and association rule filtering, all within a single instance of the software.

In this paper, we have introduced some new structures within Formal Concept Analysis that we believe are of interest in epidemiological practice. A future paper will explain the overall design and conceptual exploration capabilities of Seqer[2] and introduce ACE[3], an open source concept software project that provides many of the core functionalities of Seqer, to be posted at sourceforge.net by late 2008. We also note that the background knowledge represented by a given tag context $\mathbb{T} = (M, C, J)$ for a formal context $\mathbb{K} = (G, M, I)$ can be used to capture the dimensionality criteria and simultaneously reduce the portion of $\mathfrak{B}(\mathbb{K})$ that is visualized, through the production of a system $\mathcal{S} = \{\mathfrak{B}(\mathbb{K}_D)\}_{D \in \mathfrak{B}(\mathbb{T})}$ of concept lattices indexed by $\mathfrak{B}(\mathbb{T})$, where $\mathbb{K}_D = (G, D, I)$. The lattices $B((G, M, I)_D)$ can be considered individually, or can be used to control searches for concepts that express meets of concepts in distinct lattices in the system, thereby witnessing rules of particular forms. The foundations laid in this paper will support further development of these ideas.

Acknowledgement. The authors acknowledge support of this project by Grant Number P42ES007381 from the National Institute of Environmental Health Sciences. The content is solely the responsibility of the authors and does not necessarily represent the official views of the National Institute of Environmental

---

[2] A Formal Concept Analysis software package developed at New Mexico State University.

[3] ACE is an acronym for Analysis of Concepts for Epidemiologists, and a semordnilap for ECA, Epidemiological Concept Analysis.

Health Sciences of the National Institutes of Health. We also thank Physical Science Laboratory programmers Jesse Eyerman, Gregory Thompson, and Tim Hannan for their creation of supporting software.

References

[Bec]   Becker, P.: Numerical Analysis in Conceptual Systems with TOSCANAJ. In: Eklund, P.W. (ed.) ICFCA 2004. LNCS (LNAI), vol. 2961, pp. 96–103. Springer, Heidelberg (2004)

[BGB]   Blanchard, J., Guillet, F., Briand, H.: A User-driven and Quality-oriented Visualization for Mining Association Rules. In: Proceedings of the Third IEEE International Conference on Data Mining (ICDM 2003), p. 493 (2003)

[CFRD]   Cellier, P., Ridoux, O., Ducasse, M.: A Parametrized Algorithm for Exploring Concept Lattices. In: Kuznetsov, S.O., Schmidt, S. (eds.) ICFCA 2007. LNCS (LNAI), vol. 4390, Springer, Heidelberg (2007)

[DP]   Davey, B., Priestley, H.: Introduction to Ordered Sets and Lattices, 2nd edn. Cambridge University Press, Cambridge (2002)

[GW]   Ganter, B., Wille, R.: Formal Concept Analysis: Mathematical Foundations. Springer, Heidelberg (1999)

[HP]   Hannan, T., Pogel, A.: Spring-Based Lattice Drawing Highlighting Conceptual Similarity. In: Missaoui, R., Schmidt, J. (eds.) Formal Concept Analysis. LNCS (LNAI), vol. 3874, Springer, Heidelberg (2006)

[HLK]   Hashemi, R., Le Blanc, L., Kobayashi, T.: Formal Concept Analysis for Investigation of Normal Accidents. International Journal of General Systems 33, 469–484 (2004)

[HKB]   Hlst, P.A., Kromhout, D., Brand, R.: For debate: pet birds as an independent risk factor for lung cancer. British Medical Journal 297, 13–21 (1988)

[Im]   Imielinski, T.: Association Rule Mining of Biological Data at the DIMACS Working Group on Data Mining and Epidemiology, Rutgers, NJ (March 2004), `http://dimacs.rutgers.edu/Workshops/WGDataMining/abstracts.html`

[OPH]   Ozonoff, D., Pogel, A., Hannan, T.: Generalized Contingency Tables and Concept Lattices. In: Abello, J., Cormode, G. (eds.) Discrete Methods in Epidemiology. DIMACS Series, vol. 70, AMS, Providence (2006)

[Pea]   Pasquier, N., et al.: Efficient Mining of Association Rules Using Closed Itemset Lattices. Information Systems 24(1), 25–46 (1999)

[PHM]   Pogel, A., Hannan, T., Miller, L.: Visualization of Concept Lattices Using Weight Functions. In: Supplementary Proceedings of the International Conference on Conceptual Structures, Huntsville, AL, Shaker (2004)

[RS]   Ramsey, F.L., Schafer, D.W.: The Statistical Sleuth. Duxbury Press (2002)

[RHNV]   Rouane, M., et al.: A Proposal for Combining Formal Concept Analysis and Description Logics for Mining Relational Data. In: Kuznetsov, S.O., Schmidt, S. (eds.) ICFCA 2007. LNCS (LNAI), vol. 4390, pp. 51–65. Springer, Heidelberg (2007)

[RG]   Rothman, K.J., Greenland, S.: Modern Epidemiology. Lipponcott-Raven (1998)

[Spir]   Spirtes, P., Glymour, C., Scheines, R.: Causation, Prediction, and Search, 2nd edn. Adaptive Computation and Machine Learning. Bradford Book, Cambridge, MA (2001)

[SW]   Stumme, G., Wolff, K.E.: Numerical Aspects in the Data Model of Con-
          ceptual Information Systems. In: Kambayashi, Y., Lee, D.-L., Lim, E.-p.,
          Mohania, M., Masunaga, Y. (eds.) ER Workshops 1998. LNCS, vol. 1552,
          pp. 117–128. Springer, Heidelberg (1999)
[URL]   For data and further images related to this paper, refer to
          http://www.psl.nmsu.edu/~apogel/GeneralizedContingencyStructures
[ZO]   Zaki, M., Ogihara, M.M.: Theoretical Foundations of Associations Rules.
          In: Proceedings of 3rd SIGMOD'98 Workshop on Research Issues in Data
          Mining and Knowledge Discovery (DMKD 1998), Seattle, Washington, USA
          (June 1998)

Comparison of Dual Orderings in Time II

Vincent Duquenne[1] and John Mohr[2]

[1] CNRS-ECP6, Université Pierre et Marie Curie,
175 rue du Chevaleret, 75013 Paris, France
duquenne@math.jussieu.fr

[2] Dept. of Sociology
University of California, Santa Barbara,
CA 93106-9430
mohr@soc.ucsb.edu

Keywords: Duality practices × symbolic systems, Poverty relief in New York City 1888-1917, Progressive era. Lattices, (nested) line diagrams, (relative) canonical basis of implications, gluing decomposition, subdirect products, join / meet morphisms.

Aims and results

In the direction of researches on formalization in the social sciences [6,1,7], several papers were devoted to analyzing the dual interplay between cultural components (categories of words) and actual practices (welfare treatments, programs ...). A first analysis of *poverty in NY City in 1888 - 1917* [6] was undertaken in a joint work, with the description of *relief treatments* by *words* to investigate their institutional logic.

By making use of the abilities of lattices to analyze the duality *treatments × words*, a second note [1] refined this analysis along three directions. First, to screen the source data with the basic toolkit of FCA [10, 4 ...] and Lattice Analysis [2] (*orders on words, treatments, concept lattices ...*). Then, to make use of a second tool set for elaborating more synthetic views of the data source structures with *canonical basis of implications* [5], lattice splits generated by *transpositions / double arrows* expressing incompatibilities between words / treatments, and lattice *ungluing decompositions* [4] into intervals that expresses similarities between words or treatments and provides an objective and faithful way for dismantling the ordinal data structure. The third direction compares the findings in 1888 / 1917, and addresses the question of what was either stable, or different between these two points in time through a formal comparison using simple if not simplistic consensus by context union / intersection.

The aim of the present work is to elaborate and experiment new algorithms for pointing out more systematically what is new or unmoved concerning orders and lattice structures, as they change through time (see Fig. 1-2), and to test them on the original data set. To this end, we will mix together and make use of *specific / relative* basis of implications [3] that naturally occur when *apposition* and *subposition* of contexts have to be considered, together with *subdirect products* of lattices (see Fig. 3) that have been used in particular for *context fusion* [11], as a natural candidate for lattice consensus. The outcome is to give a simultaneous representation of the two data sets providing new ways to explore and characterize practice / cultural changes.

R. Medina and S. Obiedkov (Eds.): ICFCA 2008, LNAI 4933, pp. 321–324, 2008.
© Springer-Verlag Berlin Heidelberg 2008

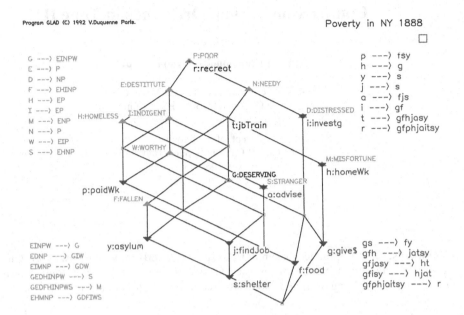

Fig. 1. The concept lattice *treatments* × *words* (1888), together with the two canonical basis of implications of implications on conjunctions of *words* (left) and *treatments* (right hand side)

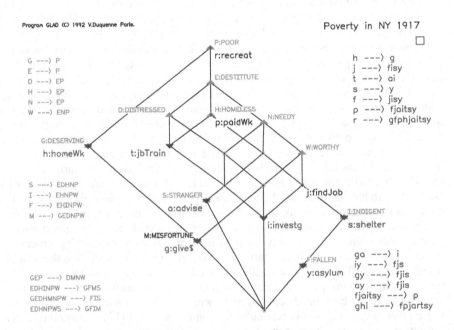

Fig. 2. The concept lattice *treatments* × *words* (1917), with its canonical basis of implications characterizing minimally discrepancy to treatment / word powersets

As for the results, both *treatment* × *word* lattices (for 1888 and 1917) are quite small as compared with the potential 2^{11} elements (26 and 18 elements respectively). This reveals a lot of implications between conjunctions of *words* (or *treatments*, dually), which are summarized by their *canonical basis* of implications (see Fig. 1-2, where the lattices are *minimally labeled* [8]). Most of their implications have a *single premise*, which means that these lattices are nearly *distributive*. Actually the intervals above *s:shelter* and *i:investigation* (in the two lattices, respectively) are distributive. Interestingly, these two lattices are decomposable in *unglued intervals* [4 §5.2] which assesses similarities [1] between *words* (/ *treatments*) respecting the global structure.

1888 reveals the *splits treatment* / *word* (*transpositions* expressed by *double arrows* in the contexts see [4]): paidWork / NEEDY, investigation / DESTITUTE, advise / INDIGENT, findJob / WORTHY, give$ / HOMELESS, food / FALLEN, and asylum / STRANGER. As shocking as it could appear now, in 1888 one gave asylum - except to strangers!-, or money −except to HOMELESS!- etc. Similarly, 1917 displays the splits: jobTrain / NEEDY, food-shelter-asylum / DISTRESSED, advise / WORTHY and give$ / HOMELESS as it was already the case in 1888. As local negations, these splits capture fundamental distinctions in systems of moral boundaries of these times.

Now a first natural idea for comparing these two lattices is to glue their contexts horizontally by taking their *apposition* (resp. vertically *subposition*), and to construct the corresponding lattice which is *join-embedded* (resp. *meet-*) in their *direct product*, as it is implicitly done with *nested line diagrams* [10], and to distinguish two *specific basis* [1] of implications going from one set to another (ex: 1888 ↔ 1917 *words*).

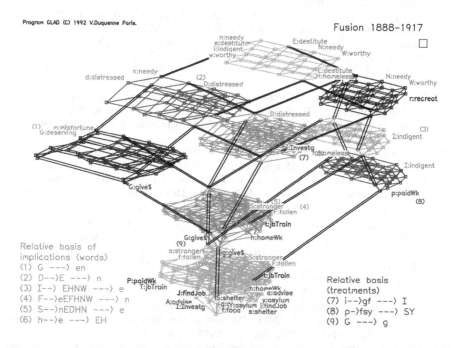

Fig. 3. The *fusion* of the 1888 & 1917 lattices is embedded in their *direct product* and is also *gluing decomposable*. The two *relative basis* of implications (1888 / 1917 = lower / upper-case letters) express discrepancy to direct product and independence.

This is specially adapted when a single set of objects is described through two different sets of attributes with a dissymmetry between object / attribute rôles. In our present case however, there is a symmetry *words / treatments* which are equally conceptual. On the other hand two pairs of different sets are needed to distinguish them for these two periods. Hence, let (T_1, W_1, I_1) and (T_2, W_2, I_2) be the 1888 / 1917 contexts and $L_1 = L(T_1, W_1, I_1)$, $L_2 = L(T_2, W_2, I_2)$ their concept lattices. The *fusion* (see [11, 4 §5.1]) of these contexts is the context generating the smallest *sublattice* of $L_1 \times L_2$ the relation of which being a superset of the relation obtained by subposition of the two appositions $(I_1 \mid I_1 \cup I_2)$, and $(I_1 \cup I_2 \mid I_2)$. This *subdirect product* construction is highly symmetric regarding the two original contexts, as well as *words & treatments*. The two *relative basis* of implications mixed together characterize minimally the discrepancy to direct product (taken as a *starting lattice* [3,9] or as *background knowledge* [8]), and the underlying *meet / join morphisms* between factors. After implementation through GLAD [2], the resulting lattice (see Fig. 3) appears to be gluing decomposable, which allows detecting attributes that are structurally similar (*m:misfortune / G:deserving,...*) or stable (*distressed, stranger, fallen...*) in time, which now requires careful screenings and further interpretations.

References

[1] Duquenne, V., Mohr, J., Le Pape, A.: Comparison of dual ordering in time. Soc. Sci. Information 37, 227–253 (1998)
[2] Duquenne, V.: Latticial structures in Data Analysis, ORDAL 96: Order and decision-making (I. Rival ed.). Theoretical Computer Science 217, 407–436 (1999)
[3] Duquenne, V., et al.: Structuration of phenotypes / genotypes through Galois lattices and implications. Applied Artificial Intelligence 17, 243–256 (2003)
[4] Ganter, B., Wille, R.: Formal Concept Analysis, Mathematical Foundations. Springer, Berlin (1999)
[5] Guigues, J.L., Duquenne, V.: Familles minimales dimplications informatives r´esultant d'un tableau de données binaires. Mathématiques & Sciences Humaines 95, 5–18 (1986) (preprint Groupe Mathématiques et Psychologie, Université Paris V-René Descartes (1984))
[6] Mohr, J., Duquenne, V.: The duality of culture and practice: Poverty relief in New-York City, 1888-1917. Theory and Society 26, 305–356 (1997)
[7] Mohr, J., Bourgeois, M., Duquenne, V.: The Logic of Opportunity: A Formal Analysis of the University of California's Outreach and Diversity Discourse. Center for Studies in Higher Education, UC Berkeley, Research and Occasional Papers Series (2004), http://cshe.berkeley.edu/publications/papers/papers.html
[8] Stumme, G.: Attribute exploration with background implications and exceptions. In: Bock, H.H., Polasek, W. (eds.) Data Analysis and Information Systems. Statistical and Conceptual approaches. Studies in classification, data analysis and knowledge organization, vol. 7, pp. 457–469. Springer, Heidelberg (1996)
[9] Valtchev, P., Duquenne, V.: Towards scalable divide-and-conquer methods for computing concepts and implications. In: SanJuan, E., et al. (eds.) Proc. of the Journées de l'Informatique Messine (JIM 2003): Knowledge Discovery and Discrete Mathematics, Metz (FR), September 3-6, 2003, pp. 3–14. INRIA (2003)
[10] Wille, R.: Restructuring lattice theory: an approach based on hierarchies of concepts. In: Rival, I. (ed.) Ordered sets, pp. 445–470. Reidel, Dordrecht, Boston (1982)
[11] Wille, R.: Sur la fusion des contextes individuels. Mathématiques & Sciences Humaines 85, 57–71 (1984)

Author Index

Lecture Notes in Artificial Intelligence (LNAI)

Vol. 4696: H.-D. Burkhard, G. Lindemann, R. Verbrugge, L.Z. Varga (Eds.), Multi-Agent Systems and Applications V. XIII, 350 pages. 2007.

Vol. 4694: B. Apolloni, R.J. Howlett, L. Jain (Eds.), Knowledge-Based Intelligent Information and Engineering Systems, Part III. XXIX, 1126 pages. 2007.

Vol. 4693: B. Apolloni, R.J. Howlett, L. Jain (Eds.), Knowledge-Based Intelligent Information and Engineering Systems, Part II. XXXII, 1380 pages. 2007.

Vol. 4692: B. Apolloni, R.J. Howlett, L. Jain (Eds.), Knowledge-Based Intelligent Information and Engineering Systems, Part I. LV, 882 pages. 2007.

Vol. 4687: P. Petta, J.P. Müller, M. Klusch, M. Georgeff (Eds.), Multiagent System Technologies. X, 207 pages. 2007.

Vol. 4682: D.-S. Huang, L. Heutte, M. Loog (Eds.), Advanced Intelligent Computing Theories and Applications. XXVII, 1373 pages. 2007.

Vol. 4676: M. Klusch, K.V. Hindriks, M.P. Papazoglou, L. Sterling (Eds.), Cooperative Information Agents XI. XI, 361 pages. 2007.

Vol. 4667: J. Hertzberg, M. Beetz, R. Englert (Eds.), KI 2007: Advances in Artificial Intelligence. IX, 516 pages. 2007.

Vol. 4660: S. Džeroski, L. Todorovski (Eds.), Computational Discovery of Scientific Knowledge. X, 327 pages. 2007.

Vol. 4659: V. Mařík, V. Vyatkin, A.W. Colombo (Eds.), Holonic and Multi-Agent Systems for Manufacturing. VIII, 456 pages. 2007.

Vol. 4651: F. Azevedo, P. Barahona, F. Fages, F. Rossi (Eds.), Recent Advances in Constraints. VIII, 185 pages. 2007.

Vol. 4648: F. Almeida e Costa, L.M. Rocha, E. Costa, I. Harvey, A. Coutinho (Eds.), Advances in Artificial Life. XVIII, 1215 pages. 2007.

Vol. 4635: B. Kokinov, D.C. Richardson, T.R. Roth-Berghofer, L. Vieu (Eds.), Modeling and Using Context. XIV, 574 pages. 2007.

Vol. 4632: R. Alhajj, H. Gao, X. Li, J. Li, O.R. Zaïane (Eds.), Advanced Data Mining and Applications. XV, 634 pages. 2007.

Vol. 4629: V. Matoušek, P. Mautner (Eds.), Text, Speech and Dialogue. XVII, 663 pages. 2007.

Vol. 4626: R.O. Weber, M.M. Richter (Eds.), Case-Based Reasoning Research and Development. XIII, 534 pages. 2007.

Vol. 4617: V. Torra, Y. Narukawa, Y. Yoshida (Eds.), Modeling Decisions for Artificial Intelligence. XII, 502 pages. 2007.

Vol. 4612: I. Miguel, W. Ruml (Eds.), Abstraction, Reformulation, and Approximation. XI, 418 pages. 2007.

Vol. 4604: U. Priss, S. Polovina, R. Hill (Eds.), Conceptual Structures: Knowledge Architectures for Smart Applications. XII, 514 pages. 2007.

Vol. 4603: F. Pfenning (Ed.), Automated Deduction – CADE-21. XII, 522 pages. 2007.

Vol. 4597: P. Perner (Ed.), Advances in Data Mining. XI, 353 pages. 2007.

Vol. 4594: R. Bellazzi, A. Abu-Hanna, J. Hunter (Eds.), Artificial Intelligence in Medicine. XVI, 509 pages. 2007.

Vol. 4585: M. Kryszkiewicz, J.F. Peters, H. Rybinski, A. Skowron (Eds.), Rough Sets and Intelligent Systems Paradigms. XIX, 836 pages. 2007.

Vol. 4578: F. Masulli, S. Mitra, G. Pasi (Eds.), Applications of Fuzzy Sets Theory. XVIII, 693 pages. 2007.

Vol. 4573: M. Kauers, M. Kerber, R. Miner, W. Windsteiger (Eds.), Towards Mechanized Mathematical Assistants. XIII, 407 pages. 2007.

Vol. 4571: P. Perner (Ed.), Machine Learning and Data Mining in Pattern Recognition. XIV, 913 pages. 2007.

Vol. 4570: H.G. Okuno, M. Ali (Eds.), New Trends in Applied Artificial Intelligence. XXI, 1194 pages. 2007.

Vol. 4565: D.D. Schmorrow, L.M. Reeves (Eds.), Foundations of Augmented Cognition. XIX, 450 pages. 2007.

Vol. 4562: D. Harris (Ed.), Engineering Psychology and Cognitive Ergonomics. XXIII, 879 pages. 2007.

Vol. 4548: N. Olivetti (Ed.), Automated Reasoning with Analytic Tableaux and Related Methods. X, 245 pages. 2007.

Vol. 4539: N.H. Bshouty, C. Gentile (Eds.), Learning Theory. XII, 634 pages. 2007.

Vol. 4529: P. Melin, O. Castillo, L.T. Aguilar, J. Kacprzyk, W. Pedrycz (Eds.), Foundations of Fuzzy Logic and Soft Computing. XIX, 830 pages. 2007.

Vol. 4520: M.V. Butz, O. Sigaud, G. Pezzulo, G. Baldassarre (Eds.), Anticipatory Behavior in Adaptive Learning Systems. X, 379 pages. 2007.

Vol. 4511: C. Conati, K. McCoy, G. Paliouras (Eds.), User Modeling 2007. XVI, 487 pages. 2007.

Vol. 4509: Z. Kobti, D. Wu (Eds.), Advances in Artificial Intelligence. XII, 552 pages. 2007.

Vol. 4496: N.T. Nguyen, A. Grzech, R.J. Howlett, L.C. Jain (Eds.), Agent and Multi-Agent Systems: Technologies and Applications. XXI, 1046 pages. 2007.

Vol. 4483: C. Baral, G. Brewka, J. Schlipf (Eds.), Logic Programming and Nonmonotonic Reasoning. IX, 327 pages. 2007.

Vol. 4482: A. An, J. Stefanowski, S. Ramanna, C.J. Butz, W. Pedrycz, G. Wang (Eds.), Rough Sets, Fuzzy Sets, Data Mining and Granular Computing. XIV, 585 pages. 2007.

Vol. 4481: J. Yao, P. Lingras, W.-Z. Wu, M.S. Szczuka, N.J. Cercone, D. Ślęzak (Eds.), Rough Sets and Knowledge Technology. XIV, 576 pages. 2007.

Vol. 4476: V. Gorodetsky, C. Zhang, V.A. Skormin, L. Cao (Eds.), Autonomous Intelligent Systems: Multi-Agents and Data Mining. XIII, 323 pages. 2007.

Vol. 4460: S. Aguzzoli, A. Ciabattoni, B. Gerla, C. Manara, V. Marra (Eds.), Algebraic and Proof-theoretic Aspects of Non-classical Logics. VIII, 309 pages. 2007.

Vol. 4457: G.M.P. O'Hare, A. Ricci, M.J. O'Grady, O. Dikenelli (Eds.), Engineering Societies in the Agents World VII. XI, 401 pages. 2007.